Scattering Amplitudes in Gauge Theory and Gravity

Providing a comprehensive, pedagogical introduction to scattering amplitudes in gauge theory and gravity, this book is ideal for graduate students and researchers. It offers a smooth transition from basic knowledge of quantum field theory to the frontier of modern research.

Building on basic quantum field theory, the book starts with an introduction to the spinor helicity formalism in the context of Feynman rules for tree-level amplitudes. The material covered includes on-shell recursion relations, superamplitudes, symmetries of $N = 4$ super Yang–Mills theory, twistors and momentum twistors, Grassmannians, and polytopes. The presentation also covers amplitudes in perturbative supergravity, 3d Chern–Simons matter theories, and color-kinematics duality and its connection to "gravity = (gauge theory)^2."

Basic knowledge of Feynman rules in scalar field theory and quantum electrodynamics is assumed, but all other tools are introduced as needed. Worked examples demonstrate the techniques discussed, and over 150 exercises help readers absorb and master the material.

Henriette Elvang is Associate Professor in the Department of Physics, University of Michigan. She has worked on various aspects of high energy theoretical physics, including black holes in string theory, scattering amplitudes, and the structure of gauge theories.

Yu-tin Huang is Assistant Professor at National Taiwan University. He is known for his work in the study of scattering amplitudes beyond four dimensions, most notably in 3-dimensional Chern–Simons matter theory.

"In recent years, a series of surprising insights and new methods have transformed the understanding of gauge and gravitational scattering amplitudes. These advances are important both for practical calculations in particle physics, and for the fundamental structure of relativistic quantum theory. Elvang and Huang have written the first comprehensive text on this subject, and their clear and pedagogical approach will make these new ideas accessible to a wide range of students."

Joseph Polchinski, *University of California, Santa Barbara*

"This book provides a much needed text covering modern techniques that have given radical new insights into the structure of quantum field theory. It gathers together a very large body of recent literature and presents it in a coherent style. The book should appeal to the wide body of researchers who wish to use quantum field theory as a tool for describing physical phenomena or who are intending to gain insight by studying its mathematical structure."

Michael B. Green, *University of Cambridge*

Scattering Amplitudes in Gauge Theory and Gravity

HENRIETTE ELVANG

University of Michigan, Ann Arbor

YU-TIN HUANG

National Taiwan University

CAMBRIDGE
UNIVERSITY PRESS

CAMBRIDGE
UNIVERSITY PRESS

University Printing House, Cambridge CB2 8BS, United Kingdom

One Liberty Plaza, 20th Floor, New York, NY 10006, USA

477 Williamstown Road, Port Melbourne, VIC 3207, Australia

314-321, 3rd Floor, Plot 3, Splendor Forum, Jasola District Centre, New Delhi - 110025, India

79 Anson Road, #06-04/06, Singapore 079906

Cambridge University Press is part of the University of Cambridge.

It furthers the University's mission by disseminating knowledge in the pursuit of education, learning and research at the highest international levels of excellence.

www.cambridge.org
Information on this title: www.cambridge.org/9781107069251

First published 2015
3rd printing 2019

A catalogue record for this publication is available from the British Library

Library of Congress Cataloging in Publication data
Elvang, Henriette, 1976– author.
Scattering amplitudes in gauge theory and gravity / Henriette Elvang, University of Michigan, Ann Arbor, Yu-tin Huang, Institute for Advanced Study and National Taiwan University.
pages   cm
Includes bibliographical references and index.
ISBN 978-1-107-06925-1 (hardback)
1. Scattering amplitude (Nuclear physics)   2. Gauge fields (Physics) – Mathematics.
I. Huang, Yu-tin, author.   II. Title.
QC20.7.S3E48   2015
539.7′58 – dc23   2014032234

ISBN 978-1-107-06925-1 Hardback

For mom, dad, and Coco.
Yu-tin

Thank you, Mormor.
Henriette

Contents

Part II Loops

Part III Topics

Preface

This book grew out of a need to have a set of easily accessible notes that introduced the basic techniques used in modern research on scattering amplitudes. In addition to the key tools, such a review should collect some of the small results and intuitions the authors had acquired from their work in the field and which had not previously been exposed in the literature. As the authors quickly realized, such an introduction would bring the reader only part of the way towards some of the most exciting topics in the field, so they decided to add "a little extra" material. While doing so – and this took quite a while – the authors remained in full and complete denial about writing a book. It was only at the end of the process that they faced their worst fears: the review was becoming a book. You now hold the result in your hands. Because the authors were not writing a book, they actually thoroughly enjoyed the work. Their hope is that you will enjoy it too and that you will find it useful.

It is a pleasure to thank our friends and collaborators who have worked with us and helped us learn the subject of scattering amplitudes: Ratin Akhoury, Nima Arkani-Hamed, Zvi Bern, Freddy Cachazo, John Joseph Carrasco, Simon Caron-Huot, Tim Cohen, Scott Davies, Tristan Dennen, Lance Dixon, Dan Freedman, David Kosower, Johannes Henn, Harald Ita, Henrik Johansson, Michael Kiermaier, Sangmin Lee, Arthur Lipstein, Thomas Lam, David McGady, Timothy Olson, Cheng Peng, Jan Plefka, Radu Roiban, Mark Srednicki, Warren Siegel, David Speyer, and Jaroslav Trnka. H.E. is grateful for the hospitality offered by Stanford/SLAC during her visit in February/March 2013 and KITP/UCSB during January–March 2014.

A few people have suffered early drafts of this book and we are indebted to them for their helpful comments/suggestions/corrections: Cindy Keeler, Timothy Olson, Sam Roland, David Speyer, Sri Suresh, Jonathan Walsh, and John Ware. Their careful readings caught multiple typos and helped us improve the presentation. We would also like to thank Michael Enciso, Karol Kampf, Joe Minahan, Stephen Naculich, and especially Stefan Theisen as well as the Harvard amplitude reading group (communications from Marat Freytsis) for detailed feedback on the manuscript.

Y.-t.H. would especially like to thank Warren Siegel, Peter van Nieuwenhuizen, and Zvi Bern for their continuing support and guidance without which his career would have taken a much different path. H.E. thanks Gary Horowitz, Dan Freedman, and Joe Polchinski for the education, support, and friendship for which she is deeply grateful. She would also like to thank her colleagues at the University of Michigan for their support for this project. Both authors have benefited tremendously from amplitude discussions with Zvi Bern, Lance Dixon, and Nima Arkani-Hamed.

The authors would also like to express their thanks to Cambridge University Press editor Simon Capelin for his support, enthusiasm, and patience throughout the project.

Yu-tin Huang and Henriette Elvang

Introduction

Scattering experiments are crucial for our understanding of the building blocks of nature. The standard model of particle physics was developed from scattering experiments, including the discovery of the weak force bosons $W^{\pm}$ and Z^0, quarks and gluons, and most recently the Higgs boson.

The key observable measured in particle scattering experiments is the ***scattering cross-section*** σ. It encodes the likelihood of a given process to take place as a function of the energy and momentum of the particles involved. A more refined version of this quantity is the ***differential cross-section*** $d\sigma/d\Omega$: it describes the dependence of the cross-section on the angles of the scattered particles.

Interpretation of data from scattering experiments relies heavily on theoretical predictions of scattering cross-sections. These are calculated in ***relativistic quantum field theory (QFT)***, which is the mathematical language for describing elementary particles and their interactions. Relativistic QFT combines special relativity with quantum physics and is a hugely successful and experimentally well-tested framework for describing elementary particles and the fundamental forces of nature. In quantum mechanics, the probability distribution $|\psi|^2 = \psi^*\psi$ for a particle is given by the norm-squared of its complex-valued wavefunction ψ. Analogously, in quantum field theory, the differential cross-section is proportional to the norm-squared of the ***scattering amplitude*** A, $d\sigma/d\Omega \propto |A|^2$. The amplitudes A are well-defined physical observables: they are the subject of this book.

Scattering amplitudes have physical relevance through their role in the scattering cross-section. Moreover, it has been realized in recent years that amplitudes themselves have a very interesting mathematical structure. Understanding this structure guides us towards more efficient methods to calculate amplitudes. It also makes it exciting to study scattering amplitudes in their own right and explore (and exploit) their connections to interesting branches of mathematics, including combinatorics and geometry.

The purpose of this book is two-fold. First, we wish to provide a pedagogical introduction to the efficient modern methods for calculating scattering amplitudes. Second, we survey several interesting mathematical properties of amplitudes and recent research advances. The hope is to bridge the gap between a standard graduate course in quantum field theory and the modern approach to scattering amplitudes as well as current research. Thus we assume readers to be familiar with the basics of quantum field theory and particle physics; however, here in the first chapter we offer an introduction that may also be helpful for a broader audience with an interest in the subject.

What is a scattering amplitude?

Let us consider a few examples of scattering processes involving electrons e^-, positrons e^+, and photons γ:

$$
\begin{aligned}
\text{Compton scattering} \quad & e^- + \gamma \to e^- + \gamma\,, \\
\text{Møller scattering} \quad & e^- + e^- \to e^- + e^-\,, \\
\text{Bhabha scattering} \quad & e^- + e^+ \to e^- + e^+\,, \\
e^- e^+ \text{ annihilation} \quad & e^- + e^+ \to \gamma + \gamma\,.
\end{aligned}
\tag{1.1}
$$

These processes are described in the quantum field theory that couples Maxwell's electromagnetism to electrons and positrons, namely quantum electrodynamics (QED). Each process is characterized by the types of particles involved in the initial and final states as well as the relativistic momentum $\vec{p}$ and energy E of each particle. This input is the ***external data*** for an amplitude: a scattering amplitude A_n involving a total of n initial and final state particles takes the list $\{E_i, \vec{p}_i\,;\, \text{type}_i\}$, $i = 1, 2, \ldots, n$, of external data and returns a complex number:

$$
n\text{-particle amplitude } A_n: \quad \{E_i, \vec{p}_i\,;\, \text{type}_i\} \to A_n\Big(\{E_i, \vec{p}_i\,;\, \text{type}_i\}\Big) \in \mathbb{C}.
\tag{1.2}
$$

"Particle type" involves more than saying which particles scatter: it also includes a specification of the appropriate quantum numbers of the initial and final states, for example the polarization of a photon or the spin-state of a fermion.

In special relativity, the energy E and momentum $\vec{p}$ of a physical particle must satisfy the relationship $E^2 = |\vec{p}|^2 c^2 + m^2 c^4$ with m the rest mass of the particle. This is simply the statement that $E^2 - |\vec{p}|^2 c^2$ is a Lorentz-invariant quantity in Minkowski space. Thus, one of the constraints on the external data for an n-particle amplitude is that the relativistic ***on-shell*** condition,

$$
E_i^2 - |\vec{p}_i|^2 c^2 = m_i^2 c^4\,,
\tag{1.3}
$$

holds for all particles $i = 1, \ldots, n$. The initial and final state electrons and positrons in the QED processes (1.1) must satisfy the on-shell condition (1.3) with m_i equal to the electron mass, $m_e = 511\,\text{keV}/c^2$, while m_i is zero for the photon since it is massless.

It is convenient to work in "natural units" where the speed of light c and Planck's constant $\hbar = h/2\pi$ are set to unity: $c = 1 = \hbar$. Proper units can always be restored by dimensional analysis. We combine the energy and momentum into a 4-momentum $p_i^\mu = (E_i, \vec{p}_i)$, with $\mu = 0, 1, 2, 3$, and write $p_i^2 = -E_i^2 + |\vec{p}_i|^2$ such that the on-shell condition (1.3) becomes

$$
p_i^2 = -m_i^2\,.
\tag{1.4}
$$

Conservation of relativistic energy and momentum requires the sum of initial momenta p_{in}^{μ} to equal the sum of final state momenta p_{out}^{μ}. It is convenient to flip the signs of all incoming momenta so that the *conservation of 4-momentum* simply reads

$$\sum_{i=1}^{n} p_i^{\mu} = 0. \tag{1.5}$$

Thus, to summarize, the external data for the amplitude involve a specification of 4-momentum and particle type for each external particle, $\{p_i^{\mu}, \text{type}_i\}$, subject to the on-shell constraints $p_i^2 = -m_i^2$ and momentum conservation (1.5). We often use the phrase *on-shell amplitude* to emphasize that the external data satisfy the kinematic constraints (1.4) and (1.5) and include the appropriate polarization vectors or fermion spin wavefunctions.

Feynman diagrams
Scattering amplitudes are typically calculated as a perturbation series in the expansion of a small dimensionless parameter that encodes the coupling of interactions between the particles. For QED processes, this dimensionless coupling is the fine structure constant[1] $\alpha \approx 1/137$. In the late 1940s, Feynman introduced a diagrammatic way to organize the perturbative calculation of the scattering amplitudes. It expresses the n-particle scattering amplitude A_n as a sum of all possible Feynman diagrams with n external legs:

$$A_n = \sum (\text{Feynman diagrams}) = \quad + \quad + \quad + \cdots \tag{1.6}$$

tree-level   1-loop   2-loop

The sum of diagrams is organized by the number of closed loops. For a given number of particles n, the loop-diagrams are suppressed by powers of the coupling, e.g. in QED an L-loop diagram is order α^{2L} compared with the tree-level. Hence the loop-expansion is a diagrammatic representation of the perturbation expansion. The leading contribution comes from the tree-diagrams; their sum is called the *tree-level amplitude*, A_n^{tree}. The next order is 1-loop; the sum of the 1-loop diagrams is the *1-loop amplitude* $A_n^{\text{1-loop}}$, etc. The full amplitude is then

$$A_n = A_n^{\text{tree}} + A_n^{\text{1-loop}} + A_n^{\text{2-loop}} + \cdots \tag{1.7}$$

It is rare that the loop-expansion is convergent; typically, the number of diagrams grows so quickly at higher-loop order that it overcomes the suppression from the higher powers in the small coupling. Amplitudes can be Borel summable, but this is not a subject we treat here. Instead, we pursue an understanding of the contributions to the amplitude order-by-order in perturbation theory.

---

[1] In SI units, the fine structure constant is $\alpha = e^2/(4\pi\epsilon_0\hbar c)$, where ϵ_0 is the vacuum permittivity.

The quantum field theories we are concerned with in this book are defined by Lagrangians that encode the particles and how they interact. The Lagrangian for QED determines a simple basic 3-particle interaction between electrons/positrons and the photon:

$$(1.8)$$

From this, one can build diagrams such as

   and $$(1.9)$$

Reading the diagrams left to right, the first diagram describes the absorption and subsequent emission of a photon (wavy line) by an electron (solid line), and as such it contributes at the leading order in perturbation theory to the Compton scattering process $e^- + \gamma \rightarrow e^- + \gamma$. The second and third diagrams in (1.9) are the two tree-level diagrams that give the leading-order contribution to the Bhabha scattering process $e^- + e^+ \rightarrow e^- + e^+$. The first of those two diagrams encodes the annihilation of an electron and a positron to a photon and the subsequent $e^+ e^-$ pair-creation. The second diagram is the exchange of a photon in the scattering of an electron and a positron.

A Feynman diagram is translated to a mathematical expression via Feynman rules. These rules are specific to the particle types and the theory that describes their interactions. A Feynman graph has the following essential parts:

- *External lines:* the Feynman diagram has an external line for each initial and final state particle. The rule for the external line depends on the particle type. For spin-1 particles, such as photons and gluons, the external line rule encodes the polarizations. For fermions, it contains information about the spin state.
- *Momentum labels:* every line in the Feynman diagram is associated with a momentum. For the external lines, the momentum is fixed by the external data. Momentum conservation is enforced at any vertex. For a tree-graph, this fully determines the momenta on all internal lines. For an L-loop graph, L momenta are undetermined and the Feynman rules state that one must integrate over all possible values of these L momenta.
- *Vertices:* the vertices describe the interactions among the particles in the theory. Vertices can in principle have any number of lines going in or out, but in many theories there are just cubic and/or quartic vertices. The Feynman rules translate each vertex into a mathematical rule; in the simplest case, this is just multiplication by a constant, but more generally the vertex rule can involve the momenta associated with the lines going in or out of the vertex.
- *Internal lines:* the Feynman rules translate every internal line into a "propagator" that depends on the momentum associated with the internal line. (A propagator is the Fourier transform of a Green's function.) The mathematical expression depends on the type of internal line, i.e. what kind of virtual particle is being exchanged.

One converts a Feynman diagram to a mathematical expression by tracing through the diagram and picking up factors, in successive order of appearance, from the rules for external and internal lines and vertices. The Feynman rules then state that $i A_n^{L\text{-loop}}$ is the sum of all possible L-loop diagrams with n external lines specified by the external data.

▷ *Example.* Consider the simplest case of a massless spin-0 scalar particle. The Feynman rules for external scalar lines are simply a factor of 1 while an internal line with momentum label P contributes $-i/P^2$; this is the massless scalar propagator. Let us assume that our scalars only interact via cubic vertices for which the Feynman rule is simply to multiply a factor of ig, where g is the coupling. A process involving four scalars then has the tree-level contribution

$$A_4^{\text{tree}}(1,2,3,4) =$$

$$= g^2 \left(\frac{1}{(p_1 + p_2)^2} + \frac{1}{(p_1 + p_3)^2} + \frac{1}{(p_1 + p_4)^2} \right). \qquad (1.10)$$

Here $(p_1 + p_2)^2 = -(p_1^0 + p_2^0)^2 + |\vec{p}_1 + \vec{p}_2|^2$ etc. ◁

Note that the amplitude (1.10) is symmetric under exchange of identical bosonic external states. This is a manifestation of Bose symmetry. Similarly, Fermi statistics requires that an amplitude must be antisymmetric under exchange of any two identical external fermions.

▷ *Example.* A 1-loop diagram in our simple scalar model takes the form

$$= g^4 \int \frac{d^4\ell}{(2\pi)^4} \frac{1}{\ell^2 (\ell - p_1)^2 (\ell - p_1 - p_2)^2 (\ell + p_4)^2}. \qquad (1.11)$$

We have used momentum conservation $p_1 + p_2 + p_3 + p_4 = 0$ to simplify the fourth propagator. The 1-loop 4-scalar amplitude in this model is obtained from the sum over inequivalent box diagrams (1.11). ◁

Loop-integrals can be divergent both in the large- and small-momentum regimes. Such "ultraviolet" and "infrared divergences" are well-understood and typically treated using a scheme such as dimensional regularization in which one replaces the measure $d^4\ell$ in the loop-integral by $d^D\ell$ with $D = 4 - 2\epsilon$. The divergences can then be cleanly extracted in the expansion of small ϵ. The treatment of ultraviolet divergences is called "renormalization"; it is a well-established method and the resulting predictions for scattering processes have been tested to high precision in particle physics experiments.

Example in QED

As an example of a tree-level process in QED, consider the annihilation of an electron–positron pair to two photons:

$$e^- + e^+ \rightarrow \gamma + \gamma \qquad\qquad\qquad$$

$$(1.12)$$

The leading-order contribution to the amplitude is a sum of two tree-level Feynman diagrams constructed from the basic QED vertex (1.8). It is

$$i A_4^{\text{tree}}\big(e^- + e^+ \rightarrow \gamma + \gamma\big) = \qquad\qquad + \qquad\qquad . \qquad (1.13)$$

The two diagrams (1.13) are translated via the QED Feynman rules to Lorentz-invariant contractions of 4-momentum vectors, photon polarizations, and fermion wavefunctions. In the standard formulation, the expression for the amplitude is not particularly illuminating. However, squaring it and subjecting it to a significant dose of index massage therapy, one finds that the scattering cross-section takes a rather simple form. In the high-energy limit, where the center of mass energy dominates the electron/positron rest mass, $E_{\text{CM}} \gg m_e c^2$, the result for the differential cross-section is

$$\frac{d\sigma}{d\Omega} = \frac{1}{64\pi^2 E_{\text{CM}}^2} \sum |A_4|^2 \xrightarrow{E_{\text{CM}} \gg m_e c^2} \alpha^2 \frac{1}{E_{\text{CM}}^2} \frac{1 + \cos^2\theta}{1 - \cos^2\theta} + O(\alpha^4). \qquad (1.14)$$

The sum indicates a spin-sum average. The cross-section (1.14) depends on the center of mass energy E_{CM}, the scattering angle θ indicated in (1.12), and the fine structure constant α. (In natural units, $\alpha = e^2/4\pi$.) The result (1.14) serves to illustrate the salient properties of the differential cross-section: the dependence on particle energies and scattering angles as well as powers of a small fundamental dimensionless coupling constant α. In particular, the expression in (1.14) is the leading-order contribution to the scattering process, and higher orders, starting at 1-loop level, are indicated by $O(\alpha^4)$.

Since the focus in this book is on scattering amplitudes, and not the cross-sections, it is worthwhile to present an expression for the amplitude of the annihilation process $e^- + e^+ \rightarrow \gamma + \gamma$. It turns out to be particularly simple in the high-energy regime $E_{\text{CM}} \gg m_e c^2$, where the masses can be neglected and the momentum vectors of the incoming electron and positron can be chosen to be light-like (i.e. null, $p_i^2 = 0$). In that case, the sum of the two diagrams (1.13) reduces to a single term that can be written compactly as

$$A_4^{\text{tree}}\big(e^- + e^+ \rightarrow \gamma + \gamma\big) = 2e^2 \frac{\langle 24 \rangle^2}{\langle 13 \rangle \langle 23 \rangle}. \qquad (1.15)$$

Here, the angle brackets $\langle ij \rangle$ are closely related to the particle 4-momenta via $|\langle ij \rangle|^2 \sim 2 p_i . p_j$, where the Lorentz-invariant dot-product is $p_i . p_j \equiv p_i^\mu p_{j\mu} = -p_i^0 p_j^0 + \vec{p}_i . \vec{p}_j$. The

4-momenta p_1 and p_2 are associated with the incoming electron and positron while p_3 and p_4 are the 4-momenta of the photons. The angle bracket notation $\langle ij \rangle$ is part of the **spinor helicity formalism** which is a powerful technical tool for describing scattering of massless particles in four spacetime dimensions. We will introduce the spinor helicity formalism in Chapter 2 and derive the expression (1.15) in full detail from the Feynman rules of QED. You may be surprised to note that the expression (1.15) is not symmetric in the momenta of the two photons. This is because the representation (1.15) selects distinct polarizations for the photons.

Using momentum conservation and the on-shell condition $p_i^2 = 0$, one can show that the norm-squared of the expression (1.15) is proportional to $(p_1.p_3)/(p_1.p_4)$. The sum in the cross-section (1.14) indicates a sum of the polarizations of the final state photons and an average over the spins of the initial state electron and positron. This procedure yields a second term with $p_3 \leftrightarrow p_4$. Hence, what goes into the formula (1.14) is

$$\sum \left| A_4^{\text{tree}}\left(e^- + e^+ \rightarrow \gamma + \gamma\right)\right|^2 = 2e^4 \left(\frac{p_1.p_3}{p_1.p_4} + \frac{p_1.p_4}{p_1.p_3}\right). \tag{1.16}$$

An explicit representation of the 4-momenta is $p_1^\mu = (E, 0, 0, E)$ and $p_2^\mu = (E, 0, 0, -E)$ for the initial states and $p_3^\mu = (E, 0, E\sin\theta, E\cos\theta)$ and $p_4^\mu = (E, 0, -E\sin\theta, -E\cos\theta)$ for the final state photons. Momentum conservation is simply $p_1 + p_2 = p_3 + p_4$. Using that the center of mass energy is $E_{\text{CM}}^2 = -(p_1 + p_2)^2 = 4E^2$, one finds that the high-energy result for the differential cross-section in (1.14) follows from (1.16).

It is typical that amplitudes involving only massless particles are remarkably simpler than those with massive particles. This can be regarded as a high-energy regime, as in our example above. Because of the simplicity, yet remarkably rich mathematical structure, the focus of many developments in the field has been on scattering amplitudes of massless particles.

The S-matrix

We have introduced the scattering amplitude as a function that takes as its input the constrained external data $\{p_i^\mu, \text{type}_i\}$ and produces a complex number that is traditionally calculated in terms of Feynman diagrams. The scattering process can also be considered as an operation that maps an initial state $|i\rangle$ to a final state $|f\rangle$, each being a collection of single-particle states characterized by momenta and particle types. The scattering matrix S, also called the **S-matrix**, is the unitary operator that "maps" the initial states to final states. In other words, the probability of an initial state $|i\rangle$ changing into a final state $|f\rangle$ is given by $|\langle f|S|i\rangle|^2$. Separating out the trivial part of the scattering process where no scattering occurs, we write

$$S = 1 + iT. \tag{1.17}$$

Then the amplitude is simply $A = \langle f|T|i\rangle$ and this is the quantity that is calculated by Feynman diagrams. Solving the S-matrix for a given theory means having a way to generate all scattering amplitudes at any order in perturbation theory.

Beyond Feynman diagrams

Armed with Feynman rules, we can, in principle, compute scattering amplitudes to our hearts' content. For instance, starting from the QED Lagrangian one can calculate the tree-level differential cross-section for the processes (1.1). It is typical for such a calculation that the starting point – the Lagrangian in its most compact form – is not too terribly complicated. And the final result for $d\sigma/d\Omega$ can be rather compact and simple too, as we have seen in (1.14). But the intermediate stages of the calculation often explode in an inferno of indices, contracted up-and-down and in all directions – providing little insight into the physics and hardly any hint of simplicity.

In a QFT course, students are exposed (hopefully!) to a lot of long character-building calculations, including the 4-particle QED processes in (1.1). But students are rarely asked to use standard Feynman rules to calculate processes that involve more than four or five particles, even at tree-level: for example, $e^- + e^+ \rightarrow e^- + e^+ + \gamma$ or $e^- + e^+ \rightarrow e^- + e^+ + \gamma + \gamma$. Why not? Well, one reason is that the number of Feynman diagrams tends to grow very quickly with the number of particles involved: for gluon scattering at *tree-level* in quantum chromodynamics (QCD) we have

$$
\begin{array}{lll}
g + g \rightarrow g + g & \text{4 diagrams} & \\
g + g \rightarrow g + g + g & \text{25 diagrams} & (1.18) \\
g + g \rightarrow g + g + g + g & \text{220 diagrams} &
\end{array}
$$

and for $g + g \rightarrow 8g$ one needs more than one million diagrams [1]. Another very important point is that the mathematical expression for each diagram becomes significantly more complicated as the number of external particles grows. So the reason students are not asked to calculate multi-gluon processes from Feynman diagrams is that it would be awful, un-insightful, and in many cases impossible.[2] There are tricks for simplifying the calculations; one is to use the spinor helicity formalism as indicated in our example for the high-energy limit of the tree-level process $e^- + e^+ \rightarrow \gamma + \gamma$. However, for a multi-gluon process even this does not directly provide a way to handle the growing number of increasingly complicated Feynman diagrams. Other methods are needed and you will learn much more about them in this book.

Thus, although visually appealing and seemingly intuitive, Feynman diagrams are not a particularly physical approach to studying amplitudes in gauge theories. The underlying reason is that the Feynman rules are non-unique: generally, individual Feynman diagrams are not physical observables. In theories, such as QED or QCD, that have gauge redundancy in their Lagrangian descriptions, we are required to fix the gauge in order to extract the Feynman rules from the Lagrangian. So the Feynman rules depend on the choice of gauge and therefore only gauge-invariant sums of diagrams are physically sensible. The on-shell amplitude is at each loop-order such a gauge-invariant sum of Feynman diagrams.

---

[2] Using computers to do the calculation can of course be very helpful, but not in all cases. Sometimes numerical evaluation of Feynman diagrams is simply so slow that it is not realistic to do. Moreover, given that there are poles that can cancel between diagrams, big numerical errors can arise in such evaluations. Therefore compact analytic expressions for the amplitudes are very useful in practical applications.

Furthermore, field redefinitions in the Lagrangian change the Feynman rules, but not the physics, hence the amplitudes are invariant. So field redefinitions and gauge choices can "move" the physical information between the diagrams in the Feynman expression for the amplitude and this can hugely obscure the appearance of its physical properties.

It turns out that despite the complications of the Feynman diagrams, the on-shell amplitudes for processes such as multi-gluon scattering $g + g \to g + g + \cdots + g$ can actually be written as remarkably simple expressions. This raises the questions: "why are the on-shell amplitudes so simple?" and "isn't there a better way to calculate amplitudes?". These are questions that have been explored in recent years and a lot of progress has been made towards improving calculational techniques and gaining insight into the underlying mathematical structure.

What do we mean by "mathematical structure" of amplitudes? At the simplest level, this means the analytic structure. For example, the amplitude (1.10) has a pole at $(p_1 + p_2)^2 = 0$ – this pole shows that a physical massless scalar particle is being exchanged in the process. Tree amplitudes are rational functions of the kinematic invariants, and understanding their pole structure is key to the derivation of on-shell recursive methods that provide a very efficient alternative to Feynman diagrams. These recursive methods allow one to construct on-shell n-particle tree amplitudes from input of on-shell amplitudes with fewer particles. The power of this approach is that gauge redundancy is eliminated since all input is manifestly on-shell and gauge invariant.

Loop amplitudes are integrals of rational functions and therefore have more complicated analytic structure, involving for example logarithmic branch cuts. The branch cuts can be exploited to reconstruct the loop amplitude from lower-loop and tree input, providing another class of efficient calculational techniques. On-shell approaches that exploit the analytic structure of amplitudes constitute a major theme in this book.

Thus, it is very useful to understand the analytic structure of amplitudes, but it is not the only mathematical structure of interest. We are also interested in the symmetries that leave the amplitudes invariant and how they constrain the result for the scattering process. Some amplitudes are invariant under symmetries that are not apparent in the Lagrangian and hence are not visible in the Feynman rules. Pursuing these symmetries and seeking to write the amplitudes in terms of variables that trivialize the constraints on the external data and simplify the action of the symmetries will often lead to new insights about the amplitudes; we will see this repeatedly as we develop the subject.

Finally, let us mention one of the highlights of what we mean by mathematical structure. It turns out that certain amplitudes naturally "live" in more abstract spaces in which they can be interpreted as volumes of geometric objects known as "polytopes"; these are higher-dimensional generalizations of polygons. The different compact representations found for a given amplitude turn out to correspond to different triangulations of the corresponding polytope. It is remarkable that a physical observable that encodes the probability for a scattering process as a function of energies and momenta has an alternative mathematical interpretation as the volume of a geometric object in a higher-dimensional abstract space! It is such surprising mathematical structures that are pursued in the third part of this book.

Some of the keywords for the topics we explore are:

1. spinor helicity formalism;
2. on-shell recursion relations (BCFW, CSW, all-line shifts, . . .);
3. on-shell superspace, superamplitudes, Ward identities;
4. twistors, zone-variables, momentum twistors;
5. dual superconformal symmetry and the Yangian;
6. generalized unitarity, maximal cuts;

7. Leading Singularities 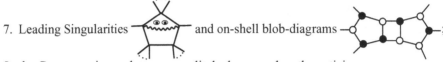 and on-shell blob-diagrams   ;

8. the Grassmannian, polytopes, amplituhedrons, and mathematicians;
9. gravity $=$ (gauge theory)^2, KLT and BCJ relations;

and much more.

The study of on-shell amplitudes may suggest a paradigm that can be phrased loosely as "avoiding the (full) Lagrangian" with all its ambiguities of field redefinitions and gauge choices, and instead focusing on how kinematics, symmetries, unitarity, and locality impact the physical observables. Or, more strongly, we may ask if the hints from the simplicity of amplitudes allow us to find another approach to perturbative quantum field theory: one might hope for a novel formulation that captures the physics of the full perturbative S-matrix. Such a new formulation could make amplitude calculations much more efficient and one might hope that it would lead to new insights even beyond amplitudes, for example for correlation functions of gauge-invariant operators and perhaps even for non-perturbative physics.

But now we are getting ahead of ourselves. The purpose of this book is to provide a practical introduction to on-shell methods for scattering amplitudes of massless particles. The choice of massless particles is because their processes are the simplest and currently the best understood. The above list of keywords will be explained and discussed in detail.

This book

Our presentation assumes a basic pre-knowledge of quantum field theory, in particular of the Dirac equation, simple scalar and QED Feynman rules and tree-level processes. Chapter 2 introduces the spinor helicity formalism in the context of tree-level Feynman rules and builds up the notation and conventions via explicit examples. Yang–Mills theory is introduced as are the needed tools for studying gluon amplitudes. Various fundamental properties of scattering amplitudes are discussed, including little group scaling and the analytic structure of tree amplitudes.

In Chapter 3 we take the first step towards modern on-shell methods for calculating scattering amplitudes, namely recursion relations for tree amplitudes. The material in Chapters 2 and 3 could be part of any modern course on quantum field theory, but traditionally it is not. So we hope that our presentation will be useful, either as part of a course or for self-study.

Chapter 4 introduces supersymmetry and superamplitudes. In Chapter 5 we explore the symmetries of superamplitudes in the most intensely studied theory in the field of amplitude-explorations, namely $\mathcal{N} = 4$ super Yang–Mills theory. We do not assume prior knowledge of supersymmetry or $\mathcal{N} = 4$ super Yang–Mills theory, but will introduce the concepts and tools as needed in Chapters 4 and 5.

Part I of the book (Chapters 2–5) focuses exclusively on tree-level processes.

Part II (Chapters 6–8) offers an upgrade to loops. Chapter 6 is an introduction to loop amplitudes and to the unitarity method, Chapter 7 discusses recursion relations for loop-integrands, and on-shell diagrams are introduced in Chapter 8.

Part III (Chapters 9–14) is devoted to a selection of topics. You may view Part III as an introduction to current research-topics on amplitudes. The subjects covered include the interesting connections between amplitudes and the Grassmannian (Chapter 9), which is the space of k-dimensional planes in n-dimensional space, and the enticing interpretation of amplitudes as volumes of polytopes (Chapter 10). While most of the book is concerned with scattering processes in four spacetime dimensions, Chapter 11 takes us on a tour of amplitudes of higher- and lower-dimensional theories. Chapter 12 explores amplitudes in perturbative gravity and supergravity while we end in Chapter 13 with the remarkable connection between scattering amplitudes of gravitons and those of gluons. Chapter 14 contains a brief outline of subjects not otherwise covered in this book and a list of suggestions for further reading.

One should keep in mind that the subject of scattering amplitudes has two main motivations. One is its practical application in particle physics: some of the on-shell methods we cover here are indeed already implemented in numerical codes for processes relevant in particle physics experiments. The other motivation is the fascinating internal mathematical beauty of the subject. The physical relevance and mathematical structure are both important, neither should be underestimated. They complement and benefit each other.

The style of presentation in this book is detailed and concrete, so that you can learn the tools. The purpose is to be pedagogical – but in this as well as other matters, there is no substitute for getting your own hands dirty. Therefore you will find many exercises scattered throughout the text. We hope you will enjoy them.

Conventions

The subject of amplitudes is often viewed as quite technical and notationally intense. We will try to avoid a long deadly-boring introduction about γ-matrix conventions and about which indices go up and down and who is dotted and who is not. Suffice it here to say that we work in four dimensions (except in Chapters 11–13), our metric convention is mostly-plus $\eta_{\mu\nu} = \text{diag}(-1, +1, +1, +1)$, and we follow the spinor and Clifford algebra conventions in Srednicki's quantum field theory textbook [2]. For easy access, and to make our presentation reasonably self-contained, some conventions are collected in the short Appendix.

PART I

TREES

Spinor helicity formalism

The spinor helicity formalism is a highly convenient and powerful notational tool for amplitudes of massless particles in 4-dimensional theories. We are going to introduce the spinor helicity formalism in the context of the basic Feynman rules that you are familiar with from Yukawa theory and QED. So we start with Dirac spinors and build up the formalism based on simple scattering problems.

2.1 Dirac spinors

The Lagrangian for a free massive 4-component Dirac field Ψ is

$$\mathcal{L} = i\overline{\Psi}\gamma^\mu \partial_\mu \Psi - m\overline{\Psi}\Psi. \tag{2.1}$$

Our conventions for the Dirac conjugate $\overline{\Psi}$ and the γ^μs are given in the Appendix. The equation of motion for $\overline{\Psi}$ gives the Dirac equation

$$(-i\slashed{\partial} + m)\Psi = 0, \tag{2.2}$$

with $\slashed{\partial} = \partial_\mu \gamma^\mu$. Multiplying the Dirac equation by $(i\slashed{\partial} + m)$ gives the Klein-Gordon equation, $(-\partial^2 + m^2)\Psi = 0$. It is solved by a plane-wave expansion

$$\Psi(x) \sim u(p)\, e^{ip.x} + v(p)\, e^{-ip.x}, \tag{2.3}$$

provided the momentum is on-shell, $p^2 \equiv p^\mu p_\mu = -m^2$. This $\Psi(x)$ will also solve the Dirac equation (2.2) if

$$(\slashed{p} + m)u(p) = 0 \quad \text{and} \quad (-\slashed{p} + m)v(p) = 0, \tag{2.4}$$

where $\slashed{p} = p_\mu \gamma^\mu$. The equations (2.4) are the momentum space form of the Dirac equation. Each of the equations in (2.4) has two independent solutions which we will label by a subscript $s = \pm$. We can now write the general free field expansion of Ψ as

$$\Psi(x) = \sum_{s=\pm} \int \widetilde{dp} \left[b_s(p)\, u_s(p)\, e^{ip.x} + d_s^\dagger(p)\, v_s(p)\, e^{-ip.x} \right], \tag{2.5}$$

where $\widetilde{dp} = \frac{d^3p}{(2\pi)^3 2E_p}$ is the 3d Lorentz-invariant momentum measure. For $\overline{\Psi}$ one finds a similar result involving $d_\pm(p)$ and $b_\pm^\dagger(p)$.

In canonical quantization, $(b_\pm^\dagger(p), d_\pm^\dagger(p))$ and $(b_\pm(p), d_\pm(p))$ become fermionic creation and annihilation operators. They take care of the Grassmann nature of $\Psi(x)$, so that $u_\pm(p)$ and $v_\pm(p)$ are *commuting* 4-component spinors that solve (2.4).

The next step in canonical quantization is to define the vacuum $|0\rangle$ such that $b_\pm(p)|0\rangle = d_\pm(p)|0\rangle = 0$. One-particle states are then defined as $|p; \pm\rangle \equiv d_\pm^\dagger(p)|0\rangle$, etc. As described in QFT textbooks, this leads to the Feynman rules for external fermions, namely that they come equipped with wavefunctions $v_\pm(p)$ for an outgoing anti-fermion (e.g. e^+) and $\overline{u}_\pm(p)$ for an outgoing fermion (e.g. e^-). The latter comes from the expansion of $\overline{\Psi}$. We can choose a basis such that in the rest-frame $u_\pm$ and $v_\pm$ are eigenstates of the z-component of the spin-matrix; then $\pm$ denotes spin up/down along the z-axis. For massless fermions, $\pm$ denotes the **helicity**, which is the projection of the spin along the momentum of the particle. It will be our interest here to study the wavefunctions $\overline{u}_\pm(p)$ and $v_\pm(p)$ further.

The wavefunction $v_\pm(p)$ solves the Dirac equation (2.4) and $\overline{u}_\pm(p)$ satisfies $\overline{u}_\pm(p)(\not{p} + m) = 0$. Starting with a momentum 4-vector $p^\mu = (p^0, p^i) = (E, p^i)$ with $p^\mu p_\mu = -m^2$, let us use the gamma-matrix conventions (A.8) in the Appendix to write

$$\not{p} = \begin{pmatrix} 0 & p_{a\dot{b}} \\ p^{\dot{a}b} & 0 \end{pmatrix}, \tag{2.6}$$

with

$$p_{a\dot{b}} \equiv p_\mu (\sigma^\mu)_{a\dot{b}} = \begin{pmatrix} -p^0 + p^3 & p^1 - ip^2 \\ p^1 + ip^2 & -p^0 - p^3 \end{pmatrix}, \tag{2.7}$$

and similarly $p^{\dot{a}b} \equiv p_\mu (\bar{\sigma}^\mu)^{\dot{a}b}$. We have $\sigma^\mu = (1, \sigma^i)$ and $\bar{\sigma}^\mu = (1, -\sigma^i)$ with $\sigma^{1,2,3}$ the Pauli-matrices (A.2). The momentum bi-spinors $p_{a\dot{b}}$ and $p^{\dot{a}b}$ can be thought of as 2×2 matrices. The determinant is Lorentz-invariant,

$$\det p = -p^\mu p_\mu = m^2. \tag{2.8}$$

In most of this book, we study scattering processes for massless particles. You can think of this as the high-energy scattering limit in which the fermion mass is neglected. So let us now specialize to the case of massless spinors.

2.2 Spinor helicity notation

When $m = 0$, the Dirac equation for the wavefunction 4-component spinors reads

$$\not{p} \, v_\pm(p) = 0, \qquad \overline{u}_\pm(p) \, \not{p} = 0. \tag{2.9}$$

We focus on $v_\pm(p)$ and $\overline{u}_\pm(p)$ as the wavefunctions associated with **outgoing** anti-fermions and fermions. As mentioned above, in the massless case, we can choose a basis such that the subscript $\pm$ indicates the helicity $h = \pm 1/2$. Crossing symmetry exchanges (incoming $\leftrightarrow$ outgoing), (fermions $\leftrightarrow$ anti-fermions) and flips the sign of the helicity, so in the massless case the wavefunctions are related as $u_\pm = v_\mp$ and $\overline{v}_\pm = \overline{u}_\mp$.

We write the two independent solutions to the Dirac equation (2.9) as

$$v_+(p) = \begin{pmatrix} |p]_a \\ 0 \end{pmatrix}, \qquad v_-(p) = \begin{pmatrix} 0 \\ |p\rangle^{\dot{a}} \end{pmatrix}, \tag{2.10}$$

and

$$\bar{u}_-(p) = \left(0 , \langle p|_{\dot{a}} \right), \qquad \bar{u}_+(p) = \left([p|^a , 0 \right). \tag{2.11}$$

The angle and square spinors are 2-component commuting spinors (think 2-component vectors) written in a very convenient Dirac bra-ket notation. By virtue of (2.6) and (2.9), they satisfy the massless Weyl equation,

$$p^{\dot{a}b}|p]_b = 0, \qquad p_{a\dot{b}}|p\rangle^{\dot{b}} = 0, \qquad [p|^b \, p_{b\dot{a}} = 0, \qquad \langle p|_{\dot{b}} \, p^{\dot{b}a} = 0. \tag{2.12}$$

Raising and lowering their indices is done with the 2-index Levi-Civita symbols defined in (A.3):

$$[p|^a = \epsilon^{ab}|p]_b, \qquad |p\rangle^{\dot{a}} = \epsilon^{\dot{a}\dot{b}}\langle p|_{\dot{b}}. \tag{2.13}$$

Now,

the angle and square spinors are the core of what we call the

"spinor helicity formalism."

As you see, these bra-kets are nothing to be scared of: they are simply 2-component commuting spinors that solve the massless Weyl equation.

It is one of the powers of the spinor helicity formalism that we do not need to find explicit representations for the angle and square-spinors; we can simply work abstractly with $|p\rangle$ and $|p]$ and later relate the results to the momentum vectors. We will see examples of how this works in this chapter. Let us now note two important properties of the spinor bras and kets.

Angle vs. square spinors: reality conditions. The spinor field $\overline{\Psi}$ is the Dirac conjugate of Ψ. Applying Dirac conjugation to the momentum space Dirac equations (2.9), we find that $\bar{u}_\mp = \bar{v}_\pm$ is related to $v_\pm$ via this conjugation *provided the momentum p^μ is real-valued,* i.e. the components of p^μ are real numbers. Thus for *real* momenta

$$p^\mu \text{ real}: \qquad [p|^a = (|p\rangle^{\dot{a}})^* \quad \text{and} \quad \langle p|_{\dot{a}} = (|p]_a)^*. \tag{2.14}$$

On the contrary, for *complex-valued momenta* p^μ, the angle and square spinors are independent.[1] It may not seem physical to take p^μ complex, but it is a very very very useful strategy. We will see this repeatedly.

---

[1] One can keep p^μ real and change the spacetime signature to $(-, +, -, +)$; in that case, the angle and square spinors are real and independent.

Spinor completeness relation. The spin-sum completeness relation with $m = 0$ reads $u_- \bar{u}_- + u_+ \bar{u}_+ = -\not{p}$. (See for example Equation (38.23) of the QFT textbook [2].) With the help of crossing symmetry $\bar{u}_\mp = \bar{v}_\pm$, this can be written in spinor helicity notation as

$$-\not{p} = |p\rangle[p| + |p]\langle p|. \tag{2.15}$$

There is a small abuse of notation in writing (2.15): the LHS is a 4×4 matrix and the RHS involves products of 2-component spinors. The relation should be read in terms of matching the appropriate L- and R-spinor indices via (2.6), namely:

$$p_{a\dot{b}} = -|p]_a \langle p|_{\dot{b}}, \qquad p^{\dot{a}b} = -|p\rangle^{\dot{a}} [p|^b. \tag{2.16}$$

The relations (2.16) may look new but they should not shock you. After all, it is taught in *some* algebra classes that if a 2×2 matrix has vanishing determinant, it can be written as a product of two 2-component vectors, say λ_a and $\tilde{\lambda}_{\dot{b}}$: i.e. $\det p = 0 \iff p_{a\dot{b}} = -\lambda_a \tilde{\lambda}_{\dot{b}}$. In fact, this is often the starting point of introductions to the spinor helicity formalism. In this presentation, we will suppress the λ_a and $\tilde{\lambda}_{\dot{b}}$ notation in favor of the more intuitive Dirac bra-kets, $\lambda_a \to |p]_a$ and $\tilde{\lambda}_{\dot{a}} \to \langle p|_{\dot{a}}$.

It is useful for keeping your feet on the ground to work out an explicit solution for $|p\rangle$ and $|p]$ for a given 4-momentum p^μ. The following exercise guides you to do just that.

▶ **Exercise 2.1**

Consider the momentum vector

$$p^\mu = (E, E \sin\theta \cos\phi, E \sin\theta \sin\phi, E \cos\theta). \tag{2.17}$$

Express $p_{a\dot{b}}$ and $p^{\dot{a}b}$ in terms of E, $\sin\frac{\theta}{2}$, $\cos\frac{\theta}{2}$ and $e^{\pm i\phi}$.

Show that the helicity spinor $|p\rangle^{\dot{a}} = \sqrt{2E} \begin{pmatrix} \cos\frac{\theta}{2} \\ \sin\frac{\theta}{2} e^{i\phi} \end{pmatrix}$ solves the massless Weyl equation. Find expressions for the spinors $\langle p|_{\dot{a}}$, $|p]_a$, and $[p|^a$ and check that they satisfy $p_{a\dot{b}} = -|p]_a\langle p|_{\dot{b}}$ and $p^{\dot{a}b} = -|p\rangle^{\dot{a}}[p|^b$.

You have probably noted that the angle and square spinors are only defined up to an overall scaling that leaves p^μ invariant. This is called the *little group scaling* and it plays a central role which we explore much more in Section 2.6.

We are now in dire need of some examples! Before we move ahead, it is convenient to summarize the external line *Feynman rules for outgoing massless (anti)fermions*:

- Outgoing fermion with $h = +1/2$: $u_+(p) \longleftrightarrow ([p|^a, 0)$;

- Outgoing fermion with $h = -1/2$: $u_-(p) \longleftrightarrow (0, \langle p|_{\dot{a}})$;

- Outgoing anti-fermion with $h = +1/2$: $\qquad v_+(p) \longleftrightarrow \begin{pmatrix} |p]_a \\ 0 \end{pmatrix}$;

- Outgoing anti-fermion with $h = -1/2$: $\qquad v_-(p) \longleftrightarrow \begin{pmatrix} 0 \\ |p\rangle^{\dot{a}} \end{pmatrix}$.

Note the useful mnemonic that *positive helicity* of an outgoing particle is associated with *square spinors* while *negative helicity* comes with *angle-spinors*. Finally, let us note that for massless fermions we usually do not bother much to distinguish fermion-anti-fermion due to the simple crossing rules. In the amplitudes, we consider all the external particles to be outgoing, so think of the rules here as the difference between the arrow on a fermion line pointing into the diagram (anti-fermion) or out of the diagram (fermion).

▶ **Exercise 2.2**

The helicity of a massless particle is the projection of the spin along the momentum 3-vector $\vec{p}$, so the helicity operator can be written $\Sigma = \mathcal{S} \cdot \vec{p}/|\vec{p}|$, where the spin $\mathcal{S}_i = \frac{1}{2}\epsilon_{ijk}S^{jk}$ $(i, j, k = 1, 2, 3)$ is defined by the spin matrix $S^{\mu\nu} = \frac{i}{4}[\gamma^\mu, \gamma^\nu]$. For simplicity, you can pick a frame where p^μ is along the z-axis. Use the results of Exercise 2.1 to show that the chiral basis (2.10), (2.11) is also a helicity basis, i.e. show that $\Sigma v_\pm = -h_\pm v_\pm$ for $h_\pm = \pm\frac{1}{2}$.

2.3 Examples from Yukawa theory

Consider a Dirac fermion interacting with a real scalar ϕ via a Yukawa coupling:

$$\mathcal{L} = i\overline{\Psi}\gamma^\mu\partial_\mu\Psi - \tfrac{1}{2}(\partial\phi)^2 + g\phi\overline{\Psi}\Psi\,. \tag{2.18}$$

The interaction term gives the simple 3-vertex Feynman rule ig. For a diagram with two outgoing Dirac fermions connecting to the rest of the particles in the process via an internal scalar line, the usual Feynman rules give

$= ig\,\overline{u}_{h_1}(p_1)v_{h_2}(p_2) \times \dfrac{-i}{(p_1 + p_2)^2} \times \text{(rest)}, \tag{2.19}$

with the spinor indices contracted and the gray blob representing the rest of the diagram. We focus on the spinor product: choosing specific examples for the helicities we find

$$\overline{u}_+(p_1)\,v_-(p_2) = \big(\lceil p_1|^a\,,\, 0\big) \begin{pmatrix} 0 \\ |p_2\rangle^{\dot{a}} \end{pmatrix} = 0, \tag{2.20}$$

$$\overline{u}_-(p_1)\,v_-(p_2) = \big(0\,,\, \langle p_1|_{\dot{a}}\big) \begin{pmatrix} 0 \\ |p_2\rangle^{\dot{a}} \end{pmatrix} = \langle p_1|_{\dot{a}}|p_2\rangle^{\dot{a}} \equiv \langle p_1 p_2\rangle\,. \tag{2.21}$$

In the first case, the diagram vanishes. In the second case, we introduce the **angle spinor bracket** $\langle p_1 p_2 \rangle$. Together with its best friend, the **square spinor bracket** $[p_1 p_2]$, it is a key ingredient for writing amplitudes in spinor helicity formalism. So let us introduce the spinor brackets properly: for two light-like vectors p^μ and q^μ, we define spinor brackets

$$\langle p\,q \rangle = \langle p|_{\dot{a}} |q\rangle^{\dot{a}}, \qquad [p\,q] = [p|^a |q]_a. \tag{2.22}$$

Since indices are raised/lowered with the antisymmetric Levi-Civita tensors (A.3), cf. (2.13), these brackets are antisymmetric:

$$\langle p\,q \rangle = -\langle q\,p \rangle, \qquad [p\,q] = -[q\,p]. \tag{2.23}$$

There are no $\langle p|q]$-brackets, because the indices cannot be contracted directly to form a Lorentz-scalar.

For real momenta, the spinor products satisfy $[p\,q]^* = \langle q\,p \rangle$.

It is a good exercise (hint: use (2.16) and (A.7)) to derive the following important relation:

$$\langle p\,q \rangle [p\,q] = 2\,p \cdot q = (p+q)^2. \tag{2.24}$$

In amplitudes with light-like momenta $p_1, p_2, \ldots$ we use the short-hand notation $|1\rangle = |p_1\rangle$, $|2\rangle = |p_2\rangle$, etc. Applying (2.24) to our Yukawa example (2.19), we find

$$g\langle 12 \rangle \times \frac{1}{2p_1 . p_2} \times (\text{rest}) = g\langle 12 \rangle \times \frac{1}{\langle 12 \rangle [12]} \times (\text{rest}) = g\frac{1}{[12]} \times (\text{rest}). \tag{2.25}$$

The cancellation of the $\langle 12 \rangle$-factors is the first tiny indication of simplifications that await us in the following.

▷ *Example.* Let us calculate the 4-fermion tree amplitude $A_4(\bar{f}^{h_1} f^{h_2} \bar{f}^{h_3} f^{h_4})$ in Yukawa theory. In our notation, f denotes an outgoing fermion and $\bar{f}$ an outgoing anti-fermion. The superscripts indicate the helicity. When specifying the helicity of each particle, we call the amplitude a ***helicity amplitude***.

The s-channel diagram for the 4-fermion process is

$$= i g\, \bar{u}_4 v_3 \times \frac{-i}{(p_1 + p_2)^2} \times i g\, \bar{u}_2 v_1. \tag{2.26}$$

Here we are using a shorthand notation $v_i = v_{h_i}(p_i)$ and $\bar{u}_i = \bar{u}_{h_i}(p_i)$. Our observations in the previous example show that the diagram (2.26) vanishes unless particles 1 and 2 have the same helicity, and 3 and 4 have the same helicity. So suppose we take particles 1 and 2 to have negative helicity and 3 and 4 positive. Then the u-channel diagram vanishes and the diagram (2.26) is the only contribution to the 4-fermion amplitude.

Translating the $\bar{u}v$-products to spinor brackets we find

$$i A_4\big(\bar{f}^- f^- \bar{f}^+ f^+\big) = ig^2[43]\frac{1}{2p_1.p_2}\langle 21\rangle = ig^2[34]\frac{1}{\langle 12\rangle[12]}\langle 12\rangle = ig^2\frac{[34]}{[12]}.$$

(2.27)

The result is a nice simple ratio of two spinor brackets. Now, it is both fun and useful to note that by momentum conservation, we have (using (2.24))

$$\langle 12\rangle[12] = 2p_1.p_2 = (p_1 + p_2)^2 = (p_3 + p_4)^2 = 2p_3.p_4 = \langle 34\rangle[34].$$

(2.28)

Using this in the second equality of (2.27) we get another expression for the same amplitude:

$$A_4\big(\bar{f}^- f^- \bar{f}^+ f^+\big) = g^2\frac{\langle 12\rangle}{\langle 34\rangle}.$$

(2.29)

This hints at another important lesson: there are various relationships among spinor brackets, implied for example by momentum conservation as in (2.28), and they allow for multiple equivalent forms of the same physical amplitude. ◁

▷ *Example.* Next, let us see what new features appear when we calculate the 4-point tree amplitude with two scalars and two fermions in Yukawa theory. Two diagrams contribute to the process:

$$i A_4\big(\phi\, \bar{f}^{h_2} f^{h_3}\phi\big) =$$

$$= (ig)^2\, \bar{u}_3\, \frac{-i(\not{p}_1 + \not{p}_2)}{(p_1 + p_2)^2}\, v_2\; + \qquad (1 \leftrightarrow 4).$$

(2.30)

If the fermions have the same helicity, each diagram has a numerator factor $\bar{u}_\pm(p_3)\gamma^\mu v_\pm(p_2) = 0$. Hence

$$A_4\big(\phi\, \bar{f}^- f^-\phi\big) = A_4\big(\phi\, \bar{f}^+ f^+\phi\big) = 0.$$

(2.31)

Thus, the fermions need to have opposite helicity to give a non-vanishing result: for example

$$\bar{u}_-(p_3)\gamma^\mu v_+(p_2) = \big(0,\, \langle 3|_{\dot{a}}\big)\begin{pmatrix} 0 & (\sigma^\mu)_{ab} \\ (\bar{\sigma}^\mu)^{\dot{a}b} & 0 \end{pmatrix}\begin{pmatrix} |2]_b \\ 0 \end{pmatrix} \equiv \langle 3|\gamma^\mu|2].$$

(2.32)

Note the abuse of notation in the definition above of the **angle-square bracket** $\langle p|\gamma^\mu|k]$: it combines the 2-component spinors with the 4×4 gamma-matrix. The meaning should be clear, though, in that the 2-component spinors project out the matching sigma-matrix from γ^μ. The spinor bracket $[p|\gamma^\mu|k\rangle$ is defined similarly. For same-helicity fermions we have $\langle p|\gamma^\mu|k\rangle = 0 = [p|\gamma^\mu|k]$. ◁

Angle-square brackets appear often, so it is useful to record the following properties:

$$[k|\gamma^\mu|p\rangle = \langle p|\gamma^\mu|k], \tag{2.33}$$

$$[k|\gamma^\mu|p\rangle^* = [p|\gamma^\mu|k\rangle \quad \text{(for real momenta)}. \tag{2.34}$$

We often use $\langle p|P|k] \equiv P_\mu \langle p|\gamma^\mu|k]$. The notation implies that p^μ and k^μ are light-like, but no assumptions are made about P^μ. However, if P^μ is also light-like, then

$$\langle p|P|k] = \langle p|_{\dot{a}} \, P^{\dot{a}b} \, |k]_b = \langle p|_{\dot{a}} \, (-|P\rangle^{\dot{a}} [P|^b) \, |k]_b = -\langle pP\rangle[Pk], \quad (P^2 = 0). \tag{2.35}$$

Finally, the Fierz identity

$$\langle 1|\gamma^\mu|2]\langle 3|\gamma_\mu|4] = 2\langle 13\rangle[24] \tag{2.36}$$

will come in handy in several applications.

▶ Exercise 2.3

Prove the Fierz identity (2.36).

▶ Exercise 2.4

Show that $\langle k|\gamma^\mu|k] = 2k^\mu$ and $\langle k|P|k] = 2\, P \cdot k$.

It is often useful to use a shorthand for the momenta "sandwiched" in the angle-square brackets, for example $\langle 1|3|4] = \langle 1|p_3|4]$ or $\langle 2|1+4|5] = \langle 2|(p_1 + p_4)|5]$.

With our new tools, we return now to the tree amplitude with two scalars and two fermions.

▷ *Example.* Choosing opposite helicities for the fermions in (2.30), we have

$$\begin{aligned}
A_4\big(\phi \, \bar{f}^+ f^- \phi\big) &= -g^2 \frac{\langle 3|1+2|2]}{(p_1 + p_2)^2} + (1 \leftrightarrow 4) \\
&= -g^2 \frac{\langle 3|1|2]}{(p_1 + p_2)^2} + (1 \leftrightarrow 4) \quad \text{(using the Weyl Eq. } p_2|2] = 0) \\
&= -g^2 \frac{-\langle 31\rangle[12]}{\langle 12\rangle[12]} + (1 \leftrightarrow 4) \quad \text{(using (2.35))} \\
&= -g^2 \frac{\langle 13\rangle}{\langle 12\rangle} + (1 \leftrightarrow 4),
\end{aligned} \tag{2.37}$$

so that the result is

$$A_4\big(\phi \, \bar{f}^+ f^- \phi\big) = -g^2 \left(\frac{\langle 13\rangle}{\langle 12\rangle} + \frac{\langle 34\rangle}{\langle 24\rangle} \right). \tag{2.38}$$

Note Bose symmetry under exchange of the scalar particle momenta. ◁

In amplitude calculations, **_momentum conservation_** is imposed on n particles as $\sum_{i=1}^{n} p_i^{\mu} = 0$ (consider all particles outgoing). This is encoded in the spinor helicity formalism as

$$\sum_{i=1}^{n} |i\rangle[i| = 0, \quad \text{i.e.} \quad \sum_{i=1}^{n} \langle qi\rangle[ik] = 0, \tag{2.39}$$

for any light-like vectors q and k. For example, you can (and should) show that for $n = 4$ momentum conservation implies $\langle 12\rangle[23] = -\langle 14\rangle[43]$. In (2.28), we already found the identity $\langle 12\rangle[12] = \langle 34\rangle[34]$ valid when $p_1 + p_2 + p_3 + p_4 = 0$.

With all momenta outgoing, the **_Mandelstam variables_** are defined as

$$s_{ij} = -(p_i + p_j)^2, \quad s_{ijk} = -(p_i + p_j + p_k)^2, \quad \text{etc.} \tag{2.40}$$

In particular, we have $s = s_{12}$, $t = s_{13}$, and $u = s_{14}$ for 4-particle processes.

To see some of the power of the spinor helicity formalism, let us now calculate the spin sum

$$\left\langle \left| A_4(\phi \, \bar{f} f \phi) \right|^2 \right\rangle = \sum_{h_2, h_3 = \pm} \left| A_4\left(\phi \, \bar{f}^{h_2} f^{h_3} \phi\right) \right|^2 \tag{2.41}$$

for the 2-scalar 2-fermion process in the previous example. To really appreciate the difference in formalism, it is educational to first do the calculation the standard way, using the spinor-completeness relations and evaluating the gamma-matrix traces:

▶ **Exercise 2.5**

Use standard techniques to show that $\langle |A_4(\phi \, \bar{f} f \phi)|^2 \rangle = 2g^4(s - t)^2/(st)$.

[Hint: This is very similar to the massless limit of the example $e^- \varphi \to e^- \varphi$ in Chapter 48 of Srednicki [2], but we include no $\frac{1}{2}$-factors from averages here.]

Having resharpened your pencils after doing this exercise, let us now do the spin sum in the spinor helicity formalism. We already know from (2.31) that the helicity amplitudes $A_4(\phi \, \bar{f}^{h_2} f^{h_3} \phi)$ vanish unless the spinors have opposite helicity, hence

$$\left\langle \left| A_4(\phi \, \bar{f} f \phi) \right|^2 \right\rangle = \left| A_4(\phi \, \bar{f}^- f^+ \phi) \right|^2 + \left| A_4(\phi \, \bar{f}^+ f^- \phi) \right|^2. \tag{2.42}$$

The first term is calculated easily using the result (2.38) for the helicity amplitude and the reality condition (2.14):

$$\left| A_4(\phi \, \bar{f}^- f^+ \phi) \right|^2 = g^4 \left(\frac{\langle 13\rangle}{\langle 12\rangle} + \frac{\langle 34\rangle}{\langle 24\rangle} \right) \left(\frac{[13]}{[12]} + \frac{[34]}{[24]} \right)$$

$$- g^4 \left(\frac{\langle 13\rangle[13]}{\langle 12\rangle[12]} + \frac{\langle 34\rangle[34]}{\langle 24\rangle[24]} + \frac{\langle 13\rangle[34]}{\langle 12\rangle[24]} + \frac{\langle 34\rangle[13]}{\langle 24\rangle[12]} \right). \tag{2.43}$$

In the first two terms, we can directly translate the spinor products to Mandelstam variables using (2.24). For the last two terms, the momentum conservation identity (2.39) comes in handy, giving $\langle 12 \rangle [24] = -\langle 13 \rangle [34]$ and $\langle 24 \rangle [12] = -\langle 34 \rangle [13]$. Thus (2.43) gives

$$\left| A_4 \big(\phi \, \bar{f}^- f^+ \phi \big) \right|^2 = g^4 \left(\frac{t}{s} + \frac{s}{t} - 2 \right) = g^4 \frac{(s-t)^2}{st} \, . \tag{2.44}$$

The second term in (2.42) gives exactly the same, so $\langle |A_4(\phi \, \bar{f} f \phi)|^2 \rangle = 2g^4 (s-t)^2/(st)$, in agreement with the result of the standard calculation – but with use of much less pencil-power!

▶ **Exercise 2.6**

Calculate the 4-fermion "all-minus" amplitude $A_4 \big(\bar{f}^- f^- \bar{f}^- f^- \big)$ in Yukawa theory.

▶ **Exercise 2.7**

Calculate the spin sum $\langle |A_4(\bar{f} f \bar{f} f)|^2 \rangle$ for the 4-fermion process in Yukawa theory.

▶ **Exercise 2.8**

Consider a model with a Weyl-fermion ψ and a complex scalar ϕ:

$$\mathcal{L} = i \psi^\dagger \bar{\sigma}^\mu \partial_\mu \psi - \partial_\mu \bar{\phi} \, \partial^\mu \phi + \tfrac{1}{2} g \, \phi \, \psi \psi + \tfrac{1}{2} g^* \, \bar{\phi} \, \psi^\dagger \psi^\dagger - \tfrac{1}{4} \lambda \, |\phi|^4 \, . \tag{2.45}$$

Show that

$$A_4 \big(\phi \phi \bar{\phi} \bar{\phi} \big) = -\lambda \, , \quad A_4 \big(\phi \, f^- f^+ \bar{\phi} \big) = -|g|^2 \frac{\langle 24 \rangle}{\langle 34 \rangle} \, , \quad A_4 \big(f^- f^- f^+ f^+ \big) = |g|^2 \frac{\langle 12 \rangle}{\langle 34 \rangle} \, . \tag{2.46}$$

We do not put a bar on the fs here because in this model the 4-component fermion field is a Majorana fermion so there is no distinction between f and $\bar{f}$. The amplitudes (2.46) serve as useful examples when we discuss supersymmetry in Chapter 4.

We end this section by discussing one more identity from the amplitudes tool-box: the **Schouten identity** is a fancy name for a rather trivial fact: three vectors in a plane cannot be linearly independent. So if we have three 2-component kets $|i\rangle$, $|j\rangle$, and $|k\rangle$, we can write one of them as a linear combination of the two others:

$$|k\rangle = a|i\rangle + b|j\rangle \qquad \text{for some } a \text{ and } b. \tag{2.47}$$

One can "dot" in spinors $\langle \cdot |$ and form antisymmetric angle brackets to solve for the coefficients a and b. Then (2.47) can be cast in the form

$$|i\rangle \langle jk \rangle + |j\rangle \langle ki \rangle + |k\rangle \langle ij \rangle = 0 \, . \tag{2.48}$$

This is the Schouten identity. It is often written with a fourth spinor $\langle r |$ "dotted-in":

$$\langle ri \rangle \langle jk \rangle + \langle rj \rangle \langle ki \rangle + \langle rk \rangle \langle ij \rangle = 0 \, . \tag{2.49}$$

A similar Schouten identity holds for the square spinors: $[ri][jk] + [rj][ki] + [rk][ij] = 0$.

▶ **Exercise 2.9**

Show that $A_5(f^- \bar{f}^- \phi\phi\phi) = g^3 \dfrac{[12][34]^2}{[13][14][23][24]} + (3 \leftrightarrow 5) + (4 \leftrightarrow 5)$ in Yukawa theory (2.18).

2.4 Massless vectors and examples from QED

The external line rule for outgoing spin-1 massless vectors is simply to "dot-in" their polarization vectors. They can be written in spinor helicity notation as follows:

$$\epsilon_-^\mu(p; q) = -\frac{\langle p | \gamma^\mu | q]}{\sqrt{2}\,[q\,p]}, \qquad \epsilon_+^\mu(p; q) = -\frac{\langle q | \gamma^\mu | p]}{\sqrt{2}\,\langle q\,p \rangle}, \qquad (2.50)$$

where $q \neq p$ denotes an arbitrary reference spinor. Note that the massless Weyl equation ensures that $p_\mu \epsilon_\pm^\mu(p) = 0$. It can be useful to write the polarizations as

$$\not{\epsilon}_-(p; q) = \frac{\sqrt{2}}{[qp]}\Big(|p\rangle[q| + |q]\langle p|\Big), \qquad \not{\epsilon}_+(p; q) = \frac{\sqrt{2}}{\langle qp \rangle}\Big(|p]\langle q| + |q\rangle[p|\Big). \quad (2.51)$$

The arbitrariness in the choice of reference spinor reflects gauge invariance, namely that one is free to shift the polarization vector with any constant times the momentum vector: $\epsilon_\pm^\mu(p) \to \epsilon_\pm^\mu(p) + C\,p^\mu$. This does not change the on-shell amplitude A_n, as encoded in the familiar Ward identity $p_\mu A_n^\mu = 0$. For each external vector boson, one has a free choice of the corresponding reference spinor $q_i \neq p_i$; however, one must stick with the same choice in each diagram of a given process. When summing over all diagrams, the final answer for the amplitude is independent of the choices of q_i.[2]

▶ **Exercise 2.10**

Consider the momentum $p^\mu = (E, E\,\sin\theta\,\cos\phi, E\,\sin\theta\,\sin\phi, E\,\cos\theta)$. In Exercise 2.1, you found the corresponding angle and square spinors $|p\rangle$ and $|p]$. In this exercise, we establish the connection between the polarization vectors (2.50) and the more familiar polarization vectors

$$\tilde{\epsilon}_\pm^\mu(p) = \pm\frac{e^{\mp i\phi}}{\sqrt{2}}\Big(0, \cos\theta\,\cos\phi \pm i\,\sin\phi, \cos\theta\,\sin\phi \mp i\,\cos\phi, -\sin\theta\Big). \qquad (2.52)$$

Note that for $\theta = \phi = 0$, we have $\tilde{\epsilon}_\pm^\mu(p) = \pm\frac{1}{\sqrt{2}}(0, 1, \mp i, 0)$.

---

[2] In practical calculations, independence of the reference spinor provides a useful check on the result for an amplitude. For example, in a numerical evaluation one has to get the same result for different choices of the polarization reference spinors.

(a) Show that $\tilde{\epsilon}_\pm(p)^2 = 0$ and $\tilde{\epsilon}_\pm(p) \cdot p = 0$.

(b) Since $\tilde{\epsilon}_\pm^\mu(p)$ is null, $(\tilde{\epsilon}_\pm^\mu(p))_{a\dot{b}} = (\sigma_\mu)_{a\dot{b}}\, \tilde{\epsilon}_\pm^\mu(p)$ can be written in as a product of a square and an angle spinor. To see this specifically, first calculate $(\tilde{\epsilon}_+^\mu(p))_{a\dot{b}}$ and then find an angle spinor $\langle r|$ such that $(\tilde{\epsilon}_+^\mu(p))_{a\dot{b}} = -|p]_a \langle r|_{\dot{b}}$.

[Hint: you should find that $\langle rp \rangle = -\sqrt{2}$.]

(c) Next, show that it follows from (2.50) that $(\epsilon_+(p;q))_{a\dot{b}} = \frac{\sqrt{2}}{\langle qp \rangle}\, |p]\langle q|$.

(d) Now suppose there is a constant c_+ such that $\epsilon_+^\mu(p;q) = \tilde{\epsilon}_+^\mu(p) + c_+\, p^\mu$. Show that this relation requires $\langle rp \rangle = -\sqrt{2}$ (as is consistent with the solution you found for $\langle r|$ in part (b)) and then show that $c_+ = -\langle rq \rangle / \langle pq \rangle$.

Since $\epsilon_+^\mu(p;q) = \tilde{\epsilon}_+^\mu(p) + c_+\, p^\mu$, the polarization vectors $\epsilon_+^\mu(p;q)$ and $\tilde{\epsilon}_+^\mu(p)$ are equivalent. You can show the same for the negative helicity polarization. It should be clear from this exercise that the arbitrariness in the reference spinors q in the polarizations (2.50) is directly related to the gauge invariance reflected in the possibility of adding any number times p^μ to the polarization vectors.

We now calculate some amplitudes in QED to illustrate the use of the spinor helicity formalism. The QED Lagrangian

$$\mathcal{L} = -\frac{1}{4}F_{\mu\nu}F^{\mu\nu} + i\overline{\Psi}\gamma^\mu(\partial_\mu - ieA_\mu)\Psi \tag{2.53}$$

describes the interaction of a massless[3] fermion with a photon via the interaction $A_\mu\overline{\Psi}\gamma^\mu\Psi$. The vertex rule is $ie\gamma^\mu$.

▷ *Example.* Consider the 3-particle QED amplitude $A_3(f^{h_1}\bar{f}^{h_2}\gamma^{h_3})$; here $f = e^-$ and $\bar{f} = e^+$ denote electron and positron outgoing states, respectively, in the massless limit. Choose, as an example, helicities $h_1 = -1/2$, $h_2 = +1/2$ and $h_3 = -1$. We then have

$$iA_3(f^-\bar{f}^+\gamma^-) = \bar{u}_-(p_1)ie\gamma_\mu v_+(p_2)\,\epsilon_-^\mu(p_3;q)$$
$$= ie\langle 1|\gamma_\mu|2]\frac{\langle 3|\gamma^\mu|q]}{\sqrt{2}[3\,q]} = \sqrt{2}ie\frac{\langle 13\rangle[2q]}{[3\,q]},$$

using in the last step the Fierz identity (2.36). Thus

$$A_3(f^-\bar{f}^+\gamma^-) = \tilde{e}\frac{\langle 13\rangle[2q]}{[3\,q]}. \tag{2.54}$$

We have absorbed the $\sqrt{2}$ into the definition of the coupling e as $\tilde{e} \equiv \sqrt{2}e$. ◁

When we introduced the polarization vectors (2.50) in spinor helicity formalism, we mentioned that the on-shell amplitude should be independent of the reference spinor q. Here, there are no other diagrams and naively it appears that (2.54) depends on $|q]$. However, it *is* actually independent of $|q]$ – and this brings us to discuss several important aspects:

---

[3] Think of this as the high-energy scattering limit in which we consider electrons/positrons massless.

- First, let us see how to eliminate $|q]$ from (2.54). Multiply (2.54) by $1 = \langle 12 \rangle / \langle 12 \rangle$. In the numerator, we then have $\langle 13 \rangle \langle 12 \rangle [2q]$. Now use (2.35), momentum conservation $p_2 = -p_1 - p_3$, and the massless Weyl equation to get

$$\langle 12 \rangle [2q] = -\langle 1|2|q] = \langle 1|1 + 3|q] = \langle 1|3|q] = \langle 13 \rangle [3q]. \tag{2.55}$$

The square bracket $[3q]$ cancels against the same factor in the denominator of (2.54), and we are left with

$$A_3\big(f^- \bar{f}^+ \gamma^-\big) = \tilde{e} \, \frac{\langle 13 \rangle^2}{\langle 12 \rangle}. \tag{2.56}$$

This is clearly independent of $|q]$.

- Second, note that the result (2.56) depends only on angle brackets, not square brackets. This is no coincidence, but a consequence of ***3-particle special kinematics***. Note that if three light-like vectors satisfy $p_1^\mu + p_2^\mu + p_3^\mu = 0$, then

$$\langle 12 \rangle [12] = 2 p_1 . p_2 = (p_1 + p_2)^2 = p_3^2 = 0, \tag{2.57}$$

so either $\langle 12 \rangle$ or $[12]$ must vanish. Suppose $\langle 12 \rangle$ is non-vanishing; then by (2.39) and the massless Weyl equation we have $\langle 12 \rangle [23] = -\langle 1|p_2|3] = \langle 1|(p_1 + p_3)|3] = 0$. So $[23] = 0$. Similarly, $[13] = 0$. Thus $[12] = [23] = [31] = 0$ which means that these three square spinors are proportional:

$$|1] \propto |2] \propto |3]. \tag{2.58}$$

Alternatively, 3-particle kinematics could hold with square brackets non-vanishing and

$$|1\rangle \propto |2\rangle \propto |3\rangle. \tag{2.59}$$

To summarize, special 3-particle kinematics is the statement that for three on-shell massless momenta satisfying momentum conservation, the associated angle and square spinors must satisfy either (2.58) or (2.59). As a consequence:

1. A non-vanishing on-shell 3-particle amplitude with only massless particles can only depend on either angle brackets or square brackets of the external momenta, never both.
2. Since for real momenta, angle and square spinors are each other's complex conjugates, *on-shell 3-particle amplitude of only massless particles can only be non-vanishing in complex momenta* (unless it is a constant, as in ϕ^3-theory).[4] Although they do not occur in nature, the massless complex momentum 3-point amplitudes are extremely useful for building up higher-point amplitudes recursively. In many cases, the on-shell 3-point amplitudes are the key building blocks. More about this in Chapter 3.

- Finally, let us comment on choices of q in (2.54). Naively, it might seem that choosing $|q] \propto |2]$ gives zero for the amplitude; this would be inconsistent with our q-independent

---

[4] Or using a $(-, +, -, +)$ spacetime signature, cf. footnote 1.

non-vanishing result (2.56). However, this choice gives $[3q] \propto [23]$, so the denominator therefore vanishes by special kinematics. One could say that the zero $[22]$ in the numerator is canceled by the zero $[23]$ in the denominator, or simply that $|q\rangle \propto |2]$ is not a legal choice since it makes the polarization vector $\epsilon_-^\mu(p_3; q)$ divergent.

At this stage it is natural to ask how, then, we know if a given 3-point amplitude of massless particles should depend on angle brackets or square brackets. This has a good answer, which we reveal in Section 2.6. For now, let us carry on exploring QED amplitudes in the spinor helicity formalism.

▷ *Example.* Consider the QED Compton scattering process: $e^- \gamma \to e^- \gamma$. By crossing symmetry, we can view this as the amplitude $A_4(\bar{f} f \gamma \gamma)$ with all particles outgoing and labeled by momenta 1,2,3,4:

$$= (ie)^2 \, \bar{u}_2 \, \epsilon\!\!\!/_4 \, \frac{-i(\not{p}_1 + \not{p}_3)}{(p_1 + p_3)^2} \, \epsilon\!\!\!/_3 \, v_1 \; + \quad\quad (3 \leftrightarrow 4) \,. \tag{2.60}$$

Note that we have an odd number of gamma-matrices sandwiched between two spinors. If $\bar{f}$ and f have the same helicity, then such spinor products vanish, e.g. $\langle 2|\gamma^\mu \gamma^\nu \gamma^\rho|1\rangle = 0$. So we need the fermions to have opposite helicity for the process to be non-vanishing.

Suppose the photons both have negative helicity. Then the first diagram in (2.60) involves $(\epsilon\!\!\!/_{3-} v_{1+}) \propto |3\rangle[q_3 1]$ using (2.51). By picking $|q_3] \propto |1]$, this diagram vanishes. Similarly, we can choose $|q_4] \propto |1]$ to make the second diagram vanish. So $A_4(\bar{f}^+ f^- \gamma^- \gamma^-) = 0$.

▶ **Exercise 2.11**

As a spinor helicity gymnastics exercise, show that $A_4(\bar{f}^+ f^- \gamma^- \gamma^-) = 0$ without making any special choices of the reference spinors q_3 and q_4.

Now consider $A_4(\bar{f}^+ f^- \gamma^+ \gamma^-)$. We have

$$A_4(\bar{f}^+ f^- \gamma^+ \gamma^-) = \frac{2e^2 \langle 24\rangle [q_4|\big(-|1]\langle 1| - |3]\langle 3|\big)|q_3\rangle [31]}{\langle 13\rangle [13]\langle q_3 3\rangle [q_4 4]}$$
$$+ \frac{2e^2 \langle 2q_3\rangle [3|\big(-|1]\langle 1| - |4]\langle 4|\big)|4\rangle [q_4 1]}{\langle 14\rangle [14]\langle q_3 3\rangle [q_4 4]} \,. \tag{2.61}$$

Let us choose $q_3 = q_4 = p_1$. Then the second diagram in (2.61) vanishes and we get

$$A_4(\bar{f}^+ f^- \gamma^+ \gamma^-) = -\tilde{e}^2 \frac{\langle 24\rangle [13]\langle 31\rangle [31]}{\langle 13\rangle [13]\langle 13\rangle [14]} = -\tilde{e}^2 \frac{\langle 24\rangle [13]}{\langle 13\rangle [14]} \,, \tag{2.62}$$

where $\tilde{e} = \sqrt{2}e$. Momentum conservation lets us rewrite this using $\langle 23\rangle[13] = -\langle 24\rangle[14]$, giving

$$A_4(\bar{f}^+ f^- \gamma^+ \gamma^-) = \tilde{e}^2 \frac{\langle 24\rangle^2}{\langle 13\rangle\langle 23\rangle}. \tag{2.63}$$

The amplitude $A_4(\bar{f}^+ f^- \gamma^- \gamma^+)$ is obtained by interchanging the momentum labels 3 and 4 in (2.63). $\lhd$

▶ **Exercise 2.12**

Show that the amplitude $A_4(\bar{f}^+ f^- \gamma^+ \gamma^-)$ is independent of q_3 and q_4 by deriving (2.63) without making a special choice for the reference spinors q_3 and q_4.

▶ **Exercise 2.13**

Via crossing symmetry, the Compton scattering process can be regarded as electron-positron annihilation $e^- + e^+ \to \gamma + \gamma$. Compare (2.63) with (1.15) and fill out the missing steps in that introductory example to obtain the differential scattering cross-section (1.14).

▶ **Exercise 2.14**

Calculate the tree-level Bhabha scattering process $e^- e^+ \to e^- e^+$ using spinor helicity formalism.

For further experience with spinor helicity formalism, consider massless **scalar-QED**: the interaction between the complex scalar φ and the gauge field A_μ is captured via the covariant derivative $D_\mu = \partial_\mu - ieA_\mu$ as

$$\begin{aligned}
\mathcal{L} &= -\frac{1}{4}F_{\mu\nu}F^{\mu\nu} - |D\varphi|^2 - \frac{1}{4}\lambda|\varphi|^4 \\
&= -\frac{1}{4}F_{\mu\nu}F^{\mu\nu} - |\partial\varphi|^2 + ieA^\mu\big[(\partial_\mu\varphi^*)\varphi - \varphi^*\partial_\mu\varphi\big] - e^2 A^\mu A_\mu \varphi^*\varphi - \frac{1}{4}\lambda|\varphi|^4. \tag{2.64}
\end{aligned}$$

The Feynman rules give a scalar–scalar–photon 3-vertex $ie(p_2 - p_1)^\mu$ (both momenta outgoing), a scalar–scalar–photon–photon 4-vertex $-2ie^2\eta_{\mu\nu}$, and a 4-scalar vertex $-i\lambda$.

We can think of φ and φ^* as the spin-0 supersymmetric partners of the electron/positron and we loosely call them *selectrons/spositrons*, though we are not assuming that our model is part of a supersymmetric theory. A process like $\varphi + \gamma \to \varphi + \gamma$ is then the spin-0 analogue of Compton scattering. Here, we consider the extreme high-energy regime in which the mass of the selectron/spositron is taken to be zero.

▶ **Exercise 2.15**

Calculate the 3-particle amplitude $A_3(\varphi \varphi^* \gamma^-)$. Show that it is independent of the reference spinor of the photon polarization vector and write the result in a form that only involves angle brackets.

Use complex conjugation to write down the amplitude $A_3(\varphi \varphi^* \gamma^+)$.

▶ **Exercise 2.16**

Consider the amplitude $A_4(\varphi\,\varphi^*\gamma\gamma)$. Show that for any choice of photon helicities, one can always pick the reference spinors in the polarizations such that the scalar–scalar–photon–photon contact term gives a vanishing contribution to the on-shell 4-point amplitude.

▶ **Exercise 2.17**

Calculate $A_4(\varphi\,\varphi^*\gamma\gamma)$ and massage the answer into a form that depends only on either angle or square brackets and is manifestly independent of the reference spinors.

▶ **Exercise 2.18**

Calculate the spin sum $\langle|A_4(\varphi\,\varphi^*\gamma\gamma)|^2\rangle$.

▶ **Exercise 2.19**

Calculate $A_4(\varphi\,\varphi^*\varphi\,\varphi^*)$. The answer can be expressed in terms of the Mandelstam variables, but show that you can bring it to the following form:

$$A_4(\varphi\,\varphi^*\varphi\,\varphi^*) = -\lambda + \tilde{e}^2\left(1 + \frac{\langle 13\rangle^2\langle 24\rangle^2}{\langle 12\rangle\langle 23\rangle\langle 34\rangle\langle 41\rangle}\right). \tag{2.65}$$

This result will resurface in Section 3.3.

2.5 Yang–Mills theory, QCD, and color-ordering

Gluons are described by the Yang–Mills Lagrangian

$$\mathcal{L} = -\frac{1}{4}\,\mathrm{Tr}\,F_{\mu\nu}F^{\mu\nu}, \tag{2.66}$$

with $F_{\mu\nu} = \partial_\mu A_\nu - \partial_\nu A_\mu - \frac{ig}{\sqrt{2}}[A_\mu, A_\nu]$ and $A_\mu = A_\mu^a T^a$. The gauge group is $G = SU(3)$ for QCD, but we will keep the number of colors N general and take $G = SU(N)$. The gluon fields are in the adjoint representation, so the color-indices run over $a, b, \ldots = 1, 2, \ldots, N^2 - 1$. The generators T^a are normalized[5] such that $\mathrm{Tr}\,T^a T^b = \delta^{ab}$ and $[T^a, T^b] = i\tilde{f}^{abc}T^c$.

To extract Feynman rules from (2.66), one needs to fix the gauge redundancy. An amplitude-friendly choice is *Gervais–Neveu gauge* for which the gauge-fixing term is $\mathcal{L}_{\mathrm{gf}} = -\frac{1}{2}\,\mathrm{Tr}(H_\mu{}^\mu)^2$ with $H_{\mu\nu} = \partial_\mu A_\nu - \frac{ig}{\sqrt{2}}A_\mu A_\nu$ [2]. In this gauge, the Lagrangian

---

[5] A more common normalization is $\mathrm{Tr}\,T^a T^b = \frac{1}{2}\delta^{ab}$ and $[T^a, T^b] = if^{abc}T^c$. So we have $\tilde{f}^{abc} = \sqrt{2}f^{abc}$, in analogue with $\tilde{e} = \sqrt{2}e$ in QED in Section 2.4. It serves the same purpose here, namely as compensation for the $\sqrt{2}$ in the polarization vectors (2.50), so the on-shell amplitudes can be written without such factors.

takes the form[6]

$$\mathcal{L} = \text{Tr}\left(-\frac{1}{2}\,\partial_\mu A_\nu \partial^\mu A^\nu - i\sqrt{2}g\,\partial^\mu A^\nu A_\nu A_\mu + \frac{g^2}{4}\,A^\mu A^\nu A_\nu A_\mu\right). \tag{2.67}$$

The Feynman rules then give a gluon propagator $\delta^{ab}\frac{\eta_{\mu\nu}}{p^2}$. The 3- and 4-gluon vertices involve $\tilde{f}^{abc}$ and $\tilde{f}^{abx}\tilde{f}^{xcd}$+perms, respectively, each dressed up with kinematic factors that we will get back to later. The amplitudes constructed from these rules can be organized into different group theory structures each dressed with a kinematic factor. For example, the **color factors** of the s-, t-, and u-channel diagram of the 4-gluon tree amplitude are

$$c_s \equiv \tilde{f}^{a_1 a_2 b}\,\tilde{f}^{b\,a_3 a_4}\,, \quad c_t \equiv \tilde{f}^{a_1 a_3 b}\,\tilde{f}^{b\,a_4 a_2}\,, \quad c_u \equiv \tilde{f}^{a_1 a_4 b}\,\tilde{f}^{b\,a_2 a_3}\,, \tag{2.68}$$

and the 4-point contact term generically gives a sum of contributions with c_s, c_t, and c_u color factors. The Jacobi identity relates the three color factors:

$$c_s + c_t + c_u = 0\,. \tag{2.69}$$

So there are only two independent color-structures for the tree-level 4-gluon amplitude. Let us now see this in terms of traces of the generators T^a. Note that

$$i\,\tilde{f}^{abc} = \text{Tr}(T^a T^b T^c) - \text{Tr}(T^b T^a T^c)\,. \tag{2.70}$$

The products of generator-traces in the amplitudes can be Fierz'ed using the completeness relation

$$(T^a)_i{}^j (T^a)_k{}^l = \delta_i{}^l \delta_k{}^j - \frac{1}{N}\delta_i{}^j \delta_k{}^l\,. \tag{2.71}$$

For example, for the 4-gluon s-channel diagram we have

$$\tilde{f}^{a_1 a_2 b}\,\tilde{f}^{b\,a_3 a_4} - \text{Tr}\left(T^{a_1} T^{a_2} T^{a_3} T^{a_4}\right) + \text{Tr}\left(T^{a_1} T^{a_2} T^{a_4} T^{a_3}\right)$$
$$= \text{Tr}\left(T^{a_1} T^{a_3} T^{a_4} T^{a_2}\right) - \text{Tr}\left(T^{a_1} T^{a_4} T^{a_3} T^{a_2}\right)\,. \tag{2.72}$$

Here we have used the cyclic property of the traces to deduce the four color-structures. Similarly, the three other diagrams contributing to the 4-gluon amplitude can also be written in terms of single-trace group theory factors. So that means that we can write the 4-gluon tree amplitude as

$$A_4^{\text{full,tree}} = g^2\left(A_4[1234]\,\text{Tr}\left(T^{a_1} T^{a_2} T^{a_3} T^{a_4}\right) + \text{perms of } (234)\right), \tag{2.73}$$

where the *partial amplitudes* $A_4[1234]$, $A_4[1243]$ etc. are called **color-ordered amplitudes**. We use the squared parenthesis in $A_4[1234]$ to distinguish the color-ordered amplitude from the non-color-ordered amplitude $A_4(1234)$. Each partial amplitude is gauge invariant.[7]

---

[6] We ignore ghosts, since our focus here is on tree-level amplitudes.

[7] This follows from a partial orthogonality property of the single-traces [1].

The color-structure generalizes to any n-point tree-level amplitude involving any particles that transform in the adjoint of the gauge group: we write

$$A_n^{\text{full,tree}} = g^{n-2} \sum_{\text{perms } \sigma} A_n[1\,\sigma(2\ldots n)] \, \text{Tr}\left(T^{a_1} T^{\sigma(a_2)} \cdots T^{a_n}\right), \qquad (2.74)$$

where the sum is over the (overcomplete) trace-basis of $(n-1)!$ elements that takes into account the cyclic nature of the traces. For loop-amplitudes, one also needs to consider multi-trace structures in addition to the simple single-trace – for more about this, see [3, 4]. We have factored out the coupling constant g to avoid carrying it along explicitly in all the color-ordered amplitudes.

In the Gervais–Neveu gauge, the Feynman vertex rules for calculating the color-ordered amplitudes are:

- 3-gluon vertex $V^{\mu_1\mu_2\mu_3}(p_1, p_2, p_3) = -\sqrt{2}\left(\eta^{\mu_1\mu_2} p_1^{\mu_3} + \eta^{\mu_2\mu_3} p_2^{\mu_1} + \eta^{\mu_3\mu_1} p_3^{\mu_2}\right)$,
- 4-gluon vertex $V^{\mu_1\mu_2\mu_3\mu_4}(p_1, p_2, p_3, p_4) = \eta^{\mu_1\mu_3}\eta^{\mu_2\mu_4}$.

The color-ordered amplitude $A_n[12\ldots n]$ is calculated in terms of diagrams with no lines crossing and the ordering of the external lines fixed as given $1, 2, 3, \ldots, n$. The gluon polarization vectors are given in (2.50), (2.51).

Let us consider the simplest case, namely the 3-gluon amplitude. From the 3-vertex rule, we get

$$A_3[1\,2\,3] = -\sqrt{2}\Big[(\epsilon_1\epsilon_2)(\epsilon_3 p_1) + (\epsilon_2\epsilon_3)(\epsilon_1 p_2) + (\epsilon_3\epsilon_1)(\epsilon_2 p_3)\Big]. \qquad (2.75)$$

Let us now pick gluons 1 and 2 to have negative helicity while gluon 3 gets to have positive helicity. Translating to spinor helicity formalism (using the Fierz identity (2.36)) we have

$$A_3\left[1^-2^-3^+\right] = -\frac{\langle 12\rangle[q_1q_2]\langle q_31\rangle[13] + \langle 2q_3\rangle[q_23]\langle 12\rangle[2q_1] + \langle q_31\rangle[3q_1]\langle 23\rangle[3q_2]}{[q_11][q_22]\langle q_33\rangle}. \qquad (2.76)$$

We must now consider 3-particle special kinematics, (2.58) or (2.59). If $|1\rangle \propto |2\rangle \propto |3\rangle$, all three terms vanish in the numerator of (2.76). So select 3-particle kinematics $|1] \propto |2] \propto |3]$. Then the first term vanishes and we are left with

$$A_3\left[1^-2^-3^+\right] = -\frac{\langle 2q_3\rangle[q_23]\langle 12\rangle[2q_1] + \langle 1q_3\rangle[q_13]\langle 23\rangle[3q_2]}{[q_11][q_22]\langle q_33\rangle}. \qquad (2.77)$$

To simplify this, first use momentum conservation to write $\langle 12\rangle[2q_1] = -\langle 13\rangle[3q_1]$. Then $[q_13][q_23]$ factors out and we get

$$A_3\left[1^-2^-3^+\right] = \frac{[q_13][q_23]\big(\langle 13\rangle\langle q_32\rangle + \langle 1q_3\rangle\langle 23\rangle\big)}{[q_11][q_22]\langle q_33\rangle}. \qquad (2.78)$$

After a quick round of Schouten'ing, this simplifies to

$$A_3\left[1^-2^-3^+\right] = \frac{[q_13][q_23]\big(-\langle 12\rangle\langle 3q_3\rangle\big)}{[q_11][q_22]\langle q_33\rangle} = \frac{\langle 12\rangle[q_13][q_23]}{[q_11][q_22]}. \qquad (2.79)$$

So we have got rid of q_3. To eliminate q_1 and q_2, use momentum conservation $[q_1 3]\langle 23 \rangle = -[q_1 1]\langle 21 \rangle$ and $[q_2 3]\langle 13 \rangle = -[q_1 1]\langle 12 \rangle$. The result for the 3-gluon amplitude is a remarkably simple ratio of angle brackets:

$$A_3[1^- 2^- 3^+] = \frac{\langle 12 \rangle^3}{\langle 23 \rangle \langle 31 \rangle}. \tag{2.80}$$

The result for the "googly" gluon amplitude $A_3[1^+ 2^+ 3^-]$ is

$$A_3[1^+ 2^+ 3^-] = \frac{[12]^3}{[23][31]}. \tag{2.81}$$

▶ **Exercise 2.20**

Fill in the details to derive the amplitude (2.81).

▶ **Exercise 2.21**

Calculate $\epsilon_-(p, q) \cdot \epsilon_-(k, q')$, $\epsilon_+(p, q) \cdot \epsilon_+(k, q')$, and $\epsilon_-(p, q) \cdot \epsilon_+(k, q')$.
Show that $\epsilon_\pm(p, q) \cdot \epsilon_\pm(k, q')$ vanishes if $q = q'$.
How can you make $\epsilon_-(p, q) \cdot \epsilon_+(k, q')$ vanish?

▶ **Exercise 2.22**

Use the previous exercise to show that for any choice of gluon helicities, it is always possible to choose the polarization vectors such that the contribution from the 4-gluon contact term to the 4-gluon amplitude vanishes.

▶ **Exercise 2.23**

Use a well-chosen set of reference spinors to show that the entire 4-gluon amplitude vanishes if all four gluons have the same helicity.

▶ **Exercise 2.24**

Calculate the color-ordered 4-gluon tree amplitude $A_4[1^- 2^- 3^+ 4^+]$ using the color-ordered Feynman rules and a smart choice of reference spinors. Show that the answer can be brought to the form

$$A_4[1^- 2^- 3^+ 4^+] = \frac{\langle 12 \rangle^4}{\langle 12 \rangle \langle 23 \rangle \langle 34 \rangle \langle 41 \rangle}. \tag{2.82}$$

Note the cyclic structure of the denominator.

The result (2.82) for the 4-gluon amplitude is an example of the famous **Parke–Taylor n-gluon tree amplitude** [5]: for the case where gluons i and j have helicity -1 and all the $n - 2$ other gluons have helicity $+1$, the Parke–Taylor formula for the gluon tree amplitude is

$$A_n\big[1^+ \ldots i^- \ldots j^- \ldots n^+\big] = \frac{\langle ij\rangle^4}{\langle 12\rangle\langle 23\rangle \cdots \langle n1\rangle}. \tag{2.83}$$

We prove this formula in Chapter 3.

It is relevant to point out that the number of Feynman diagrams that generically contribute to an n-gluon tree amplitude is[8]

$n =$	3	4	5	6	7	...
#diagrams $=$	1	3	10	38	154	...

A fun little trivia point you can impress your friends with in a bar (oh, we mean at the library), is that the number of trivalent graphs that contribute to the n-gluon tree process is counted by the Catalan numbers.

It should be clear that even though you have now learned some handy tricks for how to choose the polarization vectors to reduce the difficulty of the calculation, it would be no fun to try to compute these higher-point gluon amplitudes by brute force. But despite the complications of the many diagrams and their increased complexity, the answer is just the simple Parke–Taylor expression (2.83) for the $- - + + \cdots +$ helicity case. And that is the answer no matter which fancy field redefinitions we might subject the Lagrangian to and no matter which ugly gauge we could imagine choosing. It is precisely the point of the modern approach to amplitudes to avoid such complications and get to an answer such as (2.83) in a simple way.

▶ **Exercise 2.25**

Rewrite the expression (2.82) to show that the 4-gluon amplitude can also be written

$$A_4\big[1^- 2^- 3^+ 4^+\big] = \frac{[34]^4}{[12][23][34][41]}. \tag{2.84}$$

▶ **Exercise 2.26**

Convince yourself that in general if all helicities are flipped $h_i \to -h_i$, then the resulting amplitude $A_n[1^{h_1} 2^{h_2} \ldots n^{h_n}]$ is obtained from $A_n[1^{-h_1} 2^{-h_2} \ldots n^{-h_n}]$ by exchanging all angle and square brackets.

The color-ordered amplitudes have a number of properties worth noting:

1. *Cyclic:* It follows from the trace-structure that $A_n[12\ldots n] = A_n[2\ldots n\,1]$, etc.
2. *Reflection:* $A_n[12\ldots n] = (-1)^n A_n[n \ldots 2\,1]$. Convince yourself that this is true.
3. The $U(1)$ *decoupling identity:*

$$A_n[123\ldots n] + A_n[213\ldots n] + A_n[231\ldots n] + \cdots + A_n[23\ldots 1\,n] = 0. \tag{2.85}$$

---

[8] This can be seen by direct counting, but see also the analysis in [6].

The vanishing of this sum of $n - 1$ color-ordered amplitudes is also called the photon decoupling identity; it follows from taking one of the generators T^a proportional to the identity matrix.

▶ **Exercise 2.27**

Use (2.83) to show explicitly that (2.85) holds for $n = 4$ for the case where gluons 1 and 2 have negative helicity and 3 and 4 have positive helicity.

The trace-basis (2.74) is overcomplete and that implies that there are further linear relations among the partial *tree-level* amplitudes: these are called the ***Kleiss–Kuijf relations*** [7, 8] and they can be written [9]

$$A_n[1, \{\alpha\}, n, \{\beta\}] = (-1)^{|\beta|} \sum_{\sigma \in \text{OP}(\{\alpha\}, \{\beta^T\})} A_n[1, \sigma, n], \qquad (2.86)$$

where $\{\beta^T\}$ denotes the reverse ordering of the labels $\{\beta\}$ and the sum is over ordered permutations "OP," namely permutations of the labels in the joined set $\{\alpha\} \cup \{\beta^T\}$ such that the ordering within $\{\alpha\}$ and $\{\beta^T\}$ is preserved. The sign on the RHS is determined by the number of labels $|\beta|$ in the set $\{\beta\}$.

To make (2.86) a little less intimidating, consider the 5-point case as an example. Taking the LHS of (2.86) to be $A_5[1, \{2\}, 5, \{3, 4\}]$, we have $\{\alpha\} \cup \{\beta^T\} = \{2\} \cup \{4, 3\}$, so the sum over ordered permutations is over $\sigma = \{243\}, \{423\}, \{432\}$. Thus the Kleiss–Kuijf relation reads

$$A_5[12534] = A_5[12435] + A_5[14235] + A_5[14325]. \qquad (2.87)$$

▶ **Exercise 2.28**

Show that for $n = 4$, the Kleiss–Kuijf relation (2.86) is equivalent to the $U(1)$ decoupling relation.

▶ **Exercise 2.29**

Start with $A_5[1, \{2, 3\}, 5, \{4\}]$ to show that the Kleiss–Kuijf relation gives

$$A_5[12345] + A_5[12354] + A_5[12435] + A_5[14235] = 0. \qquad (2.88)$$

Show then that (2.88) together with the $U(1)$ decoupling relation implies that (2.87).

The Kleiss–Kuijf relations combine with the cyclic, reflection, and $U(1)$ decoupling identities to reduce the number independent n-gluon tree amplitudes to $(n - 2)!$. However, there are further linear relationships, called the (fundamental) ***BCJ relations*** – named after Bern, Carrasco, and Johansson [9] – that reduce the number of independent n-gluon color-ordered tree amplitudes to $(n - 3)!$. Examples of 4-point and 5-point BCJ amplitude relations are

$$s_{14} A_4[1234] - s_{13} A_4[1243] = 0, \qquad (2.89)$$

$$s_{12} A[21345] - s_{23} A[13245] - (s_{23} + s_{24}) A[13425] = 0. \qquad (2.90)$$

In Chapter 13, we show that the number of independent color-ordered tree amplitudes under Kleiss–Kuijf relations is $(n-2)!$ and we also discuss the origin of BCJ amplitude relations.

▶ **Exercise 2.30**

Use the Parke–Taylor formula (2.83) to verify (2.88), (2.89), and (2.90).

▶ **Exercise 2.31**

Let us get a little preview of the BCJ relations. Suppose we use the color-basis (2.68) to write the full 4-point gluon amplitude as

$$A_4^{\text{full,tree}} = \frac{n_s \, c_s}{s} + \frac{n_t \, c_t}{t} + \frac{n_u \, c_u}{u} \tag{2.91}$$

for some numerator factors n_i that in general depend on the kinematic variables and the polarizations. Write each c_i in terms of the three traces $\text{Tr}(T^{a_1} T^{a_2} T^{a_3} T^{a_4})$ and those with orderings 1243 and 1324. (Make sure to check that the Jacobi identity (2.69) holds.) Then use your expressions to convert (2.91) to a basis with those three traces.

Now use the cyclic and reflection properties of the trace and the color-ordered amplitudes to write the full amplitude $A_4^{\text{full,tree}}$ in (2.73) in terms of the traces with the same three orderings 1234, 1243, and 1324.

Comparing the resulting expressions for $A_4^{\text{full,tree}}$, read off the relationship between the numerator factors n_i and the color-ordered amplitudes. You should find

$$A_4(1234) = -\frac{n_s}{s} + \frac{n_u}{u} \tag{2.92}$$

and two similar expressions for $A_4(1243)$ and $A_4(1324)$. Show that it follows directly from these expressions that the color-ordered amplitudes satisfy the $n = 4$ photon decoupling relation (2.85).

Note that the numerator factors n_i are not unique. Suppose that there is a choice of numerator factors n_i that satisfy the same relation as the color factors c_i,

$$n_s + n_t + n_u = 0 \,. \tag{2.93}$$

Show that (2.93) implies that the color-ordered amplitudes satisfy the BCJ relation (2.89).

The existence of numerator factors n_i that satisfy the same identity (2.93) as the corresponding color factors is called *color-kinematics duality*. It has been of great interest and applicability in studies of amplitudes in both gauge theory and gravity, and we will discuss it further in Chapter 13.

We end this section with a quick look at interactions between gluons and fermions. Adding

$$\mathcal{L} = i \overline{\Psi} \gamma^\mu D_\mu \Psi = i \overline{\Psi} \gamma^\mu \partial_\mu \Psi + \frac{g}{\sqrt{2}} A_\mu \overline{\Psi} \gamma^\mu \Psi \tag{2.94}$$

to the Yang–Mills Lagrangian (2.66), we now acquire a fermion–fermion–gluon 3-vertex $i \frac{g}{\sqrt{2}} \gamma^\mu$. If the fermion represents a quark, Ψ transforms in the fundamental of the gauge group $SU(N)$. In that case the trace-structure of the amplitudes is a little different, for example for the case of scattering 2 quarks with n gluons, we get $(T^{a_1} T^{a_1} \dots T^{a_n})_i{}^j$.

If we want to study the interactions of gluons with their supersymmetric partners, the gluinos, then the fermion field must transform in the adjoint so we replace Ψ with $\lambda = \lambda^a T^a$ and include a trace in the Lagrangian. The trace-structure for gluon–gluino scattering is exactly the same as for gluon scattering.

We have by now seen enough examples of how to use spinor helicity formalism in the context of standard Feynman rules. It is about time that we get a little fancier. Therefore we postpone further discussion of Yang–Mills and super Yang–Mills amplitudes until we have developed a few more tools.

2.6 Little group scaling

We have introduced $|p\rangle$ and $|p]$ as solutions to the massless Weyl equation, $p|p\rangle = 0$ and $p|p] = 0$ for $p^2 = 0$. Their relation to p^μ was given in (2.16) as $p_{a\dot{b}} = -|p]_a \langle p|_{\dot{b}}$. It is very useful to observe that these relations are invariant under the scaling

$$|p\rangle \to t|p\rangle, \qquad |p] \to t^{-1}|p]. \qquad (2.95)$$

This is called *little group scaling*. Recall that the *little group* is the group of transformations that leave the momentum of an on-shell particle invariant. For a massless particle, we can go to a frame where $p^\mu = (E, 0, 0, E)$. Rotations in the xy-plane leave the vector invariant, so the little group representations[9] are characterized by $SO(2) = U(1)$. In the angle and square spinor representation of the momentum, the little group transformation is realized as the scaling (2.95): for real momenta, t has to be a complex phase such that $|p]^* = \langle p|$ is preserved. For complex momenta, the angle and square spinors are independent so we can be more generous and let t be any non-zero complex number.

Now let us consider *what an amplitude is made of*: each Feynman diagram consists of propagators, vertices, and external line rules. When only massless particles are involved, the amplitude can always be rewritten in terms of angle and square brackets. But note that neither propagators nor vertices can possibly scale under little group transformations. Only the *external line rules* scale under (2.95):

- The scalar rule is a constant factor 1: it does not scale under (2.95).
- The wavefunctions for (Weyl) fermions are angle ($h = -1/2$) and square spinors ($h = +1/2$): they scale as t^{-2h}.

---

[9] More precisely, the little group is $E(2)$, the group of transformations that map a 2d plane into itself. This is similar to the more familiar $SU(2)$ group, whose generators J_+ and J_- can be identified as the two translation generators of the little group and J_z can be identified with the rotation generator. Thus, just as in $SU(2)$, where representations are characterized by their J_z eigenvalue, representations of the little group $E(2)$ are characterized by their spin under the 2-dimensional rotation group $SO(2) = U(1)$.

- Polarization vectors for spin-1 bosons: you can directly check (2.50) to see that under little group scaling of $|p\rangle$ and $|p]$, the polarization vectors $\epsilon_\pm^\mu(p;q)$ scale as t^{-2h} for $h = \pm 1$. They are invariant under scaling of the reference spinor.

Thus, for an amplitude of massless particles[10] only, we have the following powerful result. Under little group scaling of each particle $i = 1, 2, \ldots, n$, the on-shell amplitude transforms homogeneously with weight $-2h_i$, where h_i is the helicity of particle i:

$$A_n\left(\{|1\rangle, |1], h_1\}, \ldots, \{t_i|i\rangle, t_i^{-1}|i], h_i\}, \ldots\right) = t_i^{-2h_i} A_n\left(\ldots \{|i\rangle, |i], h_i\} \ldots\right). \quad (2.96)$$

As an example, consider the QED amplitude (2.56), $A_3(f^- \bar{f}^+ \gamma^-) = \tilde{e} \frac{\langle 13\rangle^2}{\langle 12\rangle}$. For the negative helicity photon (particle 3) we get $t_3^2 = t_3^{-2(-1)}$. Likewise, one confirms the scaling (2.96) for the two fermions. In fact, *all massless 3-particle amplitudes are completely fixed by little group scaling!* Let us now see how.

3-particle amplitudes

Recall that by 3-particle special kinematics, (2.58) and (2.59), an on-shell 3-point amplitude with massless particles depends only on either angle or square brackets of the external momenta. Let us suppose that it depends on angle brackets only. We can then write a general Ansatz

$$A_3\left(1^{h_1} 2^{h_2} 3^{h_3}\right) = c \langle 12\rangle^{x_{12}} \langle 13\rangle^{x_{13}} \langle 23\rangle^{x_{23}}, \quad (2.97)$$

where c is some constant independent of the kinematics. Little group scaling (2.96) fixes

$$-2h_1 = x_{12} + x_{13}, \quad -2h_2 = x_{12} + x_{23}, \quad -2h_3 = x_{13} + x_{23}. \quad (2.98)$$

This system is readily solved to find $x_{12} = h_3 - h_1 - h_2$ etc. so that

$$A_3\left(1^{h_1} 2^{h_2} 3^{h_3}\right) = c \langle 12\rangle^{h_3 - h_1 - h_2} \langle 13\rangle^{h_2 - h_1 - h_3} \langle 23\rangle^{h_1 - h_2 - h_3}. \quad (2.99)$$

This means that the helicity structure uniquely fixes the 3-particle amplitude up to an overall constant! This may remind you of a closely related fact, namely that in a conformal field theory, the 3-point correlation functions are determined uniquely (up to a multiplicative constant) by the scaling dimensions of the operators.

We have already confirmed (2.99) for $A_3(f^- \bar{f}^+ \gamma^-)$. So let us do something different. Consider a 3-gluon amplitude with two negative and one positive helicity gluons. By (2.99), the kinematic structure is uniquely determined:

$$A_3\left(g_1^- g_2^- g_3^+\right) = g \frac{\langle 12\rangle^3}{\langle 13\rangle \langle 23\rangle}. \quad (2.100)$$

---

[10] For a spin-2 graviton, the helicity ± 2 states are encoded in polarizations that can be chosen as $e_\pm^{\mu\nu} = \epsilon_\pm^\mu \epsilon_\pm^\nu$; this choice automatically makes the polarizations symmetric and traceless. For a spin-3/2 gravitino, the wavefunctions can be taken to be $v_\pm \epsilon_\pm^\mu$. It follows from these external line rules that gravitons and gravitinos also obey the scaling t^{-2h} under little group scaling.

This matches our calculation (2.80). We have included here the Yang–Mills coupling g for the purpose of the discussion to follow. Note how the little group argument fixes the kinematic form of 3-gluon amplitude; this is much simpler than the direct Feynman diagram calculation we did in Section 2.5.

But – there is perhaps a small glitch in our little group scaling argument. We *assumed* that the amplitude depended only on angle brackets. What if it only depended on square brackets? Then the scaling would have been the opposite, so we would have found

$$A_3\left(g_1^- g_2^- g_3^+\right) = g' \, \frac{[13][23]}{[12]^3} \, . \tag{2.101}$$

To distinguish between (2.100) and (2.101), we use ***dimensional analysis***. From (2.24) we note that both angle and square brackets have mass-dimension 1. Thus the momentum dependence in (2.100) is $(\text{mass})^1$; this is compatible with the fact that it comes from the $AA\partial A$-interaction in $\operatorname{Tr} F_{\mu\nu}F^{\mu\nu}$. However, in (2.101), the momentum dependence has mass-dimension $(\text{mass})^{-1}$, so it would somehow have to come from an interaction of the form $g' A A \frac{\partial}{\Box} A$. Of course, we have no such interaction term in a ***local*** Lagrangian; hence we discard the expression (2.101) as unphysical.

The combination of ***little group scaling*** and ***locality*** uniquely fixes the massless 3-particle amplitudes. As we will see in Chapter 3, in *some* theories the 3-particle amplitudes can actually determine all other tree-level amplitudes!

While we are considering dimensional analysis, it is worth making a couple of other observations. First, note that while the Yang–Mills coupling g is dimensionless, the coupling g' in the $g' A A \frac{\partial}{\Box} A$ has dimension $(\text{mass})^2$. This means that the RHS of (2.101) has mass-dimension 1, just as the correct expression (2.100). This is sensible since the two amplitude-expressions had better have the same mass-dimension. In general,

> *an n-particle amplitude in $D = 4$ must have mass-dimension $4 - n$.* (2.102)

This follows from dimensional analysis since the cross-section must have dimensions of area. You can also check it by direct inspection of the Feynman diagrams.

One more comment about (2.100): you may be worried that the expression on the RHS is not Bose-symmetric in the exchange of the identical gluons 1 and 2. Fear not. The full 3-point amplitude of course comes dressed with a fully-antisymmetric group theory factor $f^{a_1 a_2 a_3}$: this restores Bose symmetry. As discussed in Section 2.5, the kinematic structure in (2.100) is exactly that of the color-ordered 3-point amplitude $A_3[1^- 2^- 3^+]$.

▶ **Exercise 2.32**

Write down spinor helicity formulas for the possible color-ordered 3-point amplitudes with two gluinos (massless spin-1/2 in the adjoint representation of the gauge group) and one gluon. The result will encode the gluino–gluino–gluon interactions in super Yang–Mills theory: work "backwards" to find out what that interaction term looks like in the Lagrangian.

▶ **Exercise 2.33**

Let us play a little game. Suppose someone gives you the following amplitudes for scattering processes involving massless particles:

(a) $\quad A_5 = g_a \, \dfrac{[13]^4}{[12][23][34][45][51]}$, \hfill (2.103)

(b) $\quad A_4 = g_b \, \dfrac{\langle 14 \rangle \langle 24 \rangle^2}{\langle 12 \rangle \langle 23 \rangle \langle 34 \rangle}$, \hfill (2.104)

(c) $\quad A_4 = g_c \, \dfrac{\langle 12 \rangle^7 [12]}{\langle 13 \rangle \langle 14 \rangle \langle 23 \rangle \langle 24 \rangle \langle 34 \rangle^2}$. \hfill (2.105)

With all particles outgoing, what are the helicities of the particles?
What is the dimension of the couplings g_i relevant to the interactions?
In each case, try to figure out which theory could produce such an amplitude.

▷ *Example.* What about a gluon amplitude with all-negative helicities? Well, let us do it. The formula (2.99) immediately tells us that

$$A_3\big(g_1^- g_2^- g_3^-\big) = a \, \langle 12 \rangle \langle 13 \rangle \langle 23 \rangle . \tag{2.106}$$

The kinematic part has mass-dimension 3 and this reveals that: (*i*) the coupling a must have mass-dimension -2 for the whole amplitude to have mass-dimension $4 - 3 = 1$; and (*ii*) this must come from a Lagrangian interaction term with three derivatives, i.e. $(\partial A)^3$. Furthermore, the kinematic terms are antisymmetric under exchanges of gluon-momenta, so Bose symmetry tells us that the couplings must be associated with antisymmetric structure constants – as is of course the case for a non-abelian gauge field. Thus, there is a natural candidate, namely the dimension-6 operator $\mathrm{Tr}\, F^\mu{}_\nu F^\nu{}_\lambda F^\lambda{}_\mu$. Indeed this operator produces the amplitude (2.106). We can also conclude that in pure Yang–Mills theory or in QED, $A_3(g_1^- g_2^- g_3^-) = 0$. ◁

▶ **Exercise 2.34**

Let us look at gravity scattering amplitudes. If we expand the Einstein–Hilbert action $\frac{1}{2\kappa^2} \int d^4x \sqrt{-g}\, R$ around flat space $g_{\mu\nu} = \eta_{\mu\nu} + \kappa\, h_{\mu\nu}$, we obtain an infinite series of 2-derivative interactions involving n fields $h_{\mu\nu}$ for any n. This makes it very complicated to calculate graviton scattering amplitudes using Feynman rules. (Gravitons are massless spin-2 particles; in 4d they have two helicity states, $h = \pm 2$.) For now just focus on the 3-point amplitude: use little group scaling to write down the result for the on-shell 3-graviton amplitudes. Check the mass-dimensions. Compare your answer with the 3-gluon amplitudes. Graviton scattering is discussed in much further detail in Chapter 12.

▶ **Exercise 2.35**

Consider in gravity an operator constructed from some contraction of the indices of three Riemann-tensors; we denote it schematically as R^3. If we linearize the metric around flat space, $g_{\mu\nu} = \eta_{\mu\nu} + \kappa\, h_{\mu\nu}$, then we can calculate graviton scattering associated with R^3. What is the mass-dimension of the coupling associated with R^3? Use little group scaling to determine $A_3(h_1^- h_2^- h_3^-)$ and $A_3(h_1^- h_2^- h_3^+)$. Compare the expressions with 3-gluon scattering processes.

▶ **Exercise 2.36**

Consider a dimension-5 Higgs-gluon fusion operator $H \operatorname{Tr} F_{\mu\nu} F^{\mu\nu}$, where H is a Higgs scalar field. Use little group scaling to determine the 3-particle amplitudes of this operator in the limit of a massless Higgs, $m_H = 0$. (For more about on-shell methods and Higgs-gluon fusion, see [10].)

▷ *Example.* Consider a 3-point amplitude with three scalars. We learn from (2.99) that there can be no momentum dependence in the amplitude, $A_3(\phi\phi\phi) = $ constant. This is of course compatible with a ϕ^3-interaction, but what about throwing in some derivatives, as in a non-linear sigma model? – something like $\phi \, \partial_\mu \phi \, \partial^\mu \phi$. The Feynman rules for this 3-scalar vertex give a vertex $\sim (p_1 \cdot p_2 + p_2 \cdot p_3 + p_3 \cdot p_1)$ which on-shell is proportional to $(s + t + u) = 0$. So the on-shell 3-point amplitude corresponding to $\phi \, \partial_\mu \phi \, \partial^\mu \phi$ vanishes. We can see this directly from the Lagrangian since the interaction term is readily rewritten as $\frac{1}{2}\partial_\mu(\phi^2) \, \partial^\mu \phi$, which by partial integration gives $-\frac{1}{2}\phi^2 \, \Box\phi$. This clearly vanishes on-shell for a massless scalar that obeys the Klein–Gordon equation $\Box\phi = 0$. ◁

▶ **Exercise 2.37**

Now your turn: why does the 3-particle on-shell amplitude for three distinct massless scalars, e.g. $\phi_1 \, \partial_\mu \phi_2 \, \partial^\mu \phi_3$, vanish? Try to see how this works both from the amplitude perspective and from the Lagrangian.

There is a lesson to learn from the above example. The expression for an amplitude can be manipulated using momentum conservation and on-shell conditions $p_i^2 = 0$. If the amplitude is the on-shell matrix element of a (local) operator in the Lagrangian, then these manipulations correspond to a rewriting of the operator by, respectively, dropping terms that are total derivatives or vanish on-shell (i.e. on the equations of motion).

2.7 MHV classification

In this section we return to the study of gluon scattering amplitudes. The Yang–Mills Lagrangian contains two types of interaction terms: schematically

$$\operatorname{Tr} F_{\mu\nu} F^{\mu\nu} \longrightarrow AA\partial A + A^4 \,. \tag{2.107}$$

In a typical gauge, such as a Feynman gauge or Neveu–Gervais gauge (2.67), this gives rise to Feynman rules with two types of interaction vertices: the cubic vertex which depends linearly on momentum and the quartic vertex which is momentum-independent. Since the

Yang–Mills coupling is dimensionless, the cubic vertex is $O(\text{mass}^1)$ and the quartic is $O(\text{mass}^0)$.

Consider tree diagrams with only cubic vertices, i.e. trivalent tree-graphs, with n external legs. If you start with a 3-point vertex ($n = 3$) you can easily convince yourself that every time you add an extra external line, you have to add both a new vertex and a new propagator to keep the graph trivalent. Hence the numbers of vertices and propagators both grow linearly with n, and it takes just a few examples to see that the number of vertices is $n - 2$ and the number of propagators is $n - 3$. Since the cubic vertices are $O(\text{mass}^1)$ and the propagators are $O(\text{mass}^{-2})$, we find that the mass-dimension of the diagrams, and hence of the tree amplitude, is

$$[A_n] \sim \frac{(\text{mass})^{n-2}}{(\text{mass}^2)^{n-3}} \sim (\text{mass})^{4-n}. \tag{2.108}$$

This confirms the statement (2.102) that n-point amplitudes in $D=4$ dimensions have mass-dimension $4 - n$. Any diagram with a mix of cubic and quartic vertices has the same mass-dimension of $(\text{mass})^{4-n}$. But note that the number of powers of momenta in the numerator cannot exceed $n - 2$; this point will be useful shortly.

Consider now the schematic form of a gluon tree amplitude:

$$A_n \sim \sum_{\text{diagrams}} \frac{\sum \left(\prod (\epsilon_i . \epsilon_j) \right) \left(\prod (\epsilon_i . k_j) \right) \left(\prod (k_i . k_j) \right)}{\prod P_I^2}. \tag{2.109}$$

The k_is stand for various sums of external momenta p_j. Each diagram has a numerator that is a polynomial in Lorentz scalar products of polarizations and momentum vectors. The denominators are products of momentum invariants from the propagators.

Perhaps you know the statement that in pure Yang–Mills theory all-plus tree gluon amplitudes vanish, $A_n(1^+ 2^+ \ldots n^+) = 0$? We have already observed this for $n = 3, 4$ in Exercise 2.23 and in the example (2.106). Let us show it for all n. First recall from Exercise 2.21 that the polarization vector dot-products are

$$\epsilon_{i+} \cdot \epsilon_{j+} \propto \langle q_i q_j \rangle, \quad \epsilon_{i-} \cdot \epsilon_{j-} \propto [q_i q_j], \quad \epsilon_{i-} \cdot \epsilon_{j+} \propto \langle i q_j \rangle [j q_i]. \tag{2.110}$$

Suppose we choose all q_i to be the same q. Then $\epsilon_{i+} \cdot \epsilon_{j+} = 0$. This means that the only non-vanishing way for the positive helicity gluon polarizations to enter in the numerator of (2.109) is as $\epsilon_{i+} \cdot k_j$. For an all-plus gluon amplitude, we need to absorb the Lorentz indices of all n polarization vectors this way, so that requires n powers of momenta in the numerator. But as we argued below (2.108), no more than $n - 2$ powers of momenta are possible in the numerator of any gluon tree diagram. Hence we conclude that $A_n(1^+ 2^+ \ldots n^+) = 0$.

If we had not known that we should write down a smart choice of the polarization vectors, but had worked with general expressions, we would have had to work very hard to prove that the sum of combinatorially many n-point tree diagrams in the all-plus amplitude adds up to zero.

Next, let us flip one of the helicities and consider an amplitude $A_n(1^- 2^+ \ldots n^+)$. This time, choose $q_2 = q_3 = \cdots = q_n = p_1$. This achieves $\epsilon_{i+} \cdot \epsilon_{j+} = 0$ and $\epsilon_{1-} \cdot \epsilon_{j+} = 0$. So again we would need n factors of $\epsilon_{i+} \cdot k_j$ in the numerators of (2.109) and as before we conclude that the tree-level amplitude vanishes: $A_n(1^- 2^+ \ldots n^+) = 0$.

We have shown that

tree-level gluon amplitudes: $A_n(1^+2^+\ldots n^+) = 0$ and $A_n(1^-2^+\ldots n^+) = 0$.

$$(2.111)$$

The same holds with all helicities flipped:

tree-level gluon amplitudes: $A_n(1^-2^-\ldots n^-) = 0$ and $A_n(1^+2^-\ldots n^-) = 0$.

$$(2.112)$$

In supersymmetric Yang–Mills theories, the results (2.111) and (2.112) hold true at any loop-order. In fact, supersymmetry provides an elegant argument for the vanishing of these so-called "helicity violating" amplitudes. We present this derivation in Chapter 4.

In pure Yang–Mills theory, the amplitudes (2.111) and (2.112) are actually non-vanishing at loop-level (and have a quite interesting structure which is briefly discussed in Section 6.1). Can you see how the argument based on (2.109) and (2.110) is changed at 1-loop level?

Let us move on and flip one more helicity: $A_n(1^-2^-3^+\ldots n^+)$. We choose the reference q_is to maximize the number of vanishing dot-products of polarization vectors. The choice $q_1 = q_2 = p_n$ and $q_3 = q_4 = \cdots = q_n = p_1$ implies that all $\epsilon_i \cdot \epsilon_j = 0$ vanish, except $\epsilon_{2-} \cdot \epsilon_{i+}$ for $i = 3, \ldots, n-1$. The polarization vector of gluon 2 can only appear once, so the terms in (2.109) can take the schematic form

$$A_n(1^-2^-3^+\ldots n^+) \sim \sum_{\text{diagrams}} \frac{\sum (\epsilon_{2-} \cdot \epsilon_{i+})(\epsilon_j \cdot k_l)^{n-2}}{\prod P_I^2}. \qquad (2.113)$$

Since only one product of ϵ_i^μs can be non-vanishing, $n-2$ factors of $(\epsilon_j \cdot k_l)$ were needed, and this exactly saturates the number of momentum vectors possible by dimensional analysis (2.108). Note also that with our choice of polarization vectors, any diagram that contributes to $A_n(1^-2^-3^+\ldots n^+)$ is trivalent.

Thus we conclude – based on dimensional analysis and thoughtful choices of the polarization vectors – that $A_n(1^-2^-3^+\ldots n^+)$ is the "first" gluon tree amplitude that can be non-vanishing, in the sense that having fewer negative helicity gluons gives a vanishing amplitude. More negative helicity states are also allowed, but one needs at least two positive helicity states to get a non-vanishing result. The only exception is for $n = 3$, as we have seen in (2.81).

The gluon amplitudes with two negative helicities and $n-2$ positive helicities are called *maximally helicity violating* – or simply **MHV** for short. The name "maximally helicity violating" comes from thinking of $2 \to (n-2)$ scattering. By crossing symmetry,

an outgoing gluon with $\begin{Bmatrix} \text{negative} \\ \text{positive} \end{Bmatrix}$ helicity is an incoming gluon with $\begin{Bmatrix} \text{positive} \\ \text{negative} \end{Bmatrix}$ helicity.

So with all outgoing particles, the process $A_n[1^+2^+3^+\ldots n^+]$ crosses over to $1^-2^- \to 3^+\ldots n^+$ in which the outgoing states all have the opposite helicity of the incoming states; it is "helicity violating." The process $1^-2^- \to 3^-4^+ \ldots n^+$ is a little less helicity violating and it crosses to $A_n[1^+2^+3^-4^+ \ldots n^+]$. We know from the above analysis that both these ***helicity violating*** processes vanish at tree-level in pure Yang–Mills theory. The process $1^-2^- \to 3^-4^-5^+ \ldots n^+$ – equivalent to $A_n[1^+2^+3^-4^-5^+ \ldots n^+]$ – is the most we can "violate" helicity and still get a non-vanishing answer at tree-level: therefore it is *maximally helicity violating*. As we have learned, the MHV gluon tree amplitudes are given by the Parke–Taylor formula,

$$A_n\big[1^+ \ldots i^- \ldots j^- \ldots n^+\big] = \frac{\langle ij \rangle^4}{\langle 12 \rangle \langle 23 \rangle \cdots \langle n1 \rangle}. \tag{2.114}$$

The MHV amplitudes play a key role in many studies of scattering amplitudes in (super) Yang–Mills theory. They are the simplest amplitudes in Yang–Mills theory. The next-to-simplest amplitudes are called **Next-to-MHV**, or **NMHV**, and this refers to the class of amplitudes with three negative helicity gluons and $n-3$ positive helicity gluons. This generalizes to the notation $\mathbf{N}^K\mathbf{MHV}$ amplitudes with $K+2$ negative helicity gluons and $n-K-2$ positive helicity gluons. When an amplitude has $(n-2)$ gluons of negative helicity and two of positive helicity, it is called **anti-MHV**. Anti-MHV amplitudes are obtained from the MHV amplitudes by flipping all helicities and exchanging angle and square brackets:

$$A_n\big[1^- \ldots i^+ \ldots j^+ \ldots n^-\big] = \frac{[ij]^4}{[12][23] \cdots [n1]}. \tag{2.115}$$

This is the anti-MHV Parke–Taylor formula.

Finally, let us mention that the MHV classification also applies to gravity amplitudes. We encountered the 4-graviton amplitude in Exercise 2.33 and saw how little group scaling and dimensional analysis fixed the 3-graviton amplitude in Exercise 2.34. In gravity, an N^KMHV graviton amplitude has $K+2$ negative helicity gravitons and $N-K-2$ positive helicity gravitons. Graviton amplitudes will show up occasionally throughout the book, but are treated more systematically in Chapter 12.

2.8 Analytic properties of tree amplitudes

A tree-level amplitude is a rational function of Lorentz invariants, such as $p_{i\mu}p_j^\mu$ and $p_{i\mu}\epsilon_j^\mu$, and it can have additional structure from color factors. In the spinor helicity formalism, a tree amplitude of massless particles is a rational function of the (complex-valued) kinematic invariants $\langle ij \rangle$ and $[ij]$, it scales homogeneously under little group scaling (2.96), and it has definite mass-dimension (2.102). But these are not the only properties of an amplitude: in addition, the physics places constraints on the analytic structure of the rational function.

Tree amplitudes must have a certain pole structure in order to legitimately describe a physical scattering process. ***Locality*** tells us that any pole of a tree-level amplitude must correspond to a propagating particle going on-shell, in other words all interactions are either local or mediated by a physical particle. This can be traced back to the locality of the Lagrangian. Its locality implies that there exists a choice of fields and a gauge such that the Feynman rules give interaction vertices that are constants or polynomial in the momenta, but which do not have any poles. Thus, in such a gauge, the only poles in the tree amplitude come from the propagators; examples are the Feynman gauge or Gervais–Neveu gauge in the standard formulation of Yang–Mills theory. Since the amplitude is gauge-independent, it must be true in *any gauge* that the only poles are those that arise from propagators going on-shell, i.e. from the exchanges of physical particles.[11] Thus the only poles that can appear in a tree amplitude of massless particles are of the form

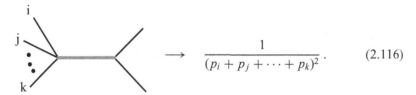

$$\longrightarrow \qquad \frac{1}{(p_i + p_j + \cdots + p_k)^2} \,. \qquad (2.116)$$

For color-ordered tree amplitudes, the propagators contain only momenta of consecutive external legs, i.e.

$$\frac{1}{(p_i + p_{i+1} + \cdots + p_{i+m})^2} \,. \qquad (2.117)$$

It may appear that the above statements hold only if we use formal polarization vectors ϵ_i^μ; for if we express them in terms of spinor variables as in (2.50), then one would invariably introduce factors such as $\langle pq \rangle$ or $[pq]$ in the denominators of the Feynman diagrams. While this is indeed true, poles from such factors depend on the choice of reference spinors q, so this means that the location of such a pole is gauge dependent. Since the amplitude is a gauge-invariant quantity, this dependence must drop out in the sum over all diagrams that contribute to the tree amplitude. In other words, the residue of such a pole must vanish in the full tree amplitude. It is an example of a ***spurious pole***, i.e. an unphysical pole that does not correspond to an exchange of a physical particle.

Field redefinitions and (obscure) gauge choices can give rise to expressions for the amplitudes in which the individual Feynman diagrams have unphysical poles. Of course, the amplitude is independent of field redefinitions and gauge choices, so the residues of such spurious poles must cancel among the diagrams to give zero. In the spinor helicity formalism, spurious poles arise from denominator factors of the form

$$\frac{1}{[i|p_i + \cdots + p_j|k\rangle} \,. \qquad (2.118)$$

In the amplitude such spurious poles must have vanishing residue.

---

[11] We encountered locality in Section 2.6 when we discussed the constraints on 3-point amplitudes from little group scaling, dimensional analysis, and locality of the Lagrangian. Also, when we discussed the helicity structure of gluon amplitudes in Section 2.7, we assumed that we were using a standard gauge, such as the Feynman gauge or Gervais–Neveu gauge, in which the only poles come from the propagators and the Feynman diagrams are of the form (2.109).

Note that the locality requirement does not exclude poles of the form $1/\langle ij \rangle$ or $1/[ij]$, where i and j are external legs. This is because we can always write $1/\langle ij \rangle = [ij]/(p_i + p_j)^2$ and $1/[ij] = \langle ij \rangle/(p_i + p_j)^2$. The exception is the 3-particle amplitudes; but special 3-particle kinematics implies that the apparent poles in, for example, the Parke–Taylor expression $\frac{\langle 12 \rangle^3}{\langle 23 \rangle \langle 31 \rangle}$ have vanishing residues. *Thus, the locality requirement for a tree amplitude is that any pole with non-vanishing residue must correspond to the propagator of a physical particle going on-shell.*

There are special limits of the external momenta in which tree amplitudes behave in a well-defined way. These are ***soft*** and ***collinear singularities***. Let us consider them in turn.

Soft singularities

Soft singularities occur when the momentum of a massless particle goes to zero. Feynman diagrams of the form

$$(2.119)$$

develop a singularity in this limit because the propagator, shown in gray, goes on-shell. As all other diagrams are finite, the amplitude becomes dominated by this subset of diagrams. With the propagator going on-shell, the leading part of the amplitude factorizes into a singular factor times an $(n-1)$-point amplitude.

As an example, consider the color-ordered amplitude $A_n[\ldots, f^+_{i-1}, i^-, f^-_{i+1}, \ldots]$ in Yang–Mills theory coupled to adjoint fermions f. Particle i is a negative-helicity gluon and $i-1$ and $i+1$ are fermions of helicity $+1/2$ and $-1/2$, respectively. In the soft-gluon limit, $p_i \to 0$, the diagrams relevant for the soft-gluon singularity are[12]

$$= -\frac{\langle i+1, i \rangle [q, i+1]}{[qi] s_{i,i+1}} A_{n-1} + \frac{[i-1, q]\langle i, i-1 \rangle}{[qi] s_{i,i-1}} A_{n-1}$$

$$= -\frac{[i+1, i-1]}{[i+1, i][i, i-1]} \times A_{n-1}. \qquad (2.120)$$

The fermion external wavefunctions for legs $i-1$ and $i+1$ in $A_{n-1}[\ldots, f^+_{i-1}, f^-_{i+1}, \ldots]$ come from the numerator of the fermion propagator. Note how the reference spinor $|q]$ of the gluon polarization vector drops out in the sum to leave the overall singular factor $\frac{[i+1,i-1]}{[i+1,i][i,i-1]}$ gauge-independent. This factor is called the ***soft function*** (or soft factor), and

---

[12] Dropping coupling constants and color factors.

it is singular since the spinor $|i]$ is near-zero. In general, for soft gluons in Yang–Mills theory, the amplitude behaves as

$$A_n\big[\dots, i-1, i^\pm, i+1, \dots\big] \xrightarrow{\;p_i \to 0\;} \mathcal{S}^\pm_{i-1,i,i+1} \times A_{n-1}\big[\dots, i-1, i+1, \dots\big],$$
(2.121)

where the soft functions $\mathcal{S}^\pm_{i-1,i,i+1}$ depend on the helicity of the soft gluon and are

$$\mathcal{S}^-_{i-1,i,i+1} = -\frac{[i+1, i-1]}{[i+1, i][i, i-1]}, \qquad \mathcal{S}^+_{i-1,i,i+1} = \frac{\langle i+1, i-1\rangle}{\langle i+1, i\rangle\langle i, i-1\rangle}.$$
(2.122)

The soft functions are independent of the helicities of the adjacent "hard" legs; the helicities of the hard particles are inherited by A_{n-1}.

▶ **Exercise 2.38**

As a test to see that the soft function is indeed independent of the helicities of the hard legs, replace the adjoint fermions with two adjoint scalars and show that the same soft function (2.122) is produced in the soft-gluon limit.

▶ **Exercise 2.39**

Show that the MHV amplitude (2.114) has the correct soft behavior (2.121).

Note that (2.121) and (2.122) were derived from Feynman rules. If you suspect that the behavior of the soft limit is not universal but depends on the theory at hand, you are certainly correct. Consider the 6-point amplitude in $\lambda\phi^4$ theory. The single soft-scalar limit does not lead to a singularity, because no internal propagator goes on-shell, as is clear from diagram (a):

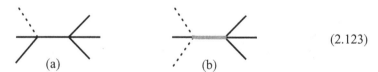

(2.123)

However, if we take a double-soft limit in which two scalars simultaneously go soft, as in diagram (b) above, one indeed finds a propagator singularity and the result is a singular factor times a residual 4-scalar amplitude.

To summarize, the tree-level amplitude must be a rational function of kinematic invariants such that only propagator-like singularities are present. For Yang–Mills theory coupled to matter fields, the amplitude must factorize as (2.121) when a gluon goes soft. Thus soft limits are useful for testing that the result of an amplitude calculation is indeed correct.

Collinear singularities

Another type of amplitude singularity arises when two external momenta become collinear. Diagrams in which the two external legs attach to a common 3-vertex are then divergent because the internal propagator goes on-shell:

$$(2.124)$$

In the collinear limit, the momentum configuration can be parameterized as

$$p_i = (1 - z)\, p_P \,, \qquad p_{i+1} = z\, p_P \,, \tag{2.125}$$

where p_P is on-shell, $p_P^2 = 0$. The parameter z cancels in the sum $p_i + p_{i+1}$, so momentum conservation for the original n on-shell momenta, $p_1 + p_2 + \cdots p_i + p_{i+1} + \cdots + p_n = 0$, becomes momentum conservation for $n - 1$ on-shell momenta $p_1 + p_2 + \cdots + p_P + \cdots + p_n = 0$, as appropriate for an $(n-1)$-point amplitude.

Let us examine the collinear limit of the MHV amplitude (2.114). Consider $A_n\big[1^-2^-3^+ \ldots n^+\big]$ and take legs 1 and n to be collinear:

$$A_n\big[1^-2^-3^+ \ldots n^+\big] = \frac{\langle 12 \rangle^3}{\langle 23 \rangle \cdots \langle n1 \rangle} \xrightarrow{\ 1 \| n\ } \frac{z^2}{\sqrt{z(z-1)}\langle n1 \rangle}\, \frac{\langle P2 \rangle^3}{\langle 23 \rangle \cdots \langle n-1, P \rangle} \,,$$

$$(2.126)$$

where $|1\rangle = \sqrt{z}|P\rangle$, $|n\rangle = \sqrt{1-z}|P\rangle$ (and similarly for the square spinors) using (2.125). Thus, in the collinear limit, the n-point MHV amplitude factorizes into the $(n-1)$-point amplitude $A_{n-1}\big[P^-2^-3^+ \ldots (n-1)^+\big]$ multiplied by a singular **splitting function**

$$\mathrm{Split}_+\big(n^+, 1^-\big) = \frac{z^2}{\sqrt{z(1-z)}\,\langle n1 \rangle} \,. \tag{2.127}$$

We write this as

$$A_n\big[1^-2^-3^+ \ldots n^+\big] \xrightarrow{\ 1 \| n\ } \mathrm{Split}_+\big(n^+, 1^-\big)\, A_{n-1}\big[P^-2^-3^+ \ldots (n-1)^+\big]. \tag{2.128}$$

This splitting function is singular because $\langle n1 \rangle$ is zero in the collinear limit. Note that – unlike the soft function – the splitting function depends on the helicities of the collinear particles as well as the helicity of the particle with momentum P.

▶ **Exercise 2.40**

For the MHV amplitude, show that the splitting function for the collinear limit with two adjacent positive helicity gluons i and $i + 1$ is

$$\mathrm{Split}_-\big(i^+, (i+1)^+\big) = \frac{1}{\sqrt{z(1-z)}\,\langle i, i+1 \rangle} \,. \tag{2.129}$$

Show that the results (2.128) and (2.129) are consistent with little group scaling.

For more general splitting functions, see the review article [11] and references therein.

At this stage, you might have noticed that all singularities of the MHV gluon amplitude precisely correspond to 2-particle collinear singularities. Why are there no multi-particle singularities of the form (2.117)? The reason is simply that the "residue" of such a pole should factorize into two amplitudes with $n \geq 4$-point, since the intermediate momentum becomes on-shell:

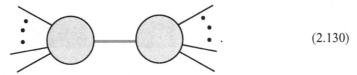

$$(2.130)$$

However, for the two gluon amplitudes in the residue to be non-vanishing, each must have two negative helicity gluons among its external states. The original n lines have two gluons with negative helicity and we can only get one from the on-shell internal line in (2.130).[13] So this means that there are not enough negative helicity gluons for the factorization channel (2.130) to have non-vanishing residue for an MHV amplitude. Hence the MHV amplitude only has the 2-particle factorization channels for the collinear limits. For N^kMHV amplitudes, multi-particle poles do appear.

The factorization property of tree-level scattering amplitudes has important and interesting consequences. Given that the residues of multi-particle poles are products of lower-point on-shell amplitudes, can one use this to set up a recursive method for calculating higher-point amplitudes? Indeed, this idea is central to the on-shell approach to calculate amplitudes. This is the next subject: go ahead to Chapter 3.

[13] A negative helicity gluon going out from the left subamplitude of the diagram (2.130) is, via crossing symmetry, a positive helicity gluon when viewed as an outgoing state from the right subamplitude.

3 On-shell recursion relations at tree-level

Recursion relations provide a method for building higher-point amplitudes from lower-point information. The 1988 *off-shell* recursion relations by Berends and Giele [12] construct n-point parton amplitudes from building blocks that are lower-point amplitudes with one leg off-shell (see the review articles [1, 3]). This off-shell method continues to be useful as an algorithm for efficient numerical evaluation of scattering amplitudes. In this review, we focus on the "modern" recursive methods whose building blocks are *on-shell* amplitudes. These *on-shell recursion relations* are elegant in that they use input only from gauge-invariant objects and they have proven very powerful for elucidating the mathematical structure of scattering amplitudes.

In the on-shell approaches, a key idea is to use the power of complex analysis to exploit the analytic properties of on-shell scattering amplitudes. The derivation of on-shell recursion relations is a great example of this, as we shall see in Section 3.1. The most famous on-shell recursion relations are the ***BCFW recursion relations*** of Britto, Cachazo, Feng, and Witten [13, 14], but there are other versions based on the same idea as BCFW, namely the use of complex deformations of the external momenta. We describe the idea here, first in a very general formulation in Section 3.1, then specialize the results to derive the BCFW recursion relations in Section 3.2. We illustrate the BCFW method with a selection of examples, including an inductive proof of the Parke–Taylor formula (2.83). Section 3.3 contains a discussion of when to expect existence of recursion relations in general local QFTs. Finally, in Section 3.4, we present the CSW construction (Cachazo-Svrcek-Witten [15]), also called the ***MHV vertex expansion***.

3.1 Complex shifts and Cauchy's theorem

An on-shell amplitude A_n is characterized by the momenta of the external particles and their type (for example a helicity label h_i for massless particles). We focus here on massless particles so $p_i^2 = 0$ for all $i = 1, 2, \ldots, n$. Of course, momentum conservation $\sum_{i=1}^n p_i^\mu = 0$ is also imposed.

Let us now introduce n complex-valued vectors r_i^μ (some of which may be zero) such that

(i) $\displaystyle\sum_{i=1}^n r_i^\mu = 0$,

(ii) $r_i \cdot r_j = 0$ for all $i, j = 1, 2, \ldots, n$. In particular $r_i^2 = 0$, and

(iii) $p_i \cdot r_i = 0$ for each i (no sum).

These vectors r_i are used to define n shifted momenta

$$\hat{p}_i^\mu \equiv p_i^\mu + z\, r_i^\mu \qquad \text{with } z \in \mathbb{C}. \tag{3.1}$$

Note that

(A) By property (i), momentum conservation holds for the shifted momenta: $\sum_{i=1}^n \hat{p}_i^\mu = 0$.

(B) By (ii) and (iii), we have $\hat{p}_i^2 = 0$, so each shifted momentum is on-shell.

(C) For a non-trivial[1] subset of generic momenta $\{p_i\}_{i \in I}$, define $P_I^\mu = \sum_{i \in I} p_i^\mu$. Then $\hat{P}_I^2$ is *linear* in z:

$$\hat{P}_I^2 = \left(\sum_{i \in I} \hat{p}_i \right)^2 = P_I^2 + z\, 2 P_I \cdot R_I \qquad \text{with} \quad R_I = \sum_{i \in I} r_i , \tag{3.2}$$

because the z^2 term vanishes by property (ii). We can write

$$\hat{P}_I^2 = -\frac{P_I^2}{z_I}(z - z_I) \qquad \text{with} \quad z_I = -\frac{P_I^2}{2 P_I \cdot R_I} . \tag{3.3}$$

As a result of (A) and (B), we can consider our amplitude A_n in terms of the shifted momenta $\hat{p}_i^\mu$ instead of the original momenta p_i^μ. In particular, it is useful to study the *shifted amplitude* as a function of z; by construction, it is a holomorphic function $\hat{A}_n(z)$. The amplitude with unshifted momenta p_i^μ is obtained by setting $z = 0$, $A_n = \hat{A}_n(z = 0)$.

We specialize to the case where A_n is a **tree-level amplitude**. In that case, the analytic structure of $\hat{A}_n(z)$ is very simple. As discussed in Section 2.8, the tree amplitude does not have any branch cuts – there are no logs, square-roots, etc. at tree-level. The tree amplitude is a rational function of the kinematic variables. As such, its analytic structure is captured by its poles and these are determined by the exchanges of physical particles. Thus, the shifted amplitude $\hat{A}_n(z)$ is a rational function of z and, for generic external momenta, its poles are all simple poles in the z-plane. To see that the poles are simple, consider the Feynman diagrams: the only way we can get poles is from the shifted propagators $1/\hat{P}_I^2$, where $\hat{P}_I$ is a sum of a non-trivial subset of the shifted momenta. By (C) above, $1/\hat{P}_I^2$ gives a simple pole at z_I, and for generic momenta, $P_I^2 \neq 0$ so $z_I \neq 0$. For generic momenta, no Feynman tree diagram can have more than one power of a given propagator $1/\hat{P}_I^2$; and poles of different propagators are at different locations in the z-plane. Hence, for generic momenta, $\hat{A}_n(z)$ only has simple poles and they are all located away from the origin $z = 0$. Note the underlying assumption of locality, i.e. that the amplitudes can be derived from some local Lagrangian so that the propagators determine the poles.

Let us now study the holomorphic function $\frac{\hat{A}_n(z)}{z}$ in the complex z-plane. Pick a contour that surrounds the simple pole at the origin. The residue at this pole is nothing but the unshifted amplitude, $A_n = \hat{A}_n(z = 0)$. Deforming the contour to surround all the other

---

[1] Non-trivial means at least two and no more than $n-2$ momenta such that $P_I^2 \neq 0$.

poles, Cauchy's theorem tells us that

$$A_n = -\sum_{z_I} \text{Res}_{z=z_I} \frac{\hat{A}_n(z)}{z} + B_n \,, \tag{3.4}$$

where B_n is the residue of the pole at $z = \infty$. By taking $z \to 1/w$ it is easily seen that B_n is the $O(z^0)$ term in the $z \to \infty$ expansion of A_n.

Now, so what? Well, at a z_I-pole the propagator $1/\hat{P}_I^2$ goes on-shell. In that limit, the shifted amplitude *factorizes* into two on-shell parts,

$$\hat{A}_n(z) \xrightarrow{z \text{ near } z_I} \hat{A}_L(z_I) \frac{1}{\hat{P}_I^2} \hat{A}_R(z_I) = -\frac{z_I}{z - z_I} \hat{A}_L(z_I) \frac{1}{P_I^2} \hat{A}_R(z_I) \,. \tag{3.5}$$

In the second step we used (3.3). This makes it easy to evaluate the residue at $z = z_I$:

$$-\text{Res}_{z=z_I} \frac{\hat{A}_n(z)}{z} = \hat{A}_L(z_I) \frac{1}{P_I^2} \hat{A}_R(z_I) = \quad \text{} \tag{3.6}$$

Note that – as opposed to Feynman diagrams – the momentum of the internal line in (3.6) is on-shell, $\hat{P}_I^2 = 0$, and the vertex-blobs represent shifted *on-shell amplitudes* evaluated at $z = z_I$; we call them *subamplitudes*. The rule for the internal line in the diagrammatic representation (3.6) is to write the scalar propagator $1/P_I^2$ of the *unshifted* momenta. Each subamplitude necessarily involves fewer than n external particles, hence all the residues at finite z can be determined in terms of on-shell amplitudes with less than n particles. This is the basis of the recursion relations.

The contribution B_n from the pole at infinity has no similar general expression in terms of lower-point amplitudes; there has been some work on how to compute B_n systematically (see for example [16, 17]), but currently there is not a general constructive method. Thus, in most applications, one assumes – or, much preferably, proves – that $B_n = 0$. This is most often justified by demonstrating the stronger statement that

$$\hat{A}_n(z) \to 0 \quad \text{for} \quad z \to \infty \,. \tag{3.7}$$

If (3.7) holds, we say that the shift (3.1) is **valid** (or **good**).

For a valid shift, the n-point on-shell amplitude is completely determined in terms of lower-point on-shell amplitudes as

$$A_n = \sum_{\text{diagrams } I} \hat{A}_L(z_I) \frac{1}{P_I^2} \hat{A}_R(z_I) = \sum_{\text{diagrams } I} \quad \text{} \tag{3.8}$$

The sum is over all possible factorization channels I. There is also implicitly a sum over all possible on-shell particle states that can be exchanged on the internal line: for example, for a gluon we have to sum the possible helicity assignments.

The recursive formula (3.8) gives a manifestly gauge-invariant construction of scattering amplitudes. Thus (3.8) is the general "prototype" of the *on-shell recursion relations for tree-level amplitudes* under a valid shift of the external momenta. We did not use any special properties of 4d spacetime, so the general derivation of the recursion relations is valid in D spacetime dimensions. In the following, we specialize to $D = 4$ again.

3.2 BCFW recursion relations

We shifted all external momenta democratically in (3.1), but with a parenthetical remark that some of the light-like shift-vectors r_i^μ might be trivial, $r_i^\mu = 0$. The BCFW shift is one in which exactly two lines, say i and j, are selected as the only ones with non-vanishing shift-vectors. In $D = 4$ spacetime dimension, the shift is implemented on angle and square spinors of the two chosen momenta:

$$|\hat{i}] = |i] + z\,|j]\,, \qquad |\hat{j}] = |j]\,, \qquad |\hat{i}\rangle = |i\rangle\,, \qquad |\hat{j}\rangle = |j\rangle - z\,|i\rangle\,. \qquad (3.9)$$

No other spinors are shifted. We call this a $[i, j\rangle$-shift. Note that $[\hat{i}k]$ and $\langle\hat{j}k\rangle$ are linear in z for $k \neq i, j$ while $\langle\hat{i}\hat{j}\rangle = \langle ij\rangle$, $[\hat{i}\hat{j}] = [ij]$, $\langle\hat{i}k\rangle = \langle ik\rangle$, and $[\hat{j}k] = [jk]$ remain unshifted.

▶ **Exercise 3.1**

Use (2.15) to calculate the shift vectors r_i^μ and r_j^μ corresponding to the shift (3.9). Then show that your shift vectors satisfy properties (i)–(iii) of Section 3.1.

With the two momenta i and j shifted according to (3.9), the BCFW recursion relation for tree amplitudes takes the form

$$A_n = \sum_{\text{diagrams } I} \hat{A}_{\mathrm{L}}(z_I)\,\frac{1}{P_I^2}\,\hat{A}_{\mathrm{R}}(z_I) = \sum_{\text{diagrams } I} \qquad\qquad\qquad . \qquad (3.10)$$

The sum is over all channels I such that the shifted lines i and j are on opposite sides of the factorization diagram in (3.10). As in the general recursion relations (3.8), there is also an implicit sum over all possible on-shell particle states that can be exchanged on the internal line.

Before diving into applications of the BCFW recursion relations (such as proving the Parke–Taylor formula), let us study the shifts a little further. As an example, consider the Parke–Taylor amplitude

$$A_n\big[1^- 2^- 3^+ \dots n^+\big] = \frac{\langle 12\rangle^4}{\langle 12\rangle\langle 23\rangle \cdots \langle n1\rangle}\,. \qquad (3.11)$$

Explore the following properties:

▶ **Exercise 3.2**

Convince yourself that for large-z the amplitude (3.11) falls off as $1/z$ under a $[-, -\rangle$-shift (i.e. choose i and j to be the two negative helicity lines). What happens under the three other types of shifts? Note the difference between shifting adjacent/non-adjacent lines.[2]

▶ **Exercise 3.3**

Consider the action of a $[1, 2\rangle$-shift of (3.11). Identify the simple pole. Calculate the residue of $\hat{A}_n(z)/z$ at this pole. Compare with (3.4). What happens if you try to repeat this for a $[1, 3\rangle$-shift?

▶ **Exercise 3.4**

The spin-1 polarization vectors (2.50) have denominators with $\langle qp \rangle$ and $[qp]$ (with q the reference spinor) which may shift under a BCFW shift (3.9) involving p. Why are there no terms in the on-shell recursion relations (3.8) corresponding to poles at $\langle q\hat{p} \rangle = 0$ or $[q\hat{p}] = 0$?

The validity of the BCFW recursion relations requires that the boundary term B_n in (3.4) is absent. The typical approach is to show that the shifted amplitude vanishes in the limit of large z, as in (3.7):

$$\hat{A}_n(z) \to 0 \quad \text{for} \quad z \to \infty. \tag{3.12}$$

In pure Yang–Mills theory, an argument [18] based on the background field method establishes the following large-z behavior of color-ordered gluon tree amplitudes under a BCFW shift of adjacent gluon lines i and j of helicity as indicated:

$$
\begin{array}{c|cccc}
[i, j\rangle & [-, -\rangle & [-, +\rangle & [+, +\rangle & [+, -\rangle \\
\hline
\hat{A}_n(z) \sim & \dfrac{1}{z} & \dfrac{1}{z} & \dfrac{1}{z} & z^3
\end{array}
. \tag{3.13}
$$

If i and j are non-adjacent, one gains an extra power $1/z$ in each case. Thus any one of the three types of shifts $[-, -\rangle$, $[-, +\rangle$, $[+, +\rangle$ gives valid recursion relations for gluon tree amplitudes.

We are now going to use the BCFW recursion relations (3.10) to construct an ***inductive proof of the Parke–Taylor formula*** (3.11). The formula (3.11) is certainly true for $n = 3$, as we saw in Section 2.5, and this establishes the base of the induction. For given n, suppose that (3.11) is true for amplitudes with less than n gluons. Then write down the recursion relation for $A_n[1^- 2^- 3^+ \ldots n^+]$ based on the valid $[1, 2\rangle$-shift: adapting from

---

[2] Of course, we cannot use the large-z behavior of formula (3.11) itself to justify the method to prove this formula! A separate argument is needed and will be discussed shortly.

(3.10), we have

$$A_n[1^-2^-3^+\ldots n^+] = \sum_{k=4}^{n}$$

$$= \sum_{k=4}^{n}\sum_{h_I=\pm} \hat{A}_{n-k+3}[\hat{1}^-,\hat{P}_I^{h_I},k^+\ldots,n^+]\frac{1}{P_I^2}$$
$$\hat{A}_{k-1}[-\hat{P}_I^{-h_I},\hat{2}^-,3^+\ldots,(k-1)^+]. \tag{3.14}$$

The internal momentum is labeled P_I, meaning that for a given $k = 4,\ldots,n$ we have $P_I = p_2 + p_3 + \cdots + p_{k-1}$ and $\hat{P}_I = \hat{p}_2 + \hat{p}_3 + \cdots + \hat{p}_{k-1}$. There are no diagrams where lines 1 and 2 belong to the same subamplitude, because in that case, the internal momentum would not be shifted and then there is no corresponding residue in (3.4). Only diagrams that preserve the color-ordering of the external states are included. Note that we are also explicitly including the sum over the possible helicity assignments for the particle exchanged on the on-shell internal line: if the exchanged gluon is outgoing from the left subamplitude and has negative helicity, then it will be a positive helicity outgoing gluon as seen from the right subamplitude.

Since one-minus amplitudes $A_n[-+\cdots+]$ vanish except for $n = 3$, (3.14) reduces to

$$A_n[1^-2^-3^+\ldots n^+] =$$

$$= \hat{A}_3[\hat{1}^-,-\hat{P}_{1n}^+,n^+]\frac{1}{P_{1n}^2}\hat{A}_{n-1}[\hat{P}_{1n}^-,\hat{2}^-,3^+\ldots(n-1)^+]$$

$$+\hat{A}_{n-1}[\hat{1}^-,\hat{P}_{23}^-,4^+\ldots,n^+]\frac{1}{P_{23}^2}\hat{A}_3[-\hat{P}_{23}^+,\hat{2}^-,3^+]. \tag{3.15}$$

It is understood here that $\hat{P}_I$ is evaluated at the residue value of $z = z_I$ such that $\hat{P}_I^2 = 0$. The notation P_{ij} means $P_{ij} = p_i + p_j$.

The next step is to implement *special kinematics* for the 3-point subamplitudes. In the first diagram of (3.15), we have a 3-point anti-MHV amplitude

$$\hat{A}_3[\hat{1}^-,-\hat{P}_{1n}^+,n^+] = \frac{[\hat{P}_{1n}\,n]^3}{[n\,\hat{1}][\hat{1}\,\hat{P}_{1n}]}. \tag{3.16}$$

Here we use the following convention for analytic continuation:

$$|-p\rangle = -|p\rangle, \quad |-p] = +|p]. \tag{3.17}$$

Since $\hat{P}_{1n}^{\mu} = \hat{p}_1^{\mu} + p_n^{\mu}$, the on-shell condition is

$$0 = \hat{P}_{1n}^2 = 2\hat{p}_1 \cdot p_n = \langle \hat{1}n \rangle [\hat{1}n] = \langle 1n \rangle [\hat{1}n]. \qquad (3.18)$$

For generic momenta, the only way for the RHS to vanish is for z to take a value such that $[\hat{1}n] = 0$. This means that the denominator in (3.16) vanishes! But so does the numerator: from

$$|\hat{P}_{1n}\rangle [\hat{P}_{1n}\, n] = -\hat{P}_{1n}|n] = -(\hat{p}_1 + p_n)|n] = |1\rangle [\hat{1}n] = 0, \qquad (3.19)$$

we conclude that $[\hat{P}_{1n}n] = 0$ since $|\hat{P}_{1n}\rangle$ is not zero. Similarly, one can show that $[\hat{1}\hat{P}_{1n}] = 0$. Thus, in the limit of imposing momentum conservation, all spinor products in (3.16) vanish; with the three powers in the numerator versus the two in the denominator, we conclude that special 3-point kinematics force $\hat{A}_3[\hat{1}^-, \hat{P}_{1n}^+, n^+] = 0$. So the contribution from the first diagram in (3.15) vanishes.

In the second diagram of (3.15), the 3-point subamplitude is also anti-MHV, but it does not vanish, since the shift of line 2 is on the angle spinor, not the square spinor. This way, the big abstract recursion formula (3.10) reduces – for the case of the $[-, -\rangle$ BCFW shift of an MHV gluon tree amplitude – to an expression with just a single non-vanishing diagram:

$$A_n\big[1^-2^-3^+ \ldots n^+\big] = \begin{array}{c} \hat{1}^- \\ n^+ \raisebox{-1ex}{\vdots} \fbox{L} \xrightarrow{\hat{P}_I} \fbox{R} \end{array} \begin{array}{c} \hat{2}^- \\ 3^+ \end{array}$$

$$= \hat{A}_{n-1}\big[\hat{1}^-, \hat{P}_{23}^-, 4^+, \ldots, n^+\big] \frac{1}{P_{23}^2} \hat{A}_3\big[-\hat{P}_{23}^+, \hat{2}^-, 3^+\big]. \quad (3.20)$$

Our inductive assumption is that (3.11) holds for $(n-1)$-point amplitudes. That, together with the result (2.81) for the 3-point anti-MHV amplitude, gives

$$A_n\big[1^-2^-3^+ \ldots n^+\big] = \frac{\langle \hat{1}\hat{P}_{23}\rangle^4}{\langle \hat{1}\hat{P}_{23}\rangle \langle \hat{P}_{23}\,4\rangle \langle 45\rangle \ldots \langle n\hat{1}\rangle} \times \frac{1}{\langle 23\rangle [23]} \times \frac{[3\hat{P}_{23}]^3}{[\hat{P}_{23}\,\hat{2}][\hat{2}3]}. \quad (3.21)$$

We could now proceed to evaluate the angle and square spinors for the shifted momenta. But it is more fun to introduce you to a nice little trick. Combine the factors from the numerator:

$$\langle \hat{1}\hat{P}_{23}\rangle [3\hat{P}_{23}] = -\langle \hat{1}\hat{P}_{23}\rangle [\hat{P}_{23}\,3] = \langle \hat{1}|\hat{P}_{23}|3] = \langle \hat{1}|(\hat{p}_2 + p_3)|3] = \langle \hat{1}|\hat{p}_2|3]$$
$$= -\langle \hat{1}\hat{2}\rangle [\hat{2}3] = -\langle 12\rangle [23]. \quad (3.22)$$

In the last step we used $\langle \hat{1}\hat{2}\rangle = \langle 12\rangle$ and $|\hat{2}] = |2]$. Playing the same game with the factors in the denominator, we find

$$\langle \hat{P}_{23}\,4\rangle [\hat{P}_{23}\,\hat{2}] = \langle 4|\hat{P}_{23}|\hat{2}] = \langle 4|3|2] = -\langle 43\rangle [32] = -\langle 34\rangle [23]. \quad (3.23)$$

Now use (3.22) and (3.23) in (3.21) to find

$$A_n\left[1^-2^-3^+\ldots n^+\right] = \frac{-\langle 12\rangle^3 [23]^3}{\left(-\langle 34\rangle [23]\right)\langle 45\rangle \cdots \langle n1\rangle \,\langle 23\rangle [23]\,[23]}$$

$$= \frac{\langle 12\rangle^4}{\langle 12\rangle\langle 23\rangle\langle 34\rangle\langle 45\rangle \cdots \langle n1\rangle}\,. \tag{3.24}$$

This completes the inductive step. With the 3-point gluon amplitude $A_3\left[1^-2^-3^+\right]$ fixed completely by little group scaling and locality, we have then proven the Parke–Taylor formula for all n. This *is* a lot easier than calculating Feynman diagrams!

You may at this point complain that we have only derived the Parke–Taylor formula recursively for the case where the negative helicity gluons are adjacent. Try your own hands on the proof for the non-adjacent case. In Chapter 4 we will use supersymmetry to derive a more general form of the tree-level gluon amplitudes: it will contain all MHV helicity arrangements in one compact expression.

We have now graduated from MHV-level and are ready to embark on the study of **NMHV amplitudes**. It is worthwhile to consider the 5-point example $A_5[1^-2^-3^-4^+5^+]$ even though this amplitude is anti-MHV: constructing it with a $[+, +\rangle$-shift is a calculation very similar to the MHV case – and that would by now be boring. So, instead, we are going to use a $[-, -\rangle$-shift to illustrate some of the manipulations used in BCFW recursion.

▷ *Example.* Consider the $[1, 2\rangle$-shift recursion relations for $A_5\left[1^-2^-3^-4^+5^+\right]$: there are two diagrams

$$A_5\left[1^-2^-3^-4^+5^+\right] = \underset{\substack{\text{diagram A}}}{\overset{\hat{1}^- \qquad \hat{P}_{15} \qquad \hat{2}^-}{\underset{5^+ \qquad\quad 4^+}{\bullet\!-\!\!\!\bullet\,3^-}}} + \underset{\substack{\text{diagram B}}}{\overset{\hat{1}^- \qquad \hat{P}_{23} \qquad \hat{2}^-}{\underset{4^+ \qquad\quad 3^-}{5^+\,\bullet\!-\!\!\!\bullet}}} . \tag{3.25}$$

We have indicated the required helicity for the gluon on the internal line. Had we chosen the opposite helicity option for the internal gluon in diagram A, the righthand subamplitude would have helicity structure $--\,-+$, so it would vanish. Diagram B also vanishes for the opposite choice of the helicity on the internal line. For the helicity choice shown, the righthand subamplitude of diagram B is MHV, $A_3[-\hat{P}_{23}^+, \hat{2}^-, 3^-]$, and since $|2\rangle$ is shifted, the special 3-particle kinematics actually makes $A_3[-\hat{P}_{23}^+, \hat{2}^-, 3^-] = 0$, just as we saw for the anti-MHV case in the discussion below (3.16). So *diagram B vanishes*, and we can focus on diagram A. Using the Parke–Taylor formula for the MHV subamplitudes, we get

$$A_5\left[1^-2^-3^-4^+5^+\right] = \frac{\langle \hat{1}\hat{P}\rangle^3}{\langle \hat{P}5\rangle\langle 5\hat{1}\rangle} \times \frac{1}{\langle 15\rangle[15]} \times \frac{\langle \hat{2}3\rangle^4}{\langle \hat{2}3\rangle\langle 34\rangle\langle 4\hat{P}\rangle\langle \hat{P}\hat{2}\rangle}\,. \tag{3.26}$$

Here $\hat{P}$ stands for $\hat{P}_{15} = \hat{p}_1 + p_5$. We have three powers of $|\hat{P}\rangle$ in the numerator and three in the denominator. A good trick to simplify such expressions is to multiply

(3.26) by $[\hat{P}X]^3/[\hat{P}X]^3$ for some useful choice of X such that $[\hat{P}X] \neq 0$. In this case, it is helpful to pick $X = 2$. Grouping terms conveniently together, we get:

- $\langle \hat{1}\hat{P}\rangle[\hat{P}2] = -\langle \hat{1}|\hat{1}+5|2] = \langle \hat{1}5\rangle[52] = -\langle 15\rangle[25]$ (since $|\hat{1}\rangle = |1\rangle$).
- $\langle 5\hat{P}\rangle[\hat{P}2] = -\langle 5|\hat{1}+5|2] = \langle 51\rangle[\hat{1}2] = \langle 51\rangle[12]$.
- $\langle 4\hat{P}\rangle[\hat{P}2] = -\langle 4|\hat{1}+5|2] = \langle 4|\hat{2}+3+4|2] = \langle 4|3|2] = -\langle 43\rangle[32] = -\langle 34\rangle[23]$.
- $\langle \hat{2}\hat{P}\rangle[\hat{P}2] = -2\,\hat{p}_2 \cdot \hat{P} = 2\,\hat{p}_2 \cdot (\hat{p}_2 + p_3 + p_4) = (\hat{p}_2 + p_3 + p_4)^2 - (p_3 + p_4)^2$
  $= \hat{P}^2 - \langle 34\rangle[34] = -\langle 34\rangle[34]$,

  since the amplitude is evaluated at z such that $\hat{P}^2 = 0$.

Using these expressions in (3.26) gives

$$A_5\big[1^-2^-3^-4^+5^+\big] = \frac{[25]^3\langle \hat{2}3\rangle^3}{[12][23][34][15]\langle 34\rangle^3}.\tag{3.27}$$

Despite the simplifications, there is some unfinished business for us to deal with: (3.27) depends on the shifted spinors via $\langle \hat{2}3\rangle$. This bracket must be evaluated at the residue value of $z = z_{15}$ which is such that $\hat{P}_{15}^2 = 0$:

$$0 = \hat{P}_{15}^2 = \langle 15\rangle[\hat{1}5], \quad \text{i.e.} \quad 0 = [\hat{1}5] = [15] + z_{15}[25], \quad \text{i.e.} \quad z_{15} = -\frac{[15]}{[25]}.\tag{3.28}$$

Use this and momentum conservation to write

$$\langle \hat{2}3\rangle = \langle 23\rangle - z_{15}\langle 13\rangle = \frac{\langle 23\rangle[25] + \langle 13\rangle[15]}{[25]} = \frac{\langle 34\rangle[45]}{[25]}.\tag{3.29}$$

Inserting this result into (3.27) we arrive at the expected anti-MHV Parke–Taylor expression

$$A_5\big[1^-2^-3^-4^+5^+\big] = \frac{[45]^4}{[12][23][34][45][51]}.\tag{3.30}$$

As noted initially, the purpose of this example was not to torture you with a difficult way to compute $A_5[1^-2^-3^-4^+5^+]$. The purpose was to illustrate the methods needed for general cases in a simple context. ◁

You may not be overly impressed with the simplicity of the manipulations needed to reduce the raw output of BCFW. Admittedly it requires some work. If you are unsatisfied, go ahead and try the calculations in this section with Feynman diagrams. Good luck.

Now you have seen the basic tricks needed to manipulate the expressions generated by BCFW. So you should get some exercise.

▶ Exercise 3.5

Let us revisit scalar-QED from the end of Section 2.4. Use little group scaling and locality to determine $A_3(\varphi\,\varphi^*\gamma^{\pm})$ and compare with your result from Exercise 2.15. Then use a $[4, 3\rangle$-shift to show that (see Exercise 2.16)

$$A_4\big(\varphi\,\varphi^*\gamma^+\gamma^-\big) = g^2\,\frac{\langle 14\rangle\langle 24\rangle}{\langle 13\rangle\langle 23\rangle}.\tag{3.31}$$

[Hint: this is not a color-ordered amplitude. See also Exercise 2.16.]
What is the large-z falloff of this amplitude under a $[4, 3\rangle$-shift?[3]

▶ Exercise 3.6

Calculate the 4-graviton amplitude $M_4(1^-2^-3^+4^+)$: first recall that little group scaling & locality fix the 3-graviton amplitudes as in Exercise 2.34. Then employ the $[1, 2\rangle$-shift BCFW recursion relations (they are valid [18, 20]).

Check the little group scaling and Bose-symmetry of your answer for $M_4(1^-2^-3^+4^+)$.

[Hint: your result should match one of the amplitudes in Exercise 2.33.]

Show that $M_4(1^-2^-3^+4^+)$ obeys the 4-point "KLT relations" [21]

$$M_4(1234) = -s_{12}\, A_4[1234]\, A_4[1243]\,, \tag{3.32}$$

where A_4 is your friend the Parke–Taylor amplitude with negative helicity states 1 and 2, and $s_{12} = -(p_1 + p_2)^2$ is a Mandelstam variable. When you are done, look up ref. [22] to see how difficult it is to do this calculation with Feynman diagrams.

Let us now take a look at some interesting aspects of BCFW for the ***split-helicity*** NMHV amplitude $A_6[1^-2^-3^-4^+5^+6^+]$. Let us first look at the recursion relations following from the $[1, 2\rangle$-shift that we are now quite familiar with. There are two non-vanishing diagrams:

$$A_6[1^-2^-3^-4^+5^+6^+] = \text{diagram A} + \text{diagram B}\,. \tag{3.33}$$

▶ Exercise 3.7

Show that the 23-channel diagram does not contribute in (3.33).

The *first thing* we want to discuss about the 6-gluon amplitude is the 3-particle poles in the expression (3.33). Diagram B involves a propagator $1/P_{156}^2$, so there is a 3-particle pole at $P_{156}^2 = 0$. By inspection of the ordering of the external states in $A_6[1^-2^-3^-4^+5^+6^+]$ there should be no distinction between the $(-++)$ 3-particle channels 561 and 345, so we would expect the amplitude to have a pole also at $P_{345}^2 = P_{126}^2 = 0$. But the $[1, 2\rangle$-shift recursion relation (3.33) does not involve any 126-channel diagram. How can it then possibly encode the correct amplitude? The answer is that it does and that the $P_{345}^2 = P_{126}^2 = 0$ pole is actually hidden in the denominator factor $\langle \hat{2}\hat{P}_{16}\rangle$ of the righthand subamplitude of diagram A in (3.33). Let us show how.

As in the 5-point example (3.26), we multiply the numerator and denominator both by three powers of $[\hat{P}_{16}\, 3]$. Then write

$$\langle \hat{2}\hat{P}_{16}\rangle[\hat{P}_{16}\, 3] = \langle 21\rangle[\hat{1}3] + \langle \hat{2}6\rangle[63]\,. \tag{3.34}$$

---

[3] Of course, we cannot use the BCFW result for the amplitude to validate the shift, only test self-consistency. The independent demonstration of the large-z falloff under the shift used here is given in [19].

It follows from $\hat{P}_{16}^2 = 0$ that $z_{16} = -[16]/[26]$, and this is then used to show that $\langle \hat{2}6 \rangle = (\langle 16 \rangle [16] + \langle 26 \rangle [26])/[26]$ and $[\hat{1}3] = [12][36]/[26]$. Plug these values into (3.34) to find

$$\langle \hat{2} \hat{P}_{16} \rangle [\hat{P}_{16}\, 3] = -\frac{[36]}{[26]}\big(\langle 12 \rangle [12] + \langle 16 \rangle [16] + \langle 26 \rangle [26]\big) = -\frac{[36]}{[26]}\, P_{126}^2 . \tag{3.35}$$

So there you have it: the 3-particle pole P_{126}^2 is indeed encoded in the BCFW result (3.33).

The *second thing* we want to show you is the actual representation for the 6-gluon NMHV tree amplitude, as it follows from (3.33):

$$A_6\big[1^- 2^- 3^- 4^+ 5^+ 6^+\big] = \frac{\langle 3|1+2|6]^3}{P_{126}^2 [21][16]\langle 34 \rangle \langle 45 \rangle \langle 5|1+6|2]}$$
$$+ \frac{\langle 1|5+6|4]^3}{P_{156}^2 [23][34]\langle 56 \rangle \langle 61 \rangle \langle 5|1+6|2]} . \tag{3.36}$$

The expression (3.36) may not look quite as delicious as the Parke–Taylor formula, but remember that it contains the same information as the sum of 38 Feynman diagrams!

▶ **Exercise 3.8**

Check the little group scaling of (3.36). Fill in the details for converting the two diagrams in (3.33) to find (3.36).

The *third thing* we would like to emphasize is that using the $[1, 2\rangle$-shift recursion relations is just one way to calculate $A_6[1^- 2^- 3^- 4^+ 5^+ 6^+]$. What happens if we use the $[2, 1\rangle$-shift? Well, now there are three non-vanishing diagrams:

$$\tag{3.37}$$

Special 3-particle kinematics force the diagram A′ to have helicity structure anti-MHV×NMHV, as opposed to the similar diagram A in (3.33) which has to be MHV×MHV. It is the first time we see a lower-point NMHV amplitude show up in the recursion relations. This is quite generic: the BCFW relations are recursive both in particle number n and in N^KMHV level K.

The two BCFW representations (3.33) and (3.37) look quite different. In order for both to describe the same amplitude, there has to be a certain identity that ensures that diagrams A+B = A′ + B′+ C′. To show that this identity holds requires a nauseating trip through Schouten identities and momentum conservation relations in order to manipulate the angle and square brackets into the right form: numerical checks can save you a lot of energy when dealing with amplitudes with more than five external lines. It turns out that the identities that guarantee the equivalence of BCFW expressions such as A+B and A′ +B′+C′ actually originate from powerful residue theorems [23] related to quite different formulations of the amplitudes. This has to do with the description of amplitudes in the Grassmannian – we get to that in Chapters 9 and 10, but wanted to give you a hint of this interesting point here.

▶ **Exercise 3.9**

Show that the BCFW recursion relations based on the $[2, 3\rangle$-shift give the following representation of the 6-point "alternating helicity" gluon amplitude:

$$A_6\left[1^+2^-3^+4^-5^+6^-\right] = \{M_2\} + \{M_4\} + \{M_6\},\qquad(3.38)$$

where

$$\{M_i\} = \frac{\langle i, i+2\rangle^4[i+3, i-1]^4}{\tilde{P}_i^2\,\langle i|\tilde{P}_i|i+3]\langle i+2|\tilde{P}_i|i-1]\langle i, i+1\rangle\langle i+1, i+2\rangle[i+3, i-2][i-2, i-1]},$$

$$(3.39)$$

and $\tilde{P}_i = P_{i,i+1,i+2}$. [Hint: $\{M_4\}$ is the value of the 12-channel diagram.]

In Chapter 9 we discover that each $\{M_i\}$ can be understood as the residue associated with a very interesting contour integral (different from the one used in the BCFW argument).

The *fourth thing* worth discussing further is the poles of scattering amplitudes. Color-ordered tree amplitudes can have physical poles only when the sum of momenta of *adjacent* external lines go on-shell. We touched on this point already in Section 2.8. There we also noted that MHV gluon amplitudes do not have multi-particle poles, only 2-particle poles. Now you have seen that 6-gluon NMHV amplitudes have both 2- and 3-particle poles. But as you stare intensely at (3.36), you will also note that there is a strange denominator-factor $\langle 5|1 + 6|2]$ in the result from each BCFW diagram. This does not correspond to a physical pole of the scattering amplitude: it is a ***spurious pole***. The residue of this unphysical pole better be zero – and it is: the spurious pole cancels in the sum of the two BCFW diagrams in (3.36). It is typical that BCFW packs the information of the amplitudes into compact expressions, but the cost is the appearance of spurious poles; this means that in the BCFW representation the locality of the underlying field theory is not manifest. Elimination of spurious poles in the representations of amplitudes leads to interesting results [24] that we discuss in Chapter 10.

Finally, for completeness, note that the color-ordered NMHV amplitudes $A_6[1^-2^-3^+4^-5^+6^+]$ and $A_6[1^-2^+3^-4^+5^-6^+]$ are inequivalent to the split-helicity amplitude $A_6[1^-2^-3^-4^+5^+6^+]$. More about this in Chapter 4.

Other comments:

1) In our study of the recursion relations, we kept insisting on "generic" momenta. However, special limits of the external momenta place useful and interesting constraints on the amplitudes: the behavior of amplitudes under ***collinear limits*** and ***soft limits*** is described in Section 2.8.

2) In some cases, the shifted amplitudes have "better than needed" large-z behavior. For example, this is the case for a BCFW shift of two non-adjacent same-helicity lines in the color-ordered Yang–Mills amplitudes: $\hat{A}_n(z) \to 1/z^2$ for large z. The vanishing of the amplitude at large z implies the validity of a recursion relation for A_n. Let us

briefly outline the reason. Start with $\oint_C \frac{\hat{A}_n(z)}{z} = 0$ with C a contour that surrounds all the simple poles. The unshifted amplitude A_n, which is the residue of the $z = 0$ pole, is therefore (minus) the sum of all the other residues. We write $A_n = \sum_I d_I$, where d_I is the factorization diagram (3.8) associated with the residue at $z = z_I$. This summarizes the derivation of the recursion relations from Section 3.1. Now, an extra power in the large-z falloff, $\hat{A}_n(z) \sim 1/z^2$, means that there is also a **bonus relation** from $\oint_C \hat{A}_n(z) = 0$ (with C as before). It gives $\sum_I d_I z_I = 0$, because there is no residue at $z = 0$. Bonus relations have practical applications, for example they have been used to demonstrate equivalence of different expressions for graviton amplitudes [25].

3.3 When does it work?

In Section 2.6 we learned that 3-point amplitudes for massless particles are uniquely determined by little group scaling, locality, and dimensional analysis. As we have just seen, with the on-shell BCFW recursion relations, we can construct all higher-point gluon tree amplitudes from the input of just the 3-point gluon amplitudes. That is a lot of information obtained from very little input! It prompts suspicion: when can we expect on-shell recursion to work? We will look at some examples now.

Yang–Mills theory and gluon scattering. The Yang–Mills Lagrangian (2.66) has two types of interaction terms, the cubic $A^2 \partial A$ and the quartic A^4. Given the former, the latter is needed for gauge invariance, and the quartic must be included along with the cubic as interaction vertices in the Feynman rules. However, existence of valid BCFW recursion relations indicates that the cubic term $A^2 \partial A$ fully captures the information needed for all on-shell gluon amplitudes, at least at tree-level, with no need for A^4. The key distinction is that the 3-vertex (and hence the Feynman rule cubic vertex) is an *off-shell* gauge non-invariant object, while the 3-point *on-shell* amplitude is gauge invariant. Since A^4 is determined from $A^2 \partial A$ by the requirement of off-shell gauge invariance of the Lagrangian, it contains no new on-shell information. In a sense, that is why the recursion relations for on-shell gluon amplitudes even have a chance to work with input only from the on-shell 3-point amplitudes.

We can rephrase the information contents of $A^2 \partial A$ in a more physical way. The actual input is then this: a 4d local theory with massless spin-1 particles (and no other dynamical states) and a dimensionless coupling constant. With valid recursion relations, this information is enough to fix the entire gluon tree-level scattering matrix!

Scalar-QED. As a second example, consider scalar-QED. The interaction between the photons and the scalar particles created/annihilated by a complex scalar field φ is encoded by the covariant derivatives $D_\mu = \partial_\mu - ieA_\mu$ in

$$\mathcal{L} \supset -|D\varphi|^2 = |\partial\varphi|^2 + ieA^\mu\big[(\partial_\mu\varphi^*)\varphi - \varphi^*\partial_\mu\varphi\big] - e^2 A^\mu A_\mu \varphi^*\varphi \,. \tag{3.40}$$

In terms of Feynman diagrams, the tree amplitude $A_4(\varphi\,\varphi^*\gamma\,\gamma)$ is constructed from the sum of two pole diagrams and the contact term from the quartic interaction (Exercise 2.17).

We have seen in Exercise 3.5 that this 4-point amplitude is constructible via BCFW. So it is clear that only the information in the 3-point vertices is needed, and the role of $A_\mu A^\mu \varphi^* \varphi$ is just to ensure off-shell gauge invariance of the Lagrangian. Thus this case is just like the Yang–Mills example above.

Thus emboldened, let us try to compute the 4-scalar tree amplitude $A_4(\varphi \varphi^* \varphi \varphi^*)$ using BCFW recursion (3.10). Using a $[1, 3\rangle$-shift, there are two diagrams and their sum simplifies to

$$A_4^{\text{BCFW}}(\varphi \varphi^* \varphi \varphi^*) = \tilde{e}^2 \frac{\langle 13 \rangle^2 \langle 24 \rangle^2}{\langle 12 \rangle \langle 23 \rangle \langle 34 \rangle \langle 41 \rangle} . \tag{3.41}$$

If, however, we calculate this amplitude using Feynman rules from the interaction terms in (3.40), we get

$$A_4^{\text{Feynman}}(\varphi \varphi^* \varphi \varphi^*) = \tilde{e}^2 \left(1 + \frac{\langle 13 \rangle^2 \langle 24 \rangle^2}{\langle 12 \rangle \langle 23 \rangle \langle 34 \rangle \langle 41 \rangle} \right). \tag{3.42}$$

Ugh! So BCFW did not compute the amplitude we expected. So what did it compute? Well, let us think about the input that BCFW knows about: 4d local theory with massless spin-1 particles and charged massless spin-0 particles (and no other dynamical states) and a dimensionless coupling constant. Note that included in this input is the possibility of a 4-scalar interaction term $\lambda |\varphi|^4$. So more generally, we should consider the scalar-QED action from (2.64):

$$\mathcal{L} = -\frac{1}{4} F_{\mu\nu} F^{\mu\nu} - |D\varphi|^2 - \frac{1}{4} \lambda |\varphi|^4$$
$$= -\frac{1}{4} F_{\mu\nu} F^{\mu\nu} - |\partial\varphi|^2 + ieA^\mu \big[(\partial_\mu \varphi^*) \varphi - \varphi^* \partial_\mu \varphi \big] - e^2 A^\mu A_\mu \varphi^* \varphi - \frac{1}{4} \lambda |\varphi|^4 . \tag{3.43}$$

In Exercise 2.19 you were asked to calculate $A_4(\varphi \varphi^* \varphi \varphi^*)$ in this model. The answer was given in (2.65): it is

$$A_4(\varphi \varphi^* \varphi \varphi^*) = -\lambda + \tilde{e}^2 \left(1 + \frac{\langle 13 \rangle^2 \langle 24 \rangle^2}{\langle 12 \rangle \langle 23 \rangle \langle 34 \rangle \langle 41 \rangle} \right). \tag{3.44}$$

So it is clear now that we have a family of scalar-QED models, labeled by λ, and that our BCFW calculation produced the very special case of $\lambda = \tilde{e}^2$. How can we understand this? Validity of the recursion relations requires the absence of the boundary term B_n (see Section 3.1). For the general family of scalar-QED models, there is a boundary term under the $[1, 3\rangle$-shift, and its value is $-\lambda + \tilde{e}^2$ (as can be seen from (3.44) by direct computation). The special choice $\lambda = \tilde{e}^2$ eliminates the boundary term, and that is then what BCFW without a boundary term computes.

The lesson is that for general λ, there is no way the 3-point interactions can know the contents of $\lambda |\varphi|^4$: it provides independent gauge-invariant information. That information needs to be supplied in order for recursion to work, so in this case one can at best expect recursion to work beyond 4-point amplitudes. The exception is of course if some symmetry, or other principle, determines the information in $\lambda |\varphi|^4$ in terms of the 3-field terms. This is what we find for $\lambda = \tilde{e}^2$. Actually, the expression (3.41) is the correct result for certain 4-scalar amplitudes in $\mathcal{N} = 2$ and $\mathcal{N} = 4$ super Yang–Mills theory (see Chapter 4), and in those cases the coupling of the 4-scalar contact term *is* fixed by the Yang–Mills coupling by supersymmetry.

Scalar theory $\lambda\phi^4$. The previous example makes us wary of $\lambda\phi^4$ interaction in the context of recursion relations – and rightly so. Suppose we consider $\lambda\phi^4$ theory with no other interactions. It is clear that one piece of input must be given to start any recursive approach, namely in this case the 4-scalar amplitude $A_4 = \lambda$. In principle, one might expect on-shell recursion to determine all tree-level A_n amplitudes with $n > 4$ from $A_4 = \lambda$ – what else could interfere? After all, this is the only interaction in the Feynman diagrams. However, given that the 6-scalar amplitude is $A_6 = \lambda^2(\frac{1}{s_{123}} + \dots)$, it is clear that any BCFW shift gives $O(z^0)$-behavior for large z and hence there are no BCFW recursion relations without boundary term for A_6 in $\lambda\phi^4$ theory. Inspection of the Feynman diagrams reveals that the $O(z^0)$-contributions are exactly the diagrams in which the two shifted lines belong to the same vertex. The sum of such diagrams equals the boundary term B_n from (3.4). Thus, in this case of $\lambda\phi^4$ theory one can reconstruct B_n recursively.[4] Hence A_4 does suffice to completely determine all tree amplitudes A_n for $n > 4$ in $\lambda\phi^4$ theory; but it is (in more than one sense) a rather trivial example.

$\mathcal{N} = 4$ **super Yang–Mills theory.** This is the favorite theory of many amplitugicians. We will review the theory in more detail in Section 4.4, for now we just comment on a few relevant aspects. The spectrum consists of 16 massless states: gluons $g^\pm$ of helicity ± 1, four gluinos λ^a and λ_a of helicity $\pm 1/2$, and six scalars S^{ab}. The indices $a, b = 1, 2, 3, 4$ are labels for the global $SU(4)$ symmetry. The Lagrangian contains standard gluon self-interactions, with standard couplings to the gluinos and the scalars. All fields transform in the adjoint of the $SU(N)$ gauge group, so we consider color-ordered tree amplitudes defined in the same way as the color-ordered gluon amplitudes. The Lagrangian includes a scalar 4-point interaction term of a schematic form $[S, S]^2$. It contains, for example, the interaction $S^{12}S^{23}S^{34}S^{41}$. The result for the corresponding color-ordered amplitude is (suppressing the gauge coupling constant):

$$A_4\big[S^{12}S^{23}S^{34}S^{41}\big] = 1. \tag{3.45}$$

Since this amplitude has no poles, it cannot be obtained via direct factorization. Actually, the amplitude (3.45) and its cousin 4-scalar amplitudes with equivalent $SU(4)$ index structures are the only tree amplitudes of $\mathcal{N} = 4$ SYM that cannot be obtained from the BCFW recursion formula (3.10); that may seem surprising, but it is true [27].

When supersymmetry is incorporated into the BCFW recursion relations, *all* tree amplitudes of $\mathcal{N} = 4$ SYM can be determined by the 3-point gluon vertex alone. The so-called super-BCFW shift mixes the external states in such a way that even the 4-scalar amplitude (3.45) can be constructed recursively. We will introduce the super-BCFW shift in Section 4.5.

---

[4] See [16]. Or avoid the term at infinity by using an all-line shift [26], to be defined in Section 3.4.

Gravity. We have already encountered the 4-point MHV amplitude $M_4(1^-2^-3^+4^+)$: you "discovered" it from little group scaling in Exercise 2.33 and constructed it with BCFW in Exercise 3.6. The validity of the BCFW recursion relations for all tree-level graviton amplitudes [18, 20] means that the entire on-shell tree-level S-matrix for gravity is determined completely by the 3-vertex interaction of three gravitons. In contrast, the expansion of the Einstein–Hilbert action $\frac{1}{2\kappa^2} \int d^4x \sqrt{-g} R$ around the flatspace Minkowski metric $g_{\mu\nu} = \eta_{\mu\nu} + \kappa\, h_{\mu\nu}$ contains *infinitely* many interaction terms. It is remarkable that all these terms are totally irrelevant from the point of view of the on-shell tree-level S-matrix; their sole purpose is to ensure diffeomorphism invariance of the off-shell Lagrangian. For on-shell (tree) amplitudes, we do not need them. Much more about gravity amplitudes in Chapter 12.

Summary. We have discussed when to expect to have recursion relations for tree-level amplitudes. The main lesson is that we do not get something for nothing: input must be given and we can only expect to recurse that input with standard BCFW when all other information in the theory is fixed by our input via gauge invariance. If another principle – such as supersymmetry – is needed to fix the interactions, then that principle should be incorporated into the recursion relations for a successful recursive approach. Further discussion of these ideas can be found in [26], mostly in the context of another recursive approach known as CSW, which we will discuss briefly next.

3.4 MHV vertex expansion (CSW)

We introduced recursion relations in Section 3.1 in the context of general shifts (3.1) satisfying the set of conditions (i)–(iii). Then we specialized to the BCFW shifts in Section 3.2. Now we would like to show you another kind of recursive structure.

Consider a shift that is implemented via a "holomorphic" square-spinor shift:

$$|\hat{i}] = |i] + z\, c_i |X] \quad \text{and} \quad |\hat{i}\rangle = |i\rangle. \tag{3.46}$$

Here $|X]$ is an arbitrary reference spinor and the coefficients c_i satisfy $\sum_{i=1}^n c_i |i\rangle = 0$.

▶ **Exercise 3.10**

Show that the square-spinor shift (3.46) gives shift-vectors r_i that fulfill requirements (i)–(iii) in Section 3.1.

The choice $c_1 = \langle 23 \rangle$, $c_2 = \langle 31 \rangle$, $c_3 = \langle 12 \rangle$, and $c_i = 0$ for $i = 4, \ldots, n$ implies that the shifted momenta satisfy momentum conservation. This particular realization of the square-spinor shift is called the Risager shift [28].

We consider here a situation where all $c_i \neq 0$ so that all momentum lines are shifted via (3.46) – this is an ***all-line shift***. It can be shown [29] that N^KMHV gluon tree amplitudes fall off as $1/z^K$ for large z under all-line shift. So this means that all gluon tree-level amplitudes can be constructed with the all-line shift recursion relations; except the MHV amplitudes ($K = 0$). It turns out that in this formulation of recursion relations, the tower of MHV amplitudes constitutes the basic building blocks for the N^KMHV amplitudes. Let us see how this works for NMHV. The recursion relations give

$$A_n^{\text{NMHV}} = \sum_{\text{diagrams } I} \text{} . \tag{3.47}$$

If you consider the possible assignments of helicity labels on the internal line, you will see that there are two options: either the diagram is anti-MHV$_3 \times$NMHV or MHV$\times$MHV. The former option vanishes by special kinematics of the 3-point anti-MHV vertex, just as in the case of the first diagram in (3.15). So all subamplitudes in (3.47) are MHV. Let us write down the example of the split-helicity NMHV 6-gluon amplitude:

$$A_n\big[1^- 2^- 3^- 4^+ 5^+ 6^+\big] = \text{} \tag{3.48}$$

All six diagrams are non-vanishing and this may look a little daunting, especially compared with the BCFW version where there were just two diagrams in the simplest version (3.36). However, the diagrams in (3.48) are easier to evaluate than the BCFW diagrams: the MHV amplitudes depend only on angle spinors, so the only way they know about the square-spinor shift is through the internal line angle spinors $|\hat{P}_I\rangle$, for example

$$\text{} = \frac{\langle 1\hat{P}_I\rangle^4}{\langle 1\hat{P}_I\rangle\langle\hat{P}_I 5\rangle\langle 56\rangle\langle 61\rangle} \frac{1}{P_{156}^2} \frac{\langle 23\rangle^4}{\langle 23\rangle\langle 34\rangle\langle 4\hat{P}_I\rangle\langle\hat{P}_I 2\rangle} . \tag{3.49}$$

We can write

$$|\hat{P}_I\rangle \frac{[\hat{P}_I X]}{[\hat{P}_I X]} = \hat{P}_I|X] \frac{1}{[\hat{P}_I X]} = P_I|X] \frac{1}{[\hat{P}_I X]} . \tag{3.50}$$

In the last step we can drop the hat, because the shift of $(\hat{P}_I)^{\dot{a}b}$ is proportional to the reference spinor $[X|^b$ of the shift (3.46). Note that the diagrams are necessarily invariant under little group scaling associated with the internal line. Therefore the factors $\frac{1}{[\hat{P}_I X]}$ in (3.50) cancel out of each diagram and we can use the prescription

$$|\hat{P}_I\rangle \to P_I|X] . \tag{3.51}$$

This gives

$$
\hat{6}^+ \underset{\hat{5}^+}{\overset{\hat{1}^-}{\bullet}} \, - \, \underset{\hat{4}^-}{\overset{\hat{2}^-}{\bullet}} \hat{3}^- = \frac{\langle 1|P_{156}|X]^4}{\langle 1|P_{156}|X]\langle 5|P_{156}|X]\langle 56\rangle\langle 61\rangle} \frac{1}{P_{156}^2} \frac{\langle 23\rangle^4}{\langle 23\rangle\langle 34\rangle\langle 4|P_{156}|X]\langle 2|P_{156}|X]}
$$

(3.52)

and similarly for the other "MHV vertex diagrams" in (3.48). Note that we can drop the indication $\hat{\ }$ of the shift on the external lines in the MHV vertex diagrams since the square-spinor shift does not affect the MHV vertices and all that is needed is the prescription (3.51) for the internal lines.

In general, each diagram depends explicitly on the reference spinor $|X]$, but of course the full tree amplitude cannot depend on an arbitrary spinor: the Cauchy theorem argument of Section 3.1 guarantees that the sum of all the diagrams will be independent of $|X]$ and reproduce the correct tree amplitude. Numerically, it is not hard to verify independence of $|X]$ and that the expressions (3.48) and (3.36) indeed produce the same scattering amplitude.

The expansion of the amplitude in terms of MHV vertex diagrams generalizes beyond the NMHV level. In general, the N^KMHV tree amplitude is written as a sum of all tree-level diagrams with precisely $K+1$ MHV vertices evaluated via the replacement rule (3.51). This construction of the amplitude is called the ***MHV vertex expansion***: it can be viewed as the closed-form solution to the all-line shift recursion relations. However, it was discovered by Cachazo, Svrcek, and Witten in 2004 [15] before the introduction of recursion relations from complex shifts. The method is therefore also known as the ***CSW expansion*** and the rule (3.51) is called the ***CSW prescription***. The first recursive derivation of the MHV vertex expansion was given by Risager [28] using the 3-line Risager shift mentioned above applied to the three negative helicity line of NMHV amplitudes. The all-line shift formulation was first presented in [29].

▶ **Exercise 3.11**

Construct $A_5[1^-2^-3^-4^+5^+]$ from the CSW expansion. Make a choice for the reference spinor $|X]$ to simplify the calculation and show that the result agrees with the anti-MHV Parke–Taylor formula (2.115).

The MHV vertex expansion was the first construction of gluon amplitudes from on-shell building blocks. The method is valid also in other cases, for example in super Yang–Mills theory [27, 29] or Higgs amplitudes with gluons and partons [30, 31]. There are also applications of the MHV vertex expansion at loop-level – for a review see [32] and references therein.

The MHV vertex expansion can also be derived directly from a Lagrangian [33]: a field redefinition and suitable light-cone gauge choice brings it to a form with an interaction term for each MHV amplitude. The N^KMHV amplitudes are then generated from the MHV vertex Lagrangian by gluing together the MHV vertices. The reference spinor $|X]$ arises from the light-cone gauge choice. There is also a twistor-action formulation of the MHV vertex expansion [34].

In the case of the BCFW shift, we have applied it to gluon as well as graviton amplitudes. A version of the MHV vertex expansion was proposed for gravity in [35] based on the Risager shift. However, the method fails for NMHV amplitudes for $n \geq 12$: under the Risager shift, $\hat{A}_n(z) \sim z^{12-n}$ for large-z, so for $n \geq 12$ there is a boundary term obstructing the recursive formula [36]. An analysis of validity of all-line shift recursion relations in general 4d QFTs can be found in [26].

At this stage, you may wonder why tree-level gluon scattering amplitudes have so many different representations: one from the MHV vertex expansion and other forms arising from BCFW applied to various pairs of external momenta. The CSW and BCFW representations reflect different aspects of the amplitudes, but they turn out to be closely related. We need more tools to learn more about this. So read on.

Supersymmetry 4

The focus of this chapter is supersymmetry in the context of scattering amplitudes. We begin with a very brief introduction to supersymmetry; it serves to provide the minimal background material we need for our amplitude studies. We then discuss *supersymmetry Ward identities* for the amplitudes and introduce *on-shell superspace* as a tool for organizing the amplitudes into *superamplitudes*. This is particularly powerful in $\mathcal{N} = 4$ super Yang–Mills theory where it allows us to solve a supersymmetric version of the BCFW recursion relations to find all tree-level superamplitudes in this theory. We had better get started.

4.1 $\mathcal{N} = 1$ supersymmetry: chiral model

Let us begin with a simple example of supersymmetry. Consider the free Lagrangian for a Weyl fermion ψ and a complex scalar field ϕ:

$$\mathcal{L}_0 = i \psi^\dagger \bar{\sigma}^\mu \partial_\mu \psi - \partial_\mu \bar{\phi} \, \partial^\mu \phi \, . \tag{4.1}$$

The bar on ϕ denotes the complex conjugate. In addition to Poincaré symmetry, $\mathcal{L}_0$ also has a symmetry that mixes fermions and bosons:

$$
\begin{aligned}
\delta_\epsilon \phi &= \epsilon \psi \, , & \delta_\epsilon \bar{\phi} &= \epsilon^\dagger \psi^\dagger \, , \\
\delta_\epsilon \psi_a &= -i \sigma^\mu_{a\dot{b}} \epsilon^{\dagger \dot{b}} \partial_\mu \phi \, , & \delta_\epsilon \psi^\dagger_{\dot{a}} &= i \partial_\mu \bar{\phi} \, \epsilon^b \sigma^\mu_{b\dot{a}} \, .
\end{aligned}
\tag{4.2}
$$

This is an example of a **supersymmetry transformation**. The anti-commuting constant spinor ϵ is the supersymmetry parameter (a fermionic analogue of the infinitesimal angle θ of a rotation transformation), and $\epsilon \psi = \epsilon^a \psi_a$ and $\epsilon^\dagger \psi^\dagger = \epsilon^\dagger_{\dot{a}} \psi^{\dagger \dot{a}}$ are the 2-component spinor products with spinor indices raised/lowered with the 2-index Levi-Civita symbol, just as in (2.13).

The conserved Noether charges associated with the supersymmetry transformations are anticommuting supercharges Q and $Q^\dagger$. They enhance the Poincaré algebra to a super-Poincaré graded Lie algebra with both bosonic and fermionic generators. Thus, in addition to the translation and rotation/boost generators P_μ and $M_{\mu\nu}$ whose algebra is

$$
\begin{aligned}
&[P^\mu, P^\nu] = 0 \, , \quad [P^\mu, M^{\rho\sigma}] = i(\eta^{\mu\rho} P^\sigma - \eta^{\mu\sigma} P^\rho) \, , \\
&[M^{\mu\nu}, M^{\rho\sigma}] = i(\eta^{\mu\rho} M^{\nu\sigma} - \eta^{\nu\rho} M^{\mu\sigma} - \eta^{\mu\sigma} M^{\nu\rho} + \eta^{\nu\sigma} M^{\mu\rho}) \, ,
\end{aligned}
\tag{4.3}
$$

the super-Poincaré algebra has

$$
[Q_a, P^\mu] = 0, \quad [Q_a, M^{\mu\nu}] = \tfrac{i}{4}\epsilon^{\dot a \dot c}\big(\sigma^\mu_{a\dot a}\sigma^\nu_{c\dot c} - \sigma^\nu_{a\dot a}\sigma^\mu_{c\dot c}\big)Q^c ,
$$
$$
\{Q_a, Q_b\} = 0, \quad \{Q_a, Q^\dagger_{\dot b}\} = -2\sigma^\mu_{a\dot b}P_\mu ,
\tag{4.4}
$$

and similarly for $Q^\dagger$. Note that the supercharges Q and $Q^\dagger$ commute with the spacetime translation operators, but not with the rotations/boosts, and that the anticommutator of Q and $Q^\dagger$ is a spacetime translation.

If you have not previously seen supersymmetry, you should promptly go ahead and do these two exercises:

▶ **Exercise 4.1**

Check that $\mathcal{L}_0$ in (4.1) is invariant under the supersymmetry variation (4.2) up to a total derivative.

▶ **Exercise 4.2**

Calculate $[\delta_{\epsilon_1}, \delta_{\epsilon_2}]$ by acting with it on the fields; first try on ϕ, then afterwards on ψ. You should find that in each case, the combination of two supersymmetry transformations is a spacetime translation.

The 2-component Weyl spinors can be combined into 4-component Majorana spinors

$$
\Psi_{\rm M} = \begin{pmatrix} \psi_a \\ \psi^{\dagger\dot a} \end{pmatrix} \quad \text{and} \quad \epsilon_{\rm M} = \begin{pmatrix} \epsilon_a \\ \epsilon^{\dagger\dot a} \end{pmatrix},
\tag{4.5}
$$

and the supersymmetry transformations are then defined with suitable L- and R-projections $P_{L,R}$ from (A.9). We write the free field expansions as

$$
\phi(x) = \int \widetilde{dp}\,\Big[a_-(p)\, e^{ip.x} + a^\dagger_+(p)\, e^{-ip.x} \Big]
$$
$$
\psi_a(x) = P_L\Psi_{\rm M}(x) = \sum_{s=\pm}\int \widetilde{dp}\,\Big[b_s(p)\, P_L u_s(p)\, e^{ip.x} + b^\dagger_s(p)\, P_L v_s(p)\, e^{-ip.x}\Big],
\tag{4.6}
$$

and similarly for $\bar\phi(x)$ and $\psi^{\dagger\dot a} = P_R\Psi_{\rm M}$. The Lorentz-invariant measure $\widetilde{dp}$ is defined below (2.5).

Upon canonical quantization, the coefficients satisfy the algebra of bosonic/fermionic creation-annihilation operators,

$$
\big[a_\pm(p), a^\dagger_\pm(p')\big] = (2\pi)^3\, 2E_p\, \delta^3(\vec p - \vec p\,'), \quad \{b_\pm(p), b^\dagger_\pm(p')\} = (2\pi)^3\, 2E_p\, \delta^3(\vec p - \vec p\,'),
\tag{4.7}
$$

with all other (anti)commutators vanishing.

For the fermions, the $\pm$-subscripts on the operators indicate the helicity $h = \pm\tfrac{1}{2}$ of the associated 1-particle states. As a matter of later convenience, we have also labeled the two sets of annihilation/creation operators associated with the complex field ϕ with

±-subscripts. The corresponding particles are of course scalars with $h = 0$, but the label indicates which spinor helicity state the scalar state is matched to via supersymmetry. Let us see how that works.

The supersymmetry transformations (4.2) transform the fields ψ and ϕ into each other, and therefore the associated annihilation/creation operators are also related. The relationship is straightforward to extract from the free field expansions (4.6). Recalling from Section 2.2 that $P_L v_s(p)$ is equal to $|p]$ for $s = +$ and vanishes for $s = -$ (and similarly for P_R), one finds

$$\delta_\epsilon a_-(p) = [\epsilon\, p]\, b_-(p)\,, \quad \delta_\epsilon b_-(p) = \langle\epsilon\, p\rangle\, a_-(p)\,,$$
$$\delta_\epsilon a_+(p) = \langle\epsilon\, p\rangle\, b_+(p)\,, \quad \delta_\epsilon b_+(p) = [\epsilon\, p]\, a_+(p)\,. \tag{4.8}$$

We have introduced anti-commuting bra-kets $\langle\epsilon|_{\dot{a}} = \epsilon_{\dot{a}}^{\dagger}$ and $|\epsilon]_a = \epsilon_a$ for the supersymmetry parameter. Using $-|p]\langle p| = p_{a\dot{b}}$, it is easy to see that $[\delta_{\epsilon_1}, \delta_{\epsilon_2}]\mathcal{O}(p) = a^\mu p_\mu \mathcal{O}(p)$ for $\mathcal{O}(p)$ any one of the creation/annihilation operators. Hence, the commutator of two supersymmetry transformations is a translation, as in the algebra (4.4). The translation parameter a^μ can be written in terms of Majorana spinors as $a^\mu = \epsilon_{\text{M},2}\gamma^\mu\epsilon_{\text{M},1}$.

The supersymmetry generators $Q_{\text{M}} = \begin{pmatrix} |Q]_a \\ |Q^{\dagger}\rangle^{\dot{a}} \end{pmatrix}$ can be deduced from

$$\delta_\epsilon\mathcal{O} = \left[\bar{\epsilon}_{\text{M}}Q_{\text{M}}, \mathcal{O}\right] = \left[[\epsilon\, Q] + \langle\epsilon\, Q\rangle, \mathcal{O}\right]. \tag{4.9}$$

Using this, one finds (and you should check it) that

$$|Q]_a = \int \widetilde{dp}\, |p]_a \left(a_+(p)\, b_+^{\dagger}(p) - b_-(p)\, a_-^{\dagger}(p)\right)\,,$$
$$|Q^{\dagger}\rangle^{\dot{a}} = \int \widetilde{dp}\, |p\rangle^{\dot{a}} \left(a_-(p)\, b_-^{\dagger}(p) - b_+(p)\, a_+^{\dagger}(p)\right)\,, \tag{4.10}$$

reproduces (4.8).

▶ **Exercise 4.3**

Show that $\{|Q]_a, \langle Q^{\dagger}|_{\dot{b}}\}$ equals $p_{a\dot{b}}$ times a sum of number operators.

The action of the supercharges (4.10) on the annihilation operators is

$$[Q, a_-(p)] = |p]\, b_-(p)\,, \quad [Q^{\dagger}, b_-(p)] = |p\rangle\, a_-(p)\,,$$
$$[Q, b_-(p)] = 0\,, \quad [Q^{\dagger}, a_-(p)] = 0\,,$$
$$[Q, b_+(p)] = |p]\, a_+(p)\,, \quad [Q^{\dagger}, a_+(p)] = |p\rangle\, b_+(p)\,, \tag{4.11}$$
$$[Q, a_+(p)] = 0\,, \quad [Q^{\dagger}, b_+(p)] = 0\,,$$

where $[.\,,.]$ is a graded bracket that is an anticommutator when both arguments are Grassmann and otherwise a commutator. The 2-component spinor-indices are suppressed. A similar set of relations holds for the creation operators.

The action of the supersymmetry generators on the creation/annihilation operators implies that the associated bosonic and fermionic 1-particle states are related – they are said to belong to the same **supermultiplet**. Note that Q lowers the helicity by $\frac{1}{2}$ and that $Q^\dagger$ raises it by $\frac{1}{2}$. Because the supersymmetry generators commute with the Hamiltonian, P^0, and the space-translation generators, P^i, it follows that states in the same supermultiplet must have the same mass.

In the chiral model (4.1), the mass of the bosons and fermions is of course zero. It follows from our discussion that the spectrum of the chiral model splits into a "negative helicity sector" and a "positive helicity sector"; CPT symmetry requires us to have both.

In a supermultiplet, the number of on-shell bosonic states equals the number of on-shell fermionic states. This is manifest in the example of the chiral model: the two real scalar states encoded in the complex scalar field are supersymmetry partners of the two fermionic states with $h = \pm 1/2$.

Interactions. Next, we would like to introduce interactions – after all, this is all about scattering amplitudes so we need something to happen! We want to study interactions that preserve supersymmetry. For our chiral model, one can introduce a "superpotential" interaction of the form

$$\mathcal{L}_I = \tfrac{1}{2} g\, \phi\, \psi\psi + \tfrac{1}{2} g^*\, \bar{\phi}\, \psi^\dagger\psi^\dagger - \tfrac{1}{4}|g|^2\, |\phi|^4 \,. \tag{4.12}$$

A small modification of the supersymmetry transformations (4.2) is now needed in the transformation rule of the fermion field:

$$\begin{aligned}
\delta_\epsilon \phi &= \epsilon\psi \,, & \delta_\epsilon \bar{\phi} &= \epsilon^\dagger\psi^\dagger \,, \\
\delta_\epsilon \psi_a &= -i\sigma^\mu_{a\dot{b}}\epsilon^{\dagger\dot{b}}\partial_\mu\phi + \tfrac{1}{2}g^*\bar{\phi}^2\epsilon_a \,, & \delta_\epsilon \psi^\dagger_{\dot{a}} &= i\,\partial_\mu\bar{\phi}\,\epsilon^b\sigma^\mu_{b\dot{a}} + \tfrac{1}{2}g\phi^2\,\epsilon^\dagger_{\dot{a}} \,.
\end{aligned} \tag{4.13}$$

Note that the coupling of the 4-scalar interaction in (4.12) is fixed in terms of the Yukawa coupling g; this is required by supersymmetry.

▶ **Exercise 4.4**

Show that (4.13) is a symmetry of $\mathcal{L} = \mathcal{L}_0 + \mathcal{L}_I$.

A linear version of the supersymmetry transformations can be given using an auxiliary field. Supersymmetric actions can be expressed compactly and conveniently using off-shell superspace formalism. You can find much more about this in textbooks such as Wess and Bagger [37].

Extended supersymmetry

So far we have focused on a very simple case of a "chiral supermultiplet" in which a spin-0 particle is partnered with a spin-$\frac{1}{2}$ particle. One can repeat the analysis for any $\mathcal{N} = 1$ supersymmetric model with particles of spin $(s, s + \frac{1}{2})$. The $\mathcal{N}$ counts the number of supersymmetry generators Q^A and $Q^\dagger_A$ with $A = 1, 2, \ldots, \mathcal{N}$. When $\mathcal{N} > 1$, the model is said to have **extended supersymmetry**. The super-Poincaré algebra with extended supersymmetry is very similar to (4.4), but with

$$\{Q_a^A, Q_{Bb}^\dagger\} = -2\sigma_{ab}^\mu P_\mu \delta^A{}_B .\tag{4.14}$$

In some cases of extended supersymmetry, a "central charge" Z^{AB} can appear in the anti-commutator $\{Q_a^A, Q_b^B\} = Z^{AB}\epsilon_{ab}$. In almost all applications of extended supersymmetry in this book, there will be no central charge.

The extended supersymmetry algebra (without central charges) has a global symmetry that rotates the supersymmetry charges: this is called an **R-symmetry**. An R-symmetry is characterized by the property that it does not commute with the supersymmetry transformations; this contrasts "ordinary" flavor-type global symmetries that do commute with supersymmetry. With $\mathcal{N}$-fold extended supersymmetry, the R-symmetry is $SU(\mathcal{N})$; this full R-symmetry may or may not be realized for a given model.

As examples of $\mathcal{N} = 1$ supermultiplets, $\mathcal{N} = 1$ super-QED has a photon and a photino with helicities ± 1 and $\pm\frac{1}{2}$. $\mathcal{N} = 1$ **super Yang–Mills theory (SYM)** with a gluon ($h = \pm 1$) and its fermionic partner, the gluino ($h = \pm\frac{1}{2}$), will be discussed in Section 4.3.

In extended supersymmetry, there are $2^{\mathcal{N}}$ states in the massless supermultiplets (and the same in the CPT conjugates). For example, for $\mathcal{N} = 2$ SYM, one supermultiplet consists of a helicity -1 gluon, two gluinos with helicity $-\frac{1}{2}$, and a scalar with helicity 0. Thus the $\mathcal{N} = 2$ gluon supermultiplet has two bosonic states and two fermionic states. The CPT conjugate multiplet contains the same types of states but with opposite signs for the helicities.

The spectrum of states is generated by acting with the supercharges on the "highest weight state"; in the case of massless particles in four dimensions, this is the highest helicity state. To avoid states with spin greater than 1, the maximal amount of supersymmetry in four dimensions is $\mathcal{N} = 4$. This large symmetry-requirement places such strong constraints on the theory that it is unique (up to choice of gauge group): this is $\mathcal{N} = 4$ super Yang–Mills theory. We are going to study the supersymmetry constraints on the $\mathcal{N} = 4$ SYM amplitudes in much further detail in Section 4.4. For now, let us content ourselves with $\mathcal{N} = 1$ supersymmetry and study the consequences it has on the scattering amplitudes in our chiral scalar model.

4.2 Amplitudes and supersymmetry Ward identities

Consider the simple supersymmetric chiral scalar model whose Lagrangian $\mathcal{L}$ is the sum of the kinetic terms (4.1) and the interaction terms (4.12). The 4-point tree amplitudes were essentially already presented in Exercise 2.8. To adapt the results (2.46), we just need to take the coupling of the 4-scalar interaction to be $\lambda = |g|^2$. The 4-point tree amplitudes are then:

$$A_4^{\text{tree}}(\phi\phi\bar\phi\bar\phi) = -|g|^2 , \quad A_4^{\text{tree}}(\phi\, f^-f^+\bar\phi) = -|g|^2 \frac{\langle 24\rangle}{\langle 34\rangle} , \quad A_4^{\text{tree}}(f^-f^-f^+f^+) = |g|^2 \frac{\langle 12\rangle}{\langle 34\rangle} .\tag{4.15}$$

By inspection of (4.15), we observe that

$$A_4^{\text{tree}}\big(\phi\, f^-\, f^+\, \bar{\phi}\big) = \frac{\langle 24\rangle}{\langle 34\rangle}\, A_4^{\text{tree}}\big(\phi\phi\bar{\phi}\bar{\phi}\big)\,, \tag{4.16}$$

$$A_4^{\text{tree}}\big(f^-\, f^-\, f^+\, f^+\big) = -\frac{\langle 12\rangle}{\langle 24\rangle}\, A_4^{\text{tree}}\big(\phi\, f^-\, f^+\, \bar{\phi}\big)\,. \tag{4.17}$$

These relations hold not just for the tree-level amplitudes, as we have seen just now; super-symmetry ensures that (4.16) and (4.17) hold at all orders in the perturbation expansion. We will now see how that comes about.

We can think of an n-point amplitude with all-outgoing particles as the S-matrix element $\langle 0|\mathcal{O}_1(p_1)\ldots\mathcal{O}_n(p_n)|0\rangle$ in which n annihilation operators $\mathcal{O}_i(p_i)$, $i = 1,\ldots,n$, act to the left on the out-vacuum. For example, $A_4\big(\phi\, f^-\, f^+\, \bar{\phi}\big) = \langle 0|a_-(p_1)b_-(p_2)b_+(p_3)a_+(p_4)|0\rangle$. Suppose the vacuum is supersymmetric: $Q|0\rangle = 0 = Q^\dagger|0\rangle$. Then for any set of n annihilation (or creation) operators, we have

$$0 = \langle 0|\big[Q^\dagger, \mathcal{O}_1(p_1)\ldots\mathcal{O}_n(p_n)\big]|0\rangle$$
$$= \sum_{i=1}^{n}(-1)^{\sum_{j<i}|\mathcal{O}_j|}\,\langle 0|\mathcal{O}_1(p_1)\cdots\big[Q^\dagger, \mathcal{O}_i(p_i)\big]\cdots\mathcal{O}_n(p_n)|0\rangle\,, \tag{4.18}$$

and similarly for Q. Here the sign-factor takes into account that a minus sign is picked up every time $Q^\dagger$ passes by a fermionic operator: so $|\mathcal{O}|$ is 0 when the operator $\mathcal{O}$ is bosonic and 1 if it is fermionic. Now using the action of the supersymmetry generators (4.11) on the free asymptotic states, the equation (4.18) describes a linear relation among scattering amplitudes whose external states are related by supersymmetry. Such relations are called **supersymmetry Ward identities** and they are valid at any order in perturbation theory. The supersymmetry Ward identities are best illustrated in explicit examples. Thus, to start with, consider:

$$0 = \langle 0|\big[Q^\dagger, a_-(p_1)b_-(p_2)a_+(p_3)a_+(p_4)\big]|0\rangle$$
$$= |2\rangle\,\langle a_-(p_1)a_-(p_2)a_+(p_3)a_+(p_4)\rangle - |3\rangle\,\langle a_-(p_1)b_-(p_2)b_+(p_3)a_+(p_4)\rangle$$
$$-|4\rangle\,\langle a_-(p_1)b_-(p_2)a_+(p_3)b_+(p_4)\rangle\,. \tag{4.19}$$

We have used (4.11) that $Q^\dagger$ annihilates $a_-(p)$. Translating (4.19) to amplitudes, we have the supersymmetry Ward identity

$$0 = |2\rangle\, A_4\big(\phi\phi\bar{\phi}\bar{\phi}\big) - |3\rangle\, A_4\big(\phi\, f^-\, f^+\, \bar{\phi}\big) - |4\rangle\, A_4\big(\phi\, f^-\, \bar{\phi}\, f^+\big)\,. \tag{4.20}$$

Each identity (4.18) encodes two relations, since $Q^\dagger$ (Q) has two components. This is also visible in our example (4.20). We project out the two independent relations by dotting in a suitable choice of bra-spinor $\langle r|$. Picking $\langle r| = \langle 4|$, we find

$$0 = \langle 42\rangle\, A_4\big(\phi\phi\bar{\phi}\bar{\phi}\big) - \langle 43\rangle\, A_4\big(\phi\, f^-\, f^+\, \bar{\phi}\big)\,. \tag{4.21}$$

At tree-level, this is precisely the relation noted in (4.16). We emphasize that supersymmetry ensures that (4.21) holds true at any loop order.

A second relation is extracted from (4.20) by choosing $\langle r| = \langle 2|$:

$$A_4\big(\phi\, f^-\bar\phi f^+\big) = -\frac{\langle 23\rangle}{\langle 24\rangle}\, A_4\big(\phi\, f^-\, f^+\bar\phi\big)\,. \tag{4.22}$$

Note how the factor $\frac{\langle 23\rangle}{\langle 24\rangle}$ compensates the different little group scaling of the two amplitudes.

▶ **Exercise 4.5**

Plug the two supersymmetry Ward identities (4.21) and (4.22) into (4.20) to show that there is no further independent information available in (4.20).

▶ **Exercise 4.6**

Derive (4.17) as a supersymmetry Ward identity.

▶ **Exercise 4.7**

Find a Q-supersymmetry Ward identity that gives $A_4(\phi\, f^-\, f^+\bar\phi) = -\frac{[13]}{[12]}\, A_4\big(\phi\phi\bar\phi\bar\phi\big)$. Show that this relation is equivalent to (4.16).

Let us take a brief look at the supersymmetry Ward identities at higher points, for example for amplitudes with 6-particles. Starting with $0 = \langle 0|[Q^\dagger, a_{1-}a_{2-}b_{3-}a_{4+}a_{5+}a_{6+}]|0\rangle$ (using a short-hand notation to indicate the momentum with a subscript) we find, after dotting in $\langle r|$,

$$\begin{aligned}
0 = {} & \langle r3\rangle\, A_6\big(\phi\phi\phi\bar\phi\bar\phi\bar\phi\big) - \langle r4\rangle\, A_6\big(\phi\phi\, f^-\, f^+\bar\phi\bar\phi\big) \\
& - \langle r5\rangle\, A_6\big(\phi\phi\, f^-\bar\phi\, f^+\bar\phi\big) - \langle r6\rangle\, A_6\big(\phi\phi\, f^-\bar\phi\bar\phi\, f^+\big)\,.
\end{aligned} \tag{4.23}$$

There are two relations, but four "unknowns" (the amplitudes), so this time the super-symmetry Ward identity does not give simple proportionality relations among pairs of amplitudes. Instead, one gets a web of linear relations. This is typical for "non-MHV type" amplitudes; we come back to this point briefly in the next two sections.

To summarize, amplitudes whose external states are related by supersymmetry are linked to each other through linear relationships called ***supersymmetry Ward identities***. They were first studied in 1977 by Grisaru, Pendleton, and van Nieuwenhuizen [38] and have since then had multiple applications.

4.3 $\mathcal{N} = 1$ **supersymmetry: gauge theory**

We have introduced supersymmetry for the simple chiral scalar model and derived the consequences of supersymmetry for the amplitudes. Let us now adapt these results to gauge theory. The Lagrangian for pure $\mathcal{N} = 1$ super Yang–Mills theory (SYM) is

$$\mathcal{L} = -\frac{1}{4}\operatorname{Tr} F_{\mu\nu}F^{\mu\nu} + i\operatorname{Tr}\lambda^\dagger\bar\sigma^\mu D_\mu\lambda\,, \tag{4.24}$$

where all fields are in the adjoint of the $SU(N)$ gauge group and the covariant derivative is $D_\mu \lambda = \partial_\mu \lambda - ig[A_\mu, \lambda]$. The on-shell spectrum consists of helicity $h = \pm 1$ states of the gluon $g^\pm$ and $h = \pm\frac{1}{2}$ states of the gluino $\lambda^\pm$. (For an abelian gauge group, we would call the states photon and photino.) We will not need the explicit supersymmetry transformation of the fields (they can be found for example in [39]), but will simply state the consequences on the states. The supercharge $\tilde{Q}$ raises the helicity by $\frac{1}{2}$ and Q lowers it by $\frac{1}{2}$, so if $c_\pm$ ($b_\pm$) are the annihilation operators for gluons of helicity ± 1 (gluinos of helicity $\pm\frac{1}{2}$), then

$$
\begin{aligned}
\left[\tilde{Q}, c_+(i)\right] &= 0, & \left[Q, c_+(i)\right] &= [i|\, b_-(i), \\
\left[\tilde{Q}, b_+(i)\right] &= |i\rangle\, c_+(i), & \left[Q, b_+(i)\right] &= 0, \\
\left[\tilde{Q}, b_-(i)\right] &= 0, & \left[Q, b_-(i)\right] &= [i|\, c_-(i), \\
\left[\tilde{Q}, c_-(i)\right] &= |i\rangle\, b_-(i), & \left[Q, c_-(i)\right] &= 0,
\end{aligned}
\tag{4.25}
$$

for $\mathcal{N} = 1$ SYM. This is derived as (4.11) for the chiral model in Section 4.1.

Supersymmetry Ward identities for $\mathcal{N} = 1$ SYM are derived from the action of the supercharges on the states in (4.25). Since $\tilde{Q}$ annihilates $c_+(i)$, the supersymmetry Ward identity $\langle 0|[\tilde{Q}, b_{1+}c_{2+} \ldots c_{n+}]|0\rangle = 0$ gives $|1\rangle A_n[g^+ g^+ g^+ \ldots g^+] = 0$. This says that the all-plus gluon amplitude vanishes at all orders in perturbation theory in $\mathcal{N} = 1$ SYM. Similarly, one can show that the gluon amplitude with exactly one negative helicity gluon also vanishes. So

$$
\text{super Yang–Mills:} \quad A_n^{L\text{-loop}}[g^+ g^+ g^+ \ldots g^+] = A_n^{L\text{-loop}}[g^- g^+ g^+ \ldots g^+] = 0. \tag{4.26}
$$

We used only one supersymmetry generator for this argument, so the statement (4.26) is true also with extended supersymmetry (as long as supersymmetry is unbroken). Moreover, as indicated, (4.26) holds *true at all orders in the loop-expansion* because it relies only on the existence of supercharges that annihilate the vacuum.

In Section 2.7, we showed that (4.26) holds at tree-level in pure non-supersymmetric Yang–Mills,

$$
\text{Yang–Mills:} \quad A_n^{\text{tree}}[g^+ g^+ g^+ \ldots g^+] = A_n^{\text{tree}}[g^- g^+ g^+ \ldots g^+] = 0. \tag{4.27}
$$

However, at loop-level (4.26) is *not true* without supersymmetry: for example in pure Yang–Mills theory, non-vanishing all-plus amplitudes are indeed generated at the 1-loop level. The reason the result (4.27) holds at tree-level in pure Yang–Mills theory is that the superpartners of the gluon couple quadratically to the gluon. So an amplitude whose external states are all gluons "sees" the superpartner states only via loops. Thus, in pure Yang–Mills the gluon amplitudes at tree-level have no choice but to obey the same supersymmetry Ward identity constraints as the gluon amplitudes in super Yang–Mills. The argument is even stronger: it says that the all-gluon amplitudes must be exactly the same at tree-level in super Yang–Mills theory as in pure Yang–Mills theory.

▶ **Exercise 4.8**

The non-vanishing of the $n = 3$ anti-MHV amplitude $A_3[1^-2^+3^+]$ escapes the Ward identity that forces $A_n[g^-g^+g^+ \ldots g^+] = 0$ for $n > 3$. Explain how.

▶ **Exercise 4.9**

Show in $\mathcal{N} = 1$ SYM that supersymmetry Ward identities give

$$A_n\big[g^-\lambda^-\lambda^+g^+ \ldots g^+\big] = \frac{\langle 13 \rangle}{\langle 12 \rangle} A_n\big[g^-g^-g^+g^+ \ldots g^+\big]. \tag{4.28}$$

Relate $A_n[\lambda^-\lambda^-\lambda^+\lambda^+g^+ \ldots g^+]$ to the MHV gluon amplitude.

Does $\mathcal{N} = 1$ supersymmetry allow you to relate gluon amplitudes such as for example $A_n[1^-2^-3^+ \ldots 4^+]$ and $A_n[1^-2^+3^- \ldots 4^+]$? Can it relate gluon MHV and NMHV amplitudes?

The $\mathcal{N} = 1$ supersymmetry Ward identities for NMHV gluon amplitude are more involved and do not result in simple proportionality relations among pairs of amplitudes, such as (4.28). For $n = 6$, the NMHV $\mathcal{N} = 1$ supersymmetry Ward identities have been solved explicitly [36, 38].

In our discussion of recursion relations in Section 3.3, we learned that in general we should not expect it to be possible to produce a 4-scalar amplitude recursively from 3-particle amplitudes because of the possibility of independent input from a 4-scalar contact term. However, in our supersymmetric chiral scalar model of Section 4.1, supersymmetry determines the coupling of the 4-scalar interaction in terms of the 3-point interactions, so one could hope that recursion relations work in that case. In particular, one might anticipate that all 4-point amplitudes can be determined by the 3-point ones. For this to be true, the recursion algorithm has to "know" which principle fixes the 4-scalar coupling, so supersymmetry must be built into the recursion relations. This is done most efficiently in two steps: first one introduces *on-shell superspace* and groups the amplitudes into *superamplitudes*. Secondly, one incorporates a shift of a Grassmann super-parameter of the on-shell superspace into the BCFW shift. Then one gets recursion relations for the superamplitudes. This is best illustrated for $\mathcal{N} = 4$ SYM, so that is what we will turn to next.

4.4 $\mathcal{N} = 4$ SYM: on-shell superspace and superamplitudes

The action for $\mathcal{N} = 4$ super Yang–Mills theory (SYM) can be written compactly as

$$S = \int d^4x \, \mathrm{Tr}\left(-\frac{1}{4}F_{\mu\nu}F^{\mu\nu} - \frac{1}{2}(D\Phi_I)^2 + \frac{i}{2}\overline{\Psi}\slashed{D}\Psi + \frac{g}{2}\overline{\Psi}\Gamma^I[\Phi_I, \Psi] + \frac{g^2}{4}[\Phi_I, \Phi_J]^2 \right). \tag{4.29}$$

Here $D_\mu = \partial_\mu - ig[A_\mu, \cdot]$ is the covariant derivative, A_μ is the vector potential field, and Φ_I are six real scalar fields with labels $I = 1, \ldots, 6$ of a global $SO(6) \sim SU(4)$ R-symmetry. All fields transform in the adjoint of the gauge group, which we take to be $SU(N)$. The commutators in (4.29) are associated with the $SU(N)$ matrix structure of the fields. The fermions are encoded in 10-dimensional Majorana–Weyl fields Ψ and the Γ^I are gamma-matrices of the 10d Clifford algebra. This description of $\mathcal{N} = 4$ SYM follows from dimensional reduction of $\mathcal{N} = 1$ SYM in 10d [40].

It is convenient to group the six real scalars Φ_I into six complex scalar fields $\varphi^{AB} = -\varphi^{BA}$, with $A, B = 1, 2, 3, 4$, that satisfy the self-duality condition $\overline{\varphi}_{AB} = \frac{1}{2}\epsilon_{ABCD} \varphi^{CD}$. Here ϵ_{ABCD} is the Levi-Civita symbol of $SU(4) \sim SO(6)$, and the scalars φ^{AB} transform in the fully antisymmetric 2-index representation of $SU(4)$. In this language, the 10d fermion fields give 4+4 gluino states λ^A ($h = +\frac{1}{2}$) and $\bar{\lambda}_A$ ($h = -\frac{1}{2}$) that transform in the (anti-)fundamental of $SU(4)$.

In a supersymmetric model, the value V_0 of the scalar potential V at the vacuum is an order parameter of supersymmetry breaking [37]: in flat space, $V_0 = 0$ is necessary for preserving supersymmetry while $V_0 > 0$ breaks supersymmetry. In $\mathcal{N} = 4$ SYM, the scalar potential is $V = \frac{g^2}{4}[\Phi_I, \Phi_J]^2$, so the theory has a *moduli space* of $\mathcal{N} = 4$ supersymmetric vacua with $[\Phi_I, \Phi_J] = 0$. At the *origin of moduli space*, where all the scalar vacuum expectation values (vevs) vanish, $\langle \varphi^{AB} \rangle = 0$, all states are massless and the theory contains no dimensionful parameters. In fact, the theory has *conformal symmetry*: the trace of the stress-tensor is zero, up to the trace anomaly. In particular, the beta-function vanishes at all orders in perturbation theory and there is no running of the coupling. The theory is invariant under an enlarged spacetime symmetry group, namely the conformal group $SO(2, 4) \sim SU(2, 2)$. Supersymmetry enhances the conformal symmetry to *superconformal symmetry* $SU(2, 2|4)$. The bosonic part of this super-group consists of the conformal group $SU(2, 2)$ and the R-symmetry group $SU(4)$. The fermionic parts are the four supersymmetry generators as well as their four superconformal partners. The symmetries of $\mathcal{N} = 4$ SYM are the subject of Chapter 5.

When the scalars acquire vevs in such a way that full supersymmetry is preserved, i.e. $[\Phi_I, \Phi_J] = 0$, the theory is said to be on the Coulomb branch. The scale introduced by the scalar vevs breaks the conformal symmetry, so the theory is not superconformal away from the origin, but it still has $\mathcal{N} = 4$ supersymmetry. We will briefly discuss scattering amplitudes on the Coulomb branch of $\mathcal{N} = 4$ SYM in Exercise 12.3, but otherwise we focus entirely on the superconformal theory at the origin of moduli space: *henceforth, when we discuss amplitudes in $\mathcal{N} = 4$ SYM this means the theory at the origin of moduli space.*

Given that the theory is conformal, we should clarify what we mean by the scattering-matrix in $\mathcal{N} = 4$ SYM. One way to think about this is to consider the theory in $4 - \epsilon$ dimensions: then the conformal symmetry is broken and the S-matrix is well-defined. This turns out to be a little inconvenient for keeping on-shell symmetries manifest, and it can therefore be better to consider the theory on the Coulomb branch and define the $\mathcal{N} = 4$ SYM S-matrix as the zero-vev limit of the Coulomb branch S-matrix. These subtleties will not affect the majority of our discussion and therefore we proceed to discuss the amplitudes of $\mathcal{N} = 4$ SYM without further hesitation.

Spectrum and supersymmetry Ward identities

The spectrum of the massless $\mathcal{N} = 4$ supermultiplet consists of 16 states:

$$\underbrace{a}_{\text{1 gluon } g^+}, \quad \underbrace{a^A}_{\text{4 gluinos } \lambda^A}, \quad \underbrace{a^{AB}}_{\text{6 scalars } S^{AB}}, \quad \underbrace{a^{ABC}}_{\text{4 gluinos } \lambda^{ABC} \sim \bar{\lambda}_D}, \quad \underbrace{a^{1234}}_{\text{1 gluon } g^-}.$$

$$\text{helicity} = \quad 1 \qquad\qquad \tfrac{1}{2} \qquad\qquad 0 \qquad\qquad -\tfrac{1}{2} \qquad\qquad -1 \,. \tag{4.30}$$

The indices $A, B, \ldots = 1, 2, 3, 4$ are labels of the global $SU(4)$ R-symmetry that rotates the four sets of supersymmetry generators Q^A and $\tilde{Q}_A \equiv Q^\dagger_A$. The helicity-$h$ states transform in fully antisymmetric $2(1 - h)$-index representations of $SU(4)$. Note the important property that the $\mathcal{N} = 4$ supermultiplet is CPT self-conjugate: in particular $\mathcal{N} = 4$ supersymmetry connects the gluon states of $h = \pm 1$.

The action of the supercharges on the annihilation operators is

$$
\begin{aligned}
\left[\tilde{Q}_A, a(i) \right] &= 0, & \left[Q^A, a(i) \right] &= [i| \, a^A(i), \\
\left[\tilde{Q}_A, a^B(i) \right] &= |i\rangle \, \delta^B_A \, a(i), & \left[Q^A, a^B(i) \right] &= [i| \, a^{AB}(i), \\
\left[\tilde{Q}_A, a^{BC}(i) \right] &= |i\rangle \, 2! \, \delta^{[B}_A \, a^{C]}(i), & \left[Q^A, a^{BC}(i) \right] &= [i| \, a^{ABC}(i), \quad (4.31) \\
\left[\tilde{Q}_A, a^{BCD}(i) \right] &= |i\rangle \, 3! \, \delta^{[B}_A \, a^{CD]}(i), & \left[Q^A, a^{BCD}(i) \right] &= [i| \, a^{ABCD}(i), \\
\left[\tilde{Q}_A, a^{BCDE}(i) \right] &= |i\rangle \, 4! \, \delta^{[B}_A \, a^{CDE]}(i), & \left[Q^A, a^{1234}(i) \right] &= 0.
\end{aligned}
$$

Note that $\tilde{Q}_A$ raises the helicity of all operators by $\frac{1}{2}$ and removes the index A (if it is not available to be removed, then the operator is annihilated). Q^A does the opposite. Truncating the spectrum to include just the gluons and gluinos, one finds that (4.31) reduces to the result (4.25) for $\mathcal{N} = 1$ SYM.

▶ **Exercise 4.10**

Show that the supersymmetry Ward identities give the following relationships among the color-ordered amplitudes in $\mathcal{N} = 4$ SYM:

$$0 = -|1\rangle A_n[\lambda^{123} g^- \lambda^4 g^+ \ldots g^+] - |2\rangle A_n[g^- \lambda^{123} \lambda^4 g^+ \ldots g^+]$$
$$+ |3\rangle A_n[g^- g^- g^+ g^+ \ldots g^+] \tag{4.32}$$

and extract

$$A_n[g^- \lambda^{123} \lambda^4 g^+ \ldots g^+] = \frac{\langle 13 \rangle}{\langle 12 \rangle} A_n[g^- g^- g^+ g^+ \ldots g^+]. \tag{4.33}$$

Then derive

$$A_n[g^- S^{12} S^{34} g^+ \ldots g^+] = \frac{\langle 13 \rangle^2}{\langle 12 \rangle^2} A_n[g^- g^- g^+ g^+ \ldots g^+]. \tag{4.34}$$

Since supersymmetry connects the entire $\mathcal{N} = 4$ SYM multiplet, there is an important $\mathcal{N} = 4$ supersymmetry Ward identity that connects color-ordered MHV gluon amplitudes with the negative helicity gluons in different positions:

$$\mathcal{N} = 4 \text{ SYM:} \quad A_n[g^+ g^+ \ldots g_i^- \ldots g_j^- \ldots g^+] = \frac{\langle ij \rangle^4}{\langle 12 \rangle^4} A_n[g^- g^- g^+ g^+ \ldots g^+].$$

$$(4.35)$$

▶ **Exercise 4.11**

Derive (4.35) by extending the strategy you used for (4.34).

Convince yourself that the supersymmetry Ward identities cannot mix gluon amplitudes with different K in the N^KMHV classification.

You were prompted in Exercise 4.9 to note that $\mathcal{N} = 1$ supersymmetry would not be sufficient to derive (4.35). We need to relate the positive and negative helicity gluons and this is only possible because the $\mathcal{N} = 4$ supermultiplet is CPT self-conjugate.

Gluon *tree* amplitudes do not see the supersymmetry partners, so they must comply with the supersymmetry Ward identities, in particular the tree amplitudes in pure Yang–Mills theory must obey (4.35). Indeed, this relation is automatic in the Parke–Taylor formula (2.114) for the tree-level MHV gluon amplitudes! Furthermore, the focus on the "split-helicity" MHV amplitude $A_n[1^- 2^- 3^+ \ldots n^+]$ in our recursive proof of the Parke–Taylor formula in Section 3.2 is now justified: the supersymmetry Ward identities (4.35) ensure that Parke–Taylor holds for an MHV tree gluon amplitude with the two negative helicity gluons in any position.

N^KMHV classification – revisited

We introduced N^KMHV amplitudes in Section 2.7 as the gluon amplitudes with $K+2$ negative helicity gluons. Now we can expand this to define the *N^KMHV sector* to be all amplitudes in $\mathcal{N} = 4$ SYM connected to the N^KMHV gluon amplitude via supersymmetry. For example, all the amplitudes in (4.33), (4.34), and (4.35) are MHV. In fact, all MHV amplitudes in $\mathcal{N} = 4$ SYM are proportional to $A_n[g^- g^- g^+ g^+ \ldots g^+]$.

The global $SU(4)$ R-symmetry implies that an amplitude vanishes unless the external states combine to an $SU(4)$ singlet. This requires that the (upper) $SU(4)$ indices appear as a combination of $(K + 2)$ sets of $\{1234\}$ for N^KMHV amplitudes. Since the supersymmetric Ward identities relate amplitudes with the same number of $SU(4)$ indices, this then provides an alternative definition of the N^KMHV sector.

On-shell superspace

It is highly convenient to introduce an on-shell[1] superspace in order to keep track of the states and the amplitudes. We introduce four Grassmann variables η_A labeled by the $SU(4)$

---

[1] There is no (known) off-shell superspace formalism for $\mathcal{N} = 4$ SYM.

index $A = 1, 2, 3, 4$.[2] This allows us to collect the 16 states into an $\mathcal{N} = 4$ on-shell chiral *superfield* (or "superwavefunction")

$$\Omega = g^+ + \eta_A \lambda^A - \frac{1}{2!} \eta_A \eta_B S^{AB} - \frac{1}{3!} \eta_A \eta_B \eta_C \lambda^{ABC} + \eta_1 \eta_2 \eta_3 \eta_4 g^- . \tag{4.36}$$

The relative signs are chosen such that the Grassmann differential operators

$$\mathcal{N} = 4 \text{ SYM:} \quad \begin{array}{c|c|c|c|c|c} \text{particle} & g^+ & \lambda^A & S^{AB} & \lambda^{ABC} & g^- = g^{1234} \\ \hline \text{operator} & 1 & \partial_i^A & \partial_i^A \partial_i^B & \partial_i^A \partial_i^B \partial_i^C & \partial_i^1 \partial_i^2 \partial_i^3 \partial_i^4 \end{array} \tag{4.37}$$

select the associated state from $\Omega(p_i)$.

In the on-shell formalism, the supercharges are represented as

$$q^{Aa} \equiv [p|^a \frac{\partial}{\partial \eta_A} , \qquad q_A^{\dagger \dot{a}} \equiv |p\rangle^{\dot{a}} \eta_A , \tag{4.38}$$

where $|p\rangle$ and $|p]$ are the spinors associated with the null momentum p of the particle.

▶ Exercise 4.12

Show that the supercharges satisfy the standard supersymmetry anticommutation relation (4.14), i.e. $\{q^{Aa}, \tilde{q}_B^{\dot{b}}\} = \delta^A{}_B |p\rangle^{\dot{b}} [p|^a = -\delta^A{}_B p^{\dot{b}a}$. The supercharges (4.38) act on the spectrum by shifting states right or left in Ω. Check that this action on Ω matches (4.31).

▷ *Example.* The purpose of this example is to clarify the relation between the *on-shell* superspace introduced here and the *off-shell* superspace formalism described in textbooks on supersymmetry (e.g. [37]). In an off-shell $\mathcal{N} = 1$ formalism, the superspace is $(x^\mu, \theta^a, \bar{\theta}^{\dot{a}})$ with Grassmann variables θ^a and $\bar{\theta}^{\dot{a}}$. The algebra of the supercharges is $\{\mathcal{Q}_a, \mathcal{Q}_b\} = 0$, $\{\bar{\mathcal{Q}}_{\dot{a}}, \bar{\mathcal{Q}}_{\dot{b}}\} = 0$, and $\{\mathcal{Q}_a, \bar{\mathcal{Q}}_{\dot{a}}\} = i(\sigma^\mu)_{a\dot{a}} \partial_\mu$. As is often convenient for studies of anti-chiral superfields, we can realize the superalgebra with an anti-chiral representation

$$\mathcal{Q}_a = \frac{\partial}{\partial \theta^a} , \qquad \bar{\mathcal{Q}}_{\dot{a}} = -\frac{\partial}{\partial \bar{\theta}^{\dot{a}}} + i\theta^a (\sigma^\mu)_{a\dot{a}} \partial_\mu . \tag{4.39}$$

The algebra (4.4) is represented faithfully, even when we replace $\bar{\mathcal{Q}}_{\dot{a}} \to i\theta^a (\sigma^\mu)_{a\dot{a}} \partial_\mu$. In momentum space, the $\mathcal{Q}\bar{\mathcal{Q}}$-anticommutator can then be written

$$\left\{ \frac{\partial}{\partial \theta^a} , \ -\theta^a (\sigma^\mu)_{a\dot{a}} p_\mu \right\} = -(\sigma^\mu)_{a\dot{a}} p_\mu . \tag{4.40}$$

---

[2] Originally, Ferber [41] introduced these variables as superpartners of bosonic twistor variables. Twistors are discussed in Chapter 5.

Let us now go on-shell by assuming that p^μ is light-like; then we can rewrite (4.40) in spinor helicity formalism as

$$\left\{ \frac{\partial}{\partial \theta^a} \, , \, \theta^a |p]_a \langle p|_{\dot{a}} \right\} = |p]_a \langle p|_{\dot{a}} \,. \tag{4.41}$$

Introduce a new Grassmann-odd variable $\eta = \theta^a |p]_a$. Then $\frac{\partial}{\partial \theta^a} = |p]_a \frac{\partial}{\partial \eta}$, so that (4.41) becomes

$$\left\{ |p]_a \frac{\partial}{\partial \eta} \, , \, \eta \langle p|_{\dot{a}} \right\} = |p]_a \langle p|_{\dot{a}} \implies \left\{ [p|^a \frac{\partial}{\partial \eta} \, , \, |p\rangle^{\dot{a}} \eta \right\} = |p\rangle^{\dot{a}} [p|^a \,. \tag{4.42}$$

The arguments, $[p|^a \frac{\partial}{\partial \eta}$ and $|p\rangle^{\dot{a}} \eta$, of the anticommutator are recognized as $\mathcal{N} = 1$ versions of our on-shell supersymmetry generators q^a and $q^{\dagger \dot{a}}$ in (4.38).

We note that dotting some arbitrary reference spinors $|\tilde{w}\rangle$ and $[w|$ (whose brackets with the p-spinors are non-vanishing) into (4.42), we find $\{\eta, \frac{\partial}{\partial \eta}\} = 1$.

Consider the consequences of the above analysis. The superspace coordinates θ and $\bar{\theta}$ have mass-dimension $(\text{mass})^{-1/2}$ and the angle and square spinors have dimension $(\text{mass})^{1/2}$. So the on-shell superspace variables η are dimensionless. Under little group scaling, the θ and $\bar{\theta}$ are inert, and therefore we have $\eta \to t^{-1} \eta$. Consequently, the on-shell superwavefunction (4.36) scales homogeneously as $\Omega \to t^{-2} \Omega$ under a little group scaling. The state-operator map (4.37) compensates this scaling when extracting component states from Ω. $\triangleleft$

Superamplitudes

We can think of $\Omega_i = \Omega(p_i)$ as a superwavefunction for the ith external particle of a (color-ordered) *superamplitude* $\mathcal{A}_n[\Omega_1, \ldots, \Omega_n]$. It depends on the on-shell momentum p_i and a set of Grassmann variables η_{iA} for each particle $i = 1, \ldots, n$. Expanding $\mathcal{A}_n[\Omega_1, \ldots, \Omega_n]$ in the Grassmann variables, we note that the $SU(4)$-symmetry requires it to be a sum of polynomials in η_{iA} of degree $4(K + 2)$. An example of a legal combination is $\eta_{i1} \eta_{i2} \eta_{i3} \eta_{i4}$; according to the map (4.37), it corresponds to particle i being a negative helicity gluon.

Using the map (4.37) to extract states from the superwavefunction of each particle $i = 1, 2, \ldots, n$, one projects out the regular amplitudes (sometimes called "component amplitudes") from the superamplitudes. For example:

$$A_n\big[1^+ \ldots i^- \ldots j^- \ldots n^+\big] = \left(\prod_{A=1}^{4} \frac{\partial}{\partial \eta_{iA}} \right) \left(\prod_{B=1}^{4} \frac{\partial}{\partial \eta_{jB}} \right) \mathcal{A}_n\big[\Omega_1, \ldots, \Omega_n\big]\bigg|_{\eta_{kC}=0}$$

$$A_n\big[S^{12} S^{34} 3^- 4^+ \ldots n^+\big] = \left(\frac{\partial}{\partial \eta_{11}} \frac{\partial}{\partial \eta_{12}} \right) \left(\frac{\partial}{\partial \eta_{23}} \frac{\partial}{\partial \eta_{24}} \right)$$
$$\times \left(\prod_{A=1}^{4} \frac{\partial}{\partial \eta_{3A}} \right) \mathcal{A}_n\big[\Omega_1, \ldots, \Omega_n\big]\bigg|_{\eta_{kC}=0} \,. \tag{4.43}$$

Grassmann differentiations can of course equally well be expressed as Grassmann integrals.

The order K of the Grassmann polynomial precisely corresponds to the N^KMHV sector. So we can organize the full superamplitude as

$$\mathcal{A}_n = \mathcal{A}_n^{\text{MHV}} + \mathcal{A}_n^{\text{NMHV}} + \mathcal{A}_n^{\text{N}^2\text{MHV}} + \cdots + \mathcal{A}_n^{\text{anti-MHV}}, \qquad (4.44)$$

where $\mathcal{A}_n^{\text{MHV}}$ has Grassmann degree 8, $\mathcal{A}_n^{\text{NMHV}}$ has Grassmann degree 12, etc. For example, expanding the MHV superamplitude in Grassmann monomials, we have

$$\mathcal{A}_n^{\text{MHV}} = A_n\big[g^- g^- g^+ \ldots g^+\big](\eta_1)^4(\eta_2)^4 + A_n\big[g^- \lambda^{123}\lambda^4 \ldots g^+\big](\eta_1)^4(\eta_{21}\eta_{22}\eta_{23})(\eta_{34}) + \cdots, \qquad (4.45)$$

where $(\eta_i)^4 \equiv \eta_{i1}\eta_{i2}\eta_{i3}\eta_{i4}$. (Quiz: why does the second term in (4.45) come with a plus?)

Next, consider the supersymmetry Ward identities. In the language of on-shell superspace, the supersymmetry Ward identities are identical to the statement that the supercharges

$$Q^A \equiv \sum_{i=1}^{n} q_i^A = \sum_{i=1}^{n} [i| \frac{\partial}{\partial \eta_{iA}}, \quad \text{and} \quad \tilde{Q}_A \equiv \sum_{i=1}^{n} q_A^\dagger = \sum_{i=1}^{n} |i\rangle \eta_{iA}, \quad A = 1,2,3,4, \qquad (4.46)$$

annihilate the superamplitude:

$$Q^A \mathcal{A}_n = 0 \quad \text{and} \quad \tilde{Q}_A \mathcal{A}_n = 0. \qquad (4.47)$$

▶ **Exercise 4.13**

It may not be totally obvious that (4.47) encodes the supersymmetry Ward identities, so the point of this exercise is to illustrate how it works. Start by writing the (relevant terms in the) MHV superamplitude as in (4.45). The supersymmetry Ward identity (4.47) says that the coefficient of each independent Grassmann monomial in $\tilde{Q}_A \mathcal{A}_n^{\text{MHV}}$ has to vanish. Pick $A = 4$ and act with $\tilde{Q}_4$ on $\mathcal{A}_n^{\text{MHV}}$ to extract all terms whose Grassmann structure is $(\eta_1)^4(\eta_2)^4(\eta_{34})$. Use that to show that the "component amplitude" supersymmetry Ward identity (4.32) follows from $\tilde{Q}_4 \mathcal{A}_n^{\text{MHV}} = 0$.

The requirement that $\{Q^A, \tilde{Q}_B\} \sim \sum_i p_i^\mu$ annihilates the superamplitude is equivalent to the statement of momentum conservation. The associated delta function $\delta^4(\sum_{i=1}^{n} p_i)$ has been left implicit throughout the early chapters, but we will begin to include it explicitly in Section 5, where it plays a central role. For example, it makes momentum conservation an operational statement that the momentum operator $P^\mu = \sum_i p_i^\mu$ annihilates the amplitude: it simply acts multiplicatively and the vanishing result follows from the fact that $x\,\delta(x)$ has support nowhere.

Note that the action of $\tilde{Q}_A$ on the superamplitude is multiplicative. We can therefore solve it easily using a Grassmann delta function $\delta^{(8)}(\tilde{Q})$ defined as

$$\delta^{(8)}(\tilde{Q}) = \frac{1}{2^4} \prod_{A=1}^{4} \tilde{Q}_{A\dot{a}} \tilde{Q}_A^{\dot{a}} = \frac{1}{2^4} \prod_{A=1}^{4} \sum_{i,j=1}^{n} \langle ij \rangle \eta_{iA} \eta_{jA} . \tag{4.48}$$

▶ **Exercise 4.14**

Show that momentum conservation ensures that Q^A annihilates $\delta^{(8)}(\tilde{Q})$.

Which property of angle spinors allows you to demonstrate explicitly that $\tilde{Q}_A \delta^{(8)}(\tilde{Q}) = 0$?

Half the supersymmetry constraints, namely $\tilde{Q}_A \mathcal{A}_n = 0$, are satisfied if we write the N^KMHV superamplitude as

$$\mathcal{A}_n^{N^K \text{MHV}} = \delta^{(8)}(\tilde{Q}) \, P_n^{(4K)} , \tag{4.49}$$

where $P_n^{(4K)}$ is a degree $4K$ polynomial in the Grassmann variables. If $P_n^{(4K)}$ is annihilated by each Q^A, then – by Exercise 4.14 – all the supersymmetry constraints are solved. The Grassmann delta function $\delta^{(8)}(\tilde{Q})$ can be viewed as conservation of supermomentum.

The Grassmann delta function is a degree 8 polynomial in the η_{iA}s. This means that for an MHV superamplitude, $\delta^{(8)}(\tilde{Q})$ fixes the η_{iA}-dependence completely and $P_n^{(0)}$ is just a normalization that depends on the momenta. Thus, the tree-level MHV superamplitude of $\mathcal{N} = 4$ SYM is

$$\mathcal{N} = 4 \text{ SYM tree-level:} \quad \mathcal{A}_n^{\text{MHV}}[123\ldots n] = \frac{\delta^{(8)}(\tilde{Q})}{\langle 12 \rangle \langle 23 \rangle \cdots \langle n1 \rangle} . \tag{4.50}$$

This object will appear repeatedly in our study of amplitudes in $\mathcal{N} = 4$ SYM, both at tree- and loop-level, so now is a good time to become familiar with it.

It is not hard to see that the MHV superamplitude (4.50) produces the Parke–Taylor gluon tree amplitudes correctly. Just use the map (4.37) to take four derivatives with respect to η_{iA} and four with respect to η_{jA} as in (4.43). Then the delta function produces the numerator-factor $\langle ij \rangle^4$ of the component amplitude $A_n[1^+ \ldots i^- \ldots j^- \ldots n^+]$. Note that one component amplitude and supersymmetry uniquely fix the form of *all* MHV amplitudes in $\mathcal{N} = 4$ at each order in perturbation theory.

▶ **Exercise 4.15**

Reproduce the three supersymmetry Ward identities (4.33), (4.34), and (4.35) from the MHV superamplitude $\mathcal{A}_n^{\text{MHV}}$ in (4.50).

▶ **Exercise 4.16**

Use the $\mathcal{N} = 4$ SYM superamplitude $\mathcal{A}_n^{\text{MHV}}$ (4.50) to calculate the 4-scalar amplitude $A_4[S^{12} S^{34} S^{12} S^{34}]$. Compare your answer to the 4-scalar amplitude (3.41).

Calculate $A_4[S^{12} S^{23} S^{34} S^{41}]$ and compare with (3.45).

In our discussion so far, we have slipped silently by the anti-MHV 3-point amplitudes $\mathcal{A}_n^{\text{anti-MHV}}$, whose supersymmetry orbit determines the anti-MHV sector with $K = -1$. Thus, the 3-point anti-MHV sector is encoded in degree-4 superamplitudes. We simply state the answer,

$$\mathcal{A}_3^{\text{anti-MHV}} = \frac{1}{[12][23][31]} \, \delta^{(4)}\big([12]\eta_3 + [23]\eta_1 + [31]\eta_2\big), \qquad (4.51)$$

with the delta function defined as

$$\delta^{(4)}\big([12]\eta_3 + [23]\eta_1 + [31]\eta_2\big) \equiv \prod_{A=1}^{4} \big([12]\eta_{3A} + [23]\eta_{1A} + [31]\eta_{2A}\big), \qquad (4.52)$$

and leave it as an exercise to show that (4.51) is annihilated by the supercharges.

▶ **Exercise 4.17**

Show that $Q^A \mathcal{A}_n^{\text{anti-MHV}} = 0$ and $\tilde{Q}_A \mathcal{A}_n^{\text{anti-MHV}} = 0$.

Let us now outline three approaches to determining the superamplitudes $\mathcal{A}_n^{\text{N}^K\text{MHV}}$ beyond the MHV level.

1. *Solution to the supersymmetry Ward identities.* The N^KMHV superamplitudes in $\mathcal{N}=4$ SYM must obey the supersymmetry Ward identities (4.47). Writing the L-loop super-amplitude $\mathcal{A}_{n,L}^{\text{N}^K\text{MHV}} = \delta^{(8)}(\tilde{Q}) \, P_{n;L}^{(4K)}$, supersymmetry requires $Q^A P_{n;L}^{(4K)} = 0$. In addition, we need $P_{n;L}^{(4K)}$ to obey the Ward identities of the global $SU(4)$ R-symmetry. As we have seen, these constraints are trivially satisfied at the MHV level where one L-loop component-amplitude suffices to fix $P_{n;L}^{(0)}$ and hence the full L-loop MHV n-point superamplitude.

   For non-MHV amplitudes, how many component-amplitudes does one need to fix the N^KMHV superamplitude? This question is answered by analyzing the constraints $Q^A P_{n;L}^{(4K)} = 0$ and R-symmetry. It turns out that for general $K > 0$, $P_{n;L}^{(4K)}$ can be written as a linear combination of L-loop "basis amplitudes" with coefficients built from the polynomials $m_{ijk,A} \equiv [ij]\eta_{kA} + [jk]\eta_{iA} + [ki]\eta_{jA}$. This is the same type of Grassmann polynomial that appears in $\mathcal{A}_3^{\text{anti-MHV}}$ given in (4.51). The solution reveals that the number of component-amplitudes sufficient to determine $\mathcal{A}_{n,L}^{\text{N}^K\text{MHV}}$ is the dimension of the irreducible representation of $SU(n-4)$ corresponding to a rectangular Young diagram with K rows and ($\mathcal{N}=$)4 columns. The independent component amplitudes are labeled by the semi-standard tableaux of this Young diagram.[3] A given particle content (gluons, fermions, scalars) of a basis amplitude corresponds to an ordered partition of the number $4K$ into n integers between 0 and 4. For each such partition, the corresponding Kostka number counts the number of independent arrangements of the $SU(4)$ R-symmetry indices. You can read more about the solution to the supersymmetry and R-symmetry Ward identities in [42, 43].

---

[3] By manipulating the color-structure, one can improve further on this count of "basis amplitudes" [42, 43].

2. **The super-MHV vertex expansion.** The tree-level superamplitudes $\mathcal{A}_n^{\mathrm{N}^K\mathrm{MHV}}$ of $\mathcal{N} = 4$ SYM can be constructed as an MHV vertex expansion in which the vertices are the MHV superamplitudes $\mathcal{A}_n^{\mathrm{MHV}}$ of (4.50). The expansion can be derived from an all-line shift, as we discussed in Section 3.4. The validity of the all-line shift for tree-level superamplitudes in $\mathcal{N} = 4$ SYM was proven in [29] (see also [44]). A new feature is that one must sum over the possible intermediate states exchanged on the internal lines of the MHV vertex diagrams. This is conveniently done by integrating over the Grassmann variables $\eta_{P_I A}$ associated with internal lines; this automatically carries out a super-sum [36]. More details about such super-sums will be offered in the next section.

3. **Super-BCFW.** Superamplitudes can be constructed with a supersymmetric version of the BCFW shift. This construction plays a central role in many developments, so we will treat it in detail in the following section.

4.5 Super-BCFW and all tree-level amplitudes in $\mathcal{N} = 4$ SYM

The BCFW shift introduced in Section 3.2 preserves the on-shell conditions $p_i^2 = 0$ and momentum conservation $\sum_{i=1}^{n} p_i = 0$. However, the shift does not preserve the condition for supermomentum conservation, $\sum_{i=1}^{n} |i\rangle \eta_{iA} = 0$. As a consequence, the shifted component amplitudes have large-z falloffs that depend on which types of particles are shifted, for example we have seen the difference between the $[-, +\rangle$ and $[+, -\rangle$ BCFW shifts of gluons in (3.13). This can be remedied by a small modification of the BCFW shift (3.9) that allows us to conserve supermomentum. We simply accompany the BCFW momentum shift by a shift in the Grassmann-variables [45–47]: for simplicity let us focus on the $[1, 2\rangle$ **super-BCFW** shift (or $[1, 2\rangle$ "supershift"), defined as

$$|\hat{1}] = |1] + z\,|2]\,, \qquad |\hat{2}\rangle = |2\rangle - z\,|1\rangle\,, \qquad \hat{\eta}_{1A} = \eta_{1A} + z\,\eta_{2A}\,. \qquad (4.53)$$

No other spinors or Grassmann variables shift.

▶ **Exercise 4.18**

Show that the supermomentum is conserved under the supershift (4.53) so that $\delta^{(8)}(\tilde{Q})$ is invariant.

It follows directly from (4.50) that the MHV superamplitudes have a $1/z$ falloff under a supershift of any adjacent lines ($1/z^2$ for non-adjacent). In fact, this falloff behavior holds true for any tree superamplitude in $\mathcal{N} = 4$ SYM, as can be shown by using supersymmetry to rotate the two shifted lines to be positive helicity gluons and using the non-supersymmetric result (3.13) for the falloff under a $[+, +\rangle$-shift [19, 47].

The tree-level recursion relations that result from the super-BCFW shift (4.53) involve diagrams with two superamplitude "vertices" connected by an internal line with on-shell momentum $\hat{P}$. As in the non-supersymmetric case, we must sum over all possible states that can be exchanged on the internal line: in this case, this includes all 16 states of $\mathcal{N} = 4$ SYM. In terms of component amplitudes, the particle exchanged on the internal line depends on the external states: if they are all gluons, then the internal line is also a gluon and one must simply sum over the helicities. The superamplitude version of this helicity sum is

$$\left(\left[\left(\prod_{A=1}^{4} \frac{\partial}{\partial \eta_{\hat{P}A}} \right) \hat{A}_{\mathrm{L}} \right] \frac{1}{P^2} \hat{A}_{\mathrm{R}} + \hat{A}_{\mathrm{L}} \frac{1}{P^2} \left[\left(\prod_{A=1}^{4} \frac{\partial}{\partial \eta_{\hat{P}A}} \right) \hat{A}_{\mathrm{R}} \right] \right) \Bigg|_{\eta_{\hat{P}A}=0} , \qquad (4.54)$$

where $\eta_{\hat{P}A}$ is the Grassmann variable associated with the internal line. In the first term of (4.54), a negative helicity gluon is projected out from the left sub-superamplitude leaving a positive helicity gluon coming out of the right sub-superamplitude. For the second term, the helicity projection is the opposite.

If a gluino can be exchanged on the internal line,[4] then we have to move one of the four Grassmann derivatives from $\hat{A}_{\mathrm{L}}$ to $\hat{A}_{\mathrm{R}}$ in the first term of (4.54) – in all four possible ways to sum over the four different gluinos labeled by the $SU(4)$-indices. And similarly for the second term. A scalar exchange means that two Grassmann derivatives act on $\hat{A}_{\mathrm{L}}$ and the two others on $\hat{A}_{\mathrm{R}}$. All in all, the entire sum over states exchanged on the internal line can be written

$$\left(\prod_{A=1}^{4} \frac{\partial}{\partial \eta_{\hat{P}A}} \right) \left[\hat{A}_{\mathrm{L}} \frac{1}{P^2} \hat{A}_{\mathrm{R}} \right] \Bigg|_{\eta_{\hat{P}A}=0} = \int d^4 \eta_{\hat{P}} \, \hat{A}_{\mathrm{L}} \frac{1}{P^2} \hat{A}_{\mathrm{R}} . \qquad (4.55)$$

Note how the product rule distributes the Grassmann derivatives $\partial/\partial \eta_{\hat{P}A}$ on the L and R superamplitudes in all possible ways to automatically carry out the state **super-sum**. In (4.55), we have rewritten the Grassmann differentiation as a Grassmann integral. Similar super-sums are used in evaluation of unitarity cuts (see Chapter 6) of loop amplitudes where one includes integration over the Grassmann variable associated with the internal line [36, 48].

Remarkably, the super-BCFW recursion relations can be solved to give closed-form expressions for *all* tree-level superamplitudes in $\mathcal{N} = 4$ SYM [49]. We are now going to show how this works. As a warm-up, we first verify that the MHV superamplitude formula (4.50) satisfies the super-BCFW recursion relations. Then we present the most essential details of the derivation of the tree-level NMHV superamplitude. Finally we comment briefly on the results for $\mathrm{N}^K\mathrm{MHV}$.

4.5.1 MHV superamplitude from super-BCFW

Consider the super-BCFW recursion relations for the MHV superamplitude. Just as in the non-supersymmetric case (3.20), there is just one non-vanishing diagram, but we must now

---

[4] This happens when there is an odd number of external gluinos on each side of the BCFW diagram.

include the super-sum (4.55) over states exchanged on the internal line:

$$\mathcal{A}_n^{\text{MHV}}[123\ldots n] = \begin{array}{c} \includegraphics \end{array}$$

$$= \int d^4\eta_{\hat{P}}\, \mathcal{A}_{n-1}\big[\hat{1},\hat{P},4,\ldots,n\big]\,\frac{1}{P^2}\,\hat{\mathcal{A}}_3\big[-\hat{P},\hat{2},3\big]$$

$$= \int d^4\eta_{\hat{P}}\,\frac{\delta^{(8)}\big(\sum_{i\in L}|\hat{i}\rangle\hat{\eta}_i\big)}{\langle 1\hat{P}\rangle\langle\hat{P}4\rangle\langle 45\rangle\cdots\langle n1\rangle}\,\frac{1}{P^2}$$

$$\times\frac{\delta^{(4)}\big([\hat{P}2]\eta_3 + [23]\eta_{\hat{P}} + [3\hat{P}]\eta_2\big)}{[23][3\hat{P}][\hat{P}2]}\,, \tag{4.56}$$

where $P = P_{23} = p_2 + p_3$ and we use (4.50) and (4.51) for the $(n-1)$-point MHV super-amplitude and 3-point anti-MHV superamplitude. We also use the analytic continuation rule (3.17) for $|-P] = |P]$.

The new feature is the Grassmann integral in (4.56). The first delta function is

$$\delta^{(8)}\left(\sum_{i\in L}|\hat{i}\rangle\hat{\eta}_i\right) = \delta^{(8)}\left(|1\rangle\hat{\eta}_1 + |\hat{P}\rangle\eta_{\hat{P}} + \sum_{i=4}^{n}|i\rangle\eta_i\right). \tag{4.57}$$

On the support of the second delta function, we can set $\eta_{\hat{P}} = -([\hat{P}2]\eta_3 + [3\hat{P}]\eta_2)/[23]$ to find

$$|1\rangle\hat{\eta}_1 + |\hat{P}\rangle\eta_{\hat{P}} = |1\rangle\hat{\eta}_1 - \frac{|\hat{P}\rangle}{[23]}\big([\hat{P}2]\eta_3 + [3\hat{P}]\eta_2\big)$$

$$= |1\rangle\hat{\eta}_1 + |3\rangle\eta_3 + |\hat{2}\rangle\eta_2$$

$$= |1\rangle\eta_1 + |3\rangle\eta_3 + |2\rangle\eta_2\,. \tag{4.58}$$

We used momentum conservation on the right-hand subamplitude to obtain the second line in (4.58) and the definition (4.53) of the supershift to get the third line. This way, the first delta function simply becomes $\delta^{(8)}(\sum_{i=1}^{n}|i\rangle\eta_i) = \delta^{(8)}(\tilde{Q})$.

Now the only $\eta_{\hat{P}}$-dependence is in $\delta^{(4)}(\ldots)$ and the Grassmann integral is easy to carry out:

$$\int d^4\eta_{\hat{P}}\,\delta^{(8)}\left(\sum_{i\in L}|\hat{i}\rangle\hat{\eta}_i\right)\delta^{(4)}\big([23]\eta_{\hat{P}} + \cdots\big) = [23]^4\,\delta^{(8)}(\tilde{Q})\,.$$

Coming back to (4.56), we then have

$$\mathcal{A}_n^{\text{MHV}}[123\ldots n] = \frac{\delta^{(8)}(\tilde{Q})}{\langle 1\hat{P}\rangle\langle\hat{P}4\rangle\langle 45\rangle\cdots\langle n1\rangle}\,\frac{1}{P^2}\,\frac{[23]^4}{[3\hat{P}][\hat{P}2][23]}\,. \tag{4.59}$$

Compare this with (3.21) and you will see that the two expressions are the same, except that $\delta^{(8)}(\tilde{Q})$ has replaced $\langle 1\hat{P}\rangle^4$ and $[23]^4$ has replaced $[3\hat{P}]^4$. Using the identities (3.22) and (3.23), we promptly recover the desired result

$$\mathcal{A}_n^{\text{MHV}}[123\ldots n] = \frac{\delta^{(8)}(\tilde{Q})}{\langle 12 \rangle \langle 23 \rangle \cdots \langle n1 \rangle}. \tag{4.60}$$

Thus we have shown that the $\mathcal{N} = 4$ SYM tree-level MHV superamplitude (4.60) satisfies the super-BCFW recursion relations. Next, we use super-BCFW to derive an important result for NMHV superamplitudes in $\mathcal{N} = 4$ SYM.

4.5.2 NMHV superamplitude and beyond

The super-BCFW recursion relation for the NMHV superamplitude involves two types of diagrams

$$\mathcal{A}_n^{\text{NMHV}}[12\ldots n] = \sum_{k=5}^{n} \quad\text{MHV diagram A}\quad + \quad\text{NMHV anti-MHV diagram B}. \tag{4.61}$$

The diagrams of type A involve two MHV sub-superamplitudes. Diagram B is present only for $n \geq 6$ because the L subamplitude is NMHV and therefore needs at least five legs; indeed, back in Section 3.2, we only had an MHV$\times$MHV diagram in the calculation (3.27) of $A_5[1^-2^-3^-4^+5^+]$.

Diagram B provides the setting for an inductive proof of the NMHV superamplitude formula we are seeking, while the diagrams of type A give an "inhomogeneous" contribution. We begin with a detailed evaluation of the type A diagrams. There will be a lot of detailed calculations in this example calculation, so if you just want the result, you are free to skip ahead to the answer in (4.82). Right after the example, we summarize the full result for the NMHV superamplitude.

▷ *Example.* Calculation of diagram A. The first step is simply to plug in the MHV superamplitudes:

$$\text{Diagram A} = \int d^4\eta_{\hat{P}} \; \frac{\delta^{(8)}(\text{L})}{\langle 1\hat{P} \rangle \langle \hat{P}\,k \rangle \langle k, k+1 \rangle \ldots \langle n1 \rangle} \frac{1}{P^2}$$

$$\times \frac{\delta^{(8)}(\text{R})}{\langle \hat{P}\,\hat{2} \rangle \langle \hat{2}3 \rangle \langle 34 \rangle \ldots \langle k-1, \hat{P} \rangle}. \tag{4.62}$$

Here $P = p_2 + p_3 + \cdots + p_{k-1}$ and (with the help of the rule (3.17)) we have

$$\delta^{(8)}(\text{L}) = \delta^{(8)}\left(-|\hat{P}\rangle\eta_{\hat{P}} + |1\rangle\hat{\eta}_1 + \sum_{r=k}^{n} |r\rangle\eta_r \right),$$

$$\delta^{(8)}(\text{R}) = \delta^{(8)}\left(|\hat{P}\rangle\eta_{\hat{P}} + |\hat{2}\rangle\eta_2 + \sum_{r=3}^{k-1} |r\rangle\eta_r \right). \tag{4.63}$$

On the support of $\delta^{(8)}(\mathrm{R})$, we can write $\delta^{(8)}(\mathrm{L}) = \delta^{(8)}(\mathrm{L} + \mathrm{R}) = \delta^{(8)}(\sum_{i=1}^{n} |i\rangle \eta_i) = \delta^{(8)}(\tilde{Q})$, which is independent of $\eta_{\hat{P}}$ and expresses the conservation of supermomentum for the n external states. Now only $\delta^{(8)}(\mathrm{R})$ remains under the state-sum integral $\int d^4 \eta_{\hat{P}}$; since $\delta^{(8)}$ enforces two conditions we can project out two separate $\delta^{(4)}$s:

$$\delta^{(8)}(\mathrm{R}) = \frac{1}{\langle 1\hat{P}\rangle^4} \, \delta^{(4)}\left(\langle 1\hat{P}\rangle \eta_{\hat{P}} + \langle 1\hat{2}\rangle \eta_2 + \sum_{r=3}^{k-1} \langle 1r\rangle \eta_r\right) \delta^{(4)}\left(\langle \hat{P}\,\hat{2}\rangle \eta_2 + \sum_{r=3}^{k-1} \langle \hat{P}\,r\rangle \eta_r\right). \tag{4.64}$$

There is $\eta_{\hat{P}}$-dependence only in the first of these two $\delta^{(4)}$s, so it is straightforward to perform the Grassmann integral: it produces a factor $\langle 1\hat{P}\rangle^4$ which cancels the normalization factor included in (4.64). All in all, we have then shown that

$$\int d^4 \eta_{\hat{P}} \, \delta^{(8)}(\mathrm{L}) \, \delta^{(8)}(\mathrm{R}) = \delta^{(8)}(\tilde{Q}) \, \delta^{(4)}\left(\langle \hat{P}\,\hat{2}\rangle \eta_2 + \sum_{r=3}^{k-1} \langle \hat{P}\,r\rangle \eta_r\right). \tag{4.65}$$

The factor $\delta^{(8)}(\tilde{Q})$ can be used to pull out an overall factor $\mathcal{A}_n^{\mathrm{MHV}}$ from diagram A in (4.62) and we can then write

$$\text{Diagram A} = \mathcal{A}_n^{\mathrm{MHV}} \, \frac{\langle 12\rangle \langle 23\rangle \langle k-1, k\rangle \, \delta^{(4)}\left(\langle \hat{P}\,\hat{2}\rangle \eta_2 + \sum_{r=3}^{k-1} \langle \hat{P}\,r\rangle \eta_r\right)}{\langle k\hat{P}\rangle \langle k-1, \hat{P}\rangle \langle \hat{2}\hat{P}\rangle \langle \hat{2}3\rangle \langle 1\hat{P}\rangle \, P^2}. \tag{4.66}$$

Next, we turn our attention to the brackets in (4.66) that involve the shifted spinors $|\hat{P}\rangle$ and $|\hat{2}\rangle$. As in the non-supersymmetric examples of Section 3.2, we manipulate these brackets by multiplying the numerator and denominator of (4.66) by $[\hat{P}\,2]^4 \langle 21\rangle^4$. We are going to encounter sums of momenta that appear in the ordering fixed by color-structure, so for convenience we introduce the shorthand notation

$$y_{ij} \equiv p_i + p_{i+1} + \cdots + p_{j-1}. \tag{4.67}$$

Further, we declare that y_{ji} with $j > i$ equals $-y_{ij}$; when we think of the p_i as cyclically ordered, this simply expresses momentum conservation. We will use the variables y_{ij} to write the result for diagram A in a form that may look slightly mysterious for now, but it has some very important features that are discussed in Chapter 5.

One type of bracket in (4.66) is $\langle r\hat{P}\rangle$ with $r \neq 1, \hat{2}$. We manipulate this as follows:

$$\langle r\hat{P}\rangle [\hat{P}\,2]\langle 21\rangle = -\langle r|(\hat{2} + 3 + \cdots + (k-1))|2]\langle 21\rangle = -\langle r|y_{3k}|2]\langle 21\rangle$$

$$= \langle r|y_{3k} \cdot y_{23}|1\rangle = \langle r|y_{3k} \cdot y_{13}|1\rangle = -\langle 1|y_{13} \cdot y_{3k}|r\rangle. \tag{4.68}$$

The double angle-bracket is defined as $\langle i|p \cdot q|j\rangle \equiv \langle i|_{\dot{a}} p^{\dot{a}b} q_{b\dot{c}}|j\rangle^{\dot{c}}$. The result (4.68) applies to the brackets $\langle r\hat{P}\rangle$, $\langle k\hat{P}\rangle$, and $\langle k-1, \hat{P}\rangle$ in (4.66).

▶ **Exercise 4.19**

To keep you actively engaged, involved, and awake, here is an exercise: show that

$$\langle i|K.K|j\rangle = -K^2 \langle ij\rangle \tag{4.69}$$

for any momentum K (light-like or not) and any spinors $\langle i|$ and $|j\rangle$.

The two brackets $\langle 1\hat{P} \rangle$ and $\langle \hat{2}\hat{P} \rangle$ from the denominator of (4.62) are dealt with as follows:

$$\langle 1\hat{P} \rangle [\hat{P} 2] = -\langle 1|y_{3k}|2] \quad \text{and} \quad \langle \hat{2}\hat{P} \rangle [\hat{P} 2] = -2\hat{p}_2 \cdot \hat{P} = -\hat{P}^2 + y_{3k}^2 = y_{3k}^2 .$$

$$(4.70)$$

In the last equality, we used the fact that the BCFW diagram is evaluated on the value of z such that $\hat{P}^2 = 0$. It is useful to note what this value of z is:

$$0 = \hat{P}^2 = \langle \hat{2}|\hat{P}|2] + y_{3k}^2 = y_{2k}^2 - z\langle 1|y_{3k}|2] \quad \Longrightarrow \quad z = \frac{y_{2k}^2}{\langle 1|y_{3k}|2]} . \qquad (4.71)$$

We use this z to evaluate $\langle \hat{2}3 \rangle$:

$$\begin{aligned}
\langle \hat{2}3 \rangle &= \langle 23 \rangle - \frac{y_{2k}^2}{\langle 1|y_{3k}|2]}\langle 13 \rangle = \frac{\langle 1|y_{3k}|2]\langle 23 \rangle - y_{2k}^2\langle 13 \rangle}{\langle 1|y_{3k}|2]} \\
&= \frac{\langle 1|y_{1k}|2]\langle 23 \rangle + \langle 1|y_{2k} \cdot y_{2k}|3 \rangle}{\langle 1|y_{3k}|2]} = \frac{-\langle 1|y_{1k} \cdot p_2|3 \rangle + \langle 1|y_{1k} \cdot y_{2k}|3 \rangle}{\langle 1|y_{3k}|2]} \\
&= \frac{\langle 1|y_{1k} \cdot (-p_2 + y_{2k})|3 \rangle}{\langle 1|y_{3k}|2]} = \frac{\langle 1|y_{1k} \cdot y_{3k}|3 \rangle}{\langle 1|y_{3k}|2]} .
\end{aligned}$$

$$(4.72)$$

In the third equality, we use the result of Exercise 4.19. Combining (4.70) and (4.72), we can write

$$\langle \hat{2}3 \rangle \langle 1\hat{P} \rangle [\hat{P} 2] = \langle 1|y_{1k} \cdot y_{k3}|3 \rangle . \qquad (4.73)$$

We have now evaluated all four angle brackets involving $|\hat{P} \rangle$ and $|\hat{2} \rangle$ in (4.66). There is, however, another manipulation that we would like to do, namely one involving the propagator $1/P^2 = 1/y_{2k}^2$. It goes like this:

$$P^2 \langle 12 \rangle = y_{2k}^2 \langle 12 \rangle = -\langle 1|y_{2k} \cdot y_{2k}|2 \rangle = -\langle 1|y_{1k} \cdot y_{3k}|2 \rangle = \langle 1|y_{1k} \cdot y_{k3}|2 \rangle . \quad (4.74)$$

And then it is time to put everything together. We can now write diagram A from (4.66) as

$$\text{Diagram A} = \mathcal{A}_n^{\text{MHV}} \frac{\langle 23 \rangle \langle k-1, k \rangle [\hat{P} 2]^4 \langle 21 \rangle^4 \, \delta^{(4)}\left(\langle \hat{P} \hat{2} \rangle \eta_2 + \sum_{r=3}^{k-1}\langle \hat{P} r \rangle \eta_r\right)}{y_{3k}^2 \langle 1|y_{13} \cdot y_{3k}|k \rangle \langle 1|y_{13} \cdot y_{3k}|k-1 \rangle \langle 1|y_{1k} \cdot y_{k3}|3 \rangle \langle 1|y_{1k} \cdot y_{k3}|2 \rangle} .$$

$$(4.75)$$

Let us examine the $\delta^{(4)}$ in the numerator. We absorb the factors $[\hat{P} 2]^4 \langle 21 \rangle^4$ into the delta function whose argument then becomes (suppressing $SU(4)$-indices)

$$\Xi \equiv \langle 12 \rangle [2\hat{P}]\langle \hat{P} \hat{2} \rangle \eta_2 + \sum_{r=3}^{k-1}\langle 12 \rangle [2\hat{P}]\langle \hat{P} r \rangle \eta_r = \langle 12 \rangle y_{3k}^2 \, \eta_2 + \sum_{r=3}^{k-1}\langle 1|y_{13}.y_{3k}|r \rangle \, \eta_r .$$

$$(4.76)$$

We are going to do a little work on this expression in order to introduce another piece of shorthand notation, namely fermionic companions of the y_{ij}s defined in (4.67):

$$|\theta_{ij,A}\rangle \equiv \sum_{r=i}^{j-1} |r\rangle\, \eta_{rA}\,. \tag{4.77}$$

Then $|\theta_{ji}\rangle = -|\theta_{ij}\rangle$ encodes supermomentum conservation.

We start by rewriting each of the terms in (4.76):

$$\langle 12\rangle\, y_{3k}^2\, \eta_2 = -\langle 1|y_{3k}\cdot y_{3k}|2\rangle\, \eta_2 = -\langle 1|y_{1k}\cdot y_{3k}|2\rangle\, \eta_2 + \langle 1|y_{23}\cdot y_{3k}|2\rangle\, \eta_2$$
$$= -\langle 1|y_{1k}\cdot y_{3k}|\theta_{13}\rangle + \langle 1|y_{1k}\cdot y_{3k}|1\rangle\, \eta_1 + \langle 1|y_{13}\cdot y_{3k}|2\rangle\, \eta_2 \tag{4.78}$$

and

$$\sum_{r=3}^{k-1} \langle 1|y_{13}\cdot y_{3k}|r\rangle\, \eta_r = \langle 1|y_{13}\cdot y_{3k}|\theta_{1k}\rangle - \langle 1|y_{13}\cdot y_{3k}|1\rangle\, \eta_1 - \langle 1|y_{13}\cdot y_{3k}|2\rangle\, \eta_2\,. \tag{4.79}$$

Adding (4.78) and (4.79) to get Ξ, the two extra η_2-terms cancel and the η_1-terms combine to

$$\langle 1|y_{1k}\cdot y_{3k}|1\rangle\, \eta_1 - \langle 1|y_{13}\cdot y_{3k}|1\rangle\, \eta_1 = \langle 1|(y_{1k} - y_{13})\cdot y_{3k}|1\rangle\, \eta_1$$
$$= \langle 1|y_{3k}\cdot y_{3k}|1\rangle\, \eta_1 = 0 \tag{4.80}$$

by (4.69). Thus we have

$$\Xi = -\langle 1|y_{1k}\cdot y_{k3}|\theta_{31}\rangle - \langle 1|y_{13}.y_{3k}|\theta_{k1}\rangle\,. \tag{4.81}$$

Our work brings us to the following form of a diagram of type A:

$$\text{Diagram A} = \mathcal{A}_n^{\text{MHV}}\, \frac{\langle 23\rangle\langle k-1,k\rangle\,\delta^{(4)}\Big(\langle 1|y_{1k}\cdot y_{k3}|\theta_{31}\rangle + \langle 1|y_{13}\cdot y_{3k}|\theta_{k1}\rangle\Big)}{y_{3k}^2\,\langle 1|y_{13}\cdot y_{3k}|k\rangle\langle 1|y_{13}\cdot y_{3k}|k-1\rangle\langle 1|y_{1k}\cdot y_{k3}|3\rangle\langle 1|y_{1k}\cdot y_{k3}|2\rangle}\,. \tag{4.82}$$

This completes our calculation of diagram A in the super-BCFW recursion relation (4.61). Next, we discuss what the full NMHV superamplitude looks like. ◁

The result (4.82) for diagram A is often written $\mathcal{A}_n^{\text{MHV}}\, R_{13k}$, where the so-called **R-invariants**[5] are defined as

$$R_{1jk} = \frac{\langle j-1,j\rangle\langle k-1,k\rangle\,\delta^{(4)}\big(\Xi_{1jk}\big)}{y_{jk}^2\,\langle 1|y_{1j}\cdot y_{jk}|k\rangle\langle 1|y_{1j}\cdot y_{jk}|k-1\rangle\langle 1|y_{1k}\cdot y_{kj}|j\rangle\langle 1|y_{1k}\cdot y_{kj}|j-1\rangle}\,, \tag{4.83}$$

where

$$\Xi_{1jk,A} = \langle 1|y_{1k}\cdot y_{kj}|\theta_{j1,A}\rangle + \langle 1|y_{1j}\cdot y_{jk}|\theta_{k1,A}\rangle\,. \tag{4.84}$$

---

[5] We discuss in Chapter 5 under which symmetries R_{ijk} is invariant.

For completeness, we repeat here the definitions (4.67) and (4.77) of the variables $y_{ij} = -y_{ji}$ and $\theta_{ij,A} = -\theta_{ji,A}$:

$$y_{ij} \equiv p_i + p_{i+1} + \cdots + p_{j-1} \quad \text{and} \quad |\theta_{ij,A}\rangle \equiv \sum_{r=i}^{j-1} |r\rangle \eta_{rA}. \qquad (4.85)$$

Note the structure of (4.83) and (4.84): neighbor indices match up with each other. This is an important feature.

For $n = 5$, diagram B vanishes and diagram A with $k = 5$ is the complete result:

$$\mathcal{A}_5^{\text{NMHV}} = \mathcal{A}_5^{\text{MHV}} R_{135}. \qquad (4.86)$$

For $n > 5$, the diagrams of type A contribute $\mathcal{A}_n^{\text{MHV}} \sum_{k=5}^{n} R_{13k}$. Diagram B recurses this form (we will not present the details, see [49]) and the result is a sum of R-invariants:

$$\text{Diagram B} = \mathcal{A}_n^{\text{MHV}} \sum_{j=4}^{n-2} \sum_{k=j+2}^{n} R_{1jk}. \qquad (4.87)$$

Hence, the entire NMHV superamplitude can be expressed in terms of the R-invariants as

$$\mathcal{A}_n^{\text{NMHV}} = \mathcal{A}_n^{\text{MHV}} \sum_{j=3}^{n-2} \sum_{k=j+2}^{n} R_{1jk}. \qquad (4.88)$$

For example,

$$\mathcal{A}_6^{\text{NMHV}} = \mathcal{A}_6^{\text{MHV}} \left(R_{135} + R_{136} + R_{146} \right), \qquad (4.89)$$

and

$$\mathcal{A}_7^{\text{NMHV}} = \mathcal{A}_7^{\text{MHV}} \left(R_{135} + R_{136} + R_{137} + R_{146} + R_{147} + R_{157} \right). \qquad (4.90)$$

A generic Feynman diagram calculation of the 7-gluon NMHV component amplitude requires 154 diagrams. It is a dramatic simplification that reduces the sum of Feynman diagrams down to the six terms in (4.90). Moreover, note that it is not just the Feynman diagrams for gluon scattering that are encoded in the sums of R_{ijk}s, rather the relatively simple expression (4.88) encodes all tree-level NMHV scattering amplitudes of the entire $\mathcal{N} = 4$ SYM spectrum.

The remarkably simple result (4.88) for the all-n tree-level NMHV superamplitudes in $\mathcal{N} = 4$ SYM was first found in studies of loop amplitudes [50]. Later, the NMHV formula was constructed as we did here using super-BCFW [49].

One final comment about the R-invariants (4.83) is that we defined them here "anchored" at momentum line 1; this came about because we used a $[1, 2\rangle$-supershift. However, by cyclic symmetry all pairs of adjacent lines in the superamplitude are on equal footing, so an $[i, i + 1\rangle$-supershift would have resulted in an NMHV formula anchored at i, giving R-invariants R_{ijk}. These are defined by replacing all momentum labels 1 in (4.83) by i. In the literature you will often find the expressions for the NMHV superamplitude given with n as the anchor; i.e. $\mathcal{A}_n^{\text{NMHV}} = \mathcal{A}_n^{\text{MHV}} \sum_{j=2}^{n-3} \sum_{k=j+2}^{n-1} R_{njk}$.

▶ **Exercise 4.20**

Write down the NMHV superamplitude formula that results from a $[2, 3\rangle$-supershift. Then project out the gluon amplitude $A_6[1^+2^-3^+4^-5^+6^-]$. Can you match your result to the 3-term expression in Exercise 3.9? Can you guess what expression you get from a $[3, 2\rangle$-supershift after projecting out $A_6[1^+2^-3^+4^-5^+6^-]$? The answer is given in Chapter 10.

Formulas for tree-level N^KMHV superamplitudes have also been derived – they take the form of sums of generalized R-invariants [49]. We refer you to the original paper [49], the discussions in [46, 47, 51], and the review [52] for further details.

The superamplitudes of $\mathcal{N} = 4$ SYM are invariant under a large symmetry group. In particular, there are both "ordinary" and "hidden" symmetries, and they are realized for the tree-level superamplitudes. The structure of the R-invariants is essential for this. We will discuss these symmetries next in Chapter 5.

This chapter is dedicated to a detailed description of the symmetries of superamplitudes in $\mathcal{N} = 4$ SYM. As mentioned in Section 4.4, $\mathcal{N} = 4$ SYM is a superconformal theory.[1] In four dimensions, the conformal group is $SO(4, 2) \sim SU(2, 2)$, and with $\mathcal{N} = 4$ supersymmetry, it is enhanced to the superconformal group $SU(2, 2|4)$. Its bosonic symmetries consist of the conformal group $SO(4, 2)$ and the R-symmetry group $SU(4)$. Our first task is to show that the generators of $SU(2, 2|4)$ annihilate the $\mathcal{N} = 4$ SYM superamplitudes; this shows that the superamplitudes respect $SU(2, 2|4)$ symmetry. In Section 5.1, we demonstrate this property using the spinor helicity formalism and the on-shell superspace from Section 4.4. And in Section 5.2 we introduce the – for this purpose – more suitable language of *twistors*.

The superconformal symmetry $SU(2, 2|4)$ can be called the "obvious" symmetry of $\mathcal{N} = 4$ SYM. This is said to contrast that there is also a "hidden" symmetry: it turns out that the tree-level superamplitudes of $\mathcal{N} = 4$ SYM enjoy another symmetry, namely ***dual superconformal symmetry*** $SU(2, 2|4)$.[2] We treat this in Section 5.3, where we also consider the consequences of having both "ordinary" and dual superconformal symmetry. The natural variables for dual superconformal symmetry are ***momentum twistors*** and they are introduced in Section 5.4 for our enjoyment now and later.

5.1 Superconformal symmetry of $\mathcal{N} = 4$ SYM

All theories studied in this book are Poincaré invariant. In four dimensions, the ten Poincaré symmetry generators, P^μ and $M^{\mu\nu}$, can be written in spinor helicity formalism. This is achieved by converting P^μ and $M^{\mu\nu}$ to momentum space and then contracting the Lorentz indices with $(\bar{\sigma}^\mu)^{\dot{a}b}$ and $(\sigma^\mu \bar{\sigma}^\nu - \sigma^\nu \bar{\sigma}^\mu)_a{}^b$ (and its conjugate). The action of the Poincaré generators on the scattering amplitudes can then be realized by the operators

$$P^{\dot{a}b} = -\sum_i |i\rangle^{\dot{a}} [i|^b \,, \qquad M_{ab} = \sum_i |i]_{(a} \partial_{[i|^b)} \,, \qquad M_{\dot{a}\dot{b}} = \sum_i \langle i|_{(\dot{a}} \partial_{|i)_{\dot{b})}} \,, \qquad (5.1)$$

where the sum is over external particle labels $i = 1, 2, \ldots, n$ and $(..)$ indicates symmetrization of the enclosed indices. The operators (5.1) satisfy the Poincaré commutator algebra (4.3) without imposing momentum conservation on the n momenta.

---

[1] At the origin of moduli space

[2] Dual conformal symmetry at loop-level is discussed in Sections 6.4 and 6.5.

▶ **Exercise 5.1**

Show that

$$\sum_i \langle i|_{\dot{a}}\, \partial_{|i\rangle^{\dot{b}}} \langle jk \rangle = \epsilon_{\dot{a}\dot{b}} \langle jk \rangle \tag{5.2}$$

and hence $M_{\dot{a}\dot{b}}$ annihilates angle brackets. M_{ab} trivially gives zero on any angle brackets. The equivalent conclusions hold for square brackets.
[Hint: use $A^{ab} - A^{ba} = -A^c{}_c\, \epsilon^{ab}$ which is valid for any 2×2 matrix.]

In the earlier chapters, we have imposed momentum conservation on the external momenta "by hand." For the purpose of studying the symmetries of the amplitudes it is necessary to include the momentum conservation delta function $\delta^4(\sum_i p_i)$ explicitly in the amplitudes. For example, for the MHV gluon amplitude

$$A_n\big[1^+ \ldots i^- \ldots j^- \ldots n^+\big] = \frac{\langle ij \rangle^4}{\langle 12 \rangle \langle 23 \rangle \cdots \langle n1 \rangle}\, \delta^4(P). \tag{5.3}$$

The translation generator $P^{\dot{a}a}$ acts multiplicatively and annihilates an amplitude in the distributional sense $P^{\dot{a}a}\, \delta^4(P) = 0$. The action of the rotations/boosts follows from the following useful identity:

▷ *Example.* Calculate

$$\sum_i |i\rangle^{\dot{a}}\, \partial_{|i\rangle^{\dot{b}}}\, \delta^4(P) = \sum_i |i\rangle^{\dot{a}}\, \frac{\partial P^{\dot{c}d}}{\partial |i\rangle^{\dot{b}}}\, \frac{\partial}{\partial P^{\dot{c}d}}\, \delta^4(P) = P^{\dot{a}d}\, \frac{\partial}{\partial P^{\dot{b}d}}\, \delta^4(P) = -2\delta^{\dot{a}}{}_{\dot{b}}\, \delta^4(P), \tag{5.4}$$

where the last equality holds as a distribution since $\int x\, f(x)\, \partial_x\, \delta(x) = -\int f(x)\, \delta(x)$ and $\frac{\partial P^{\dot{a}d}}{\partial P^{\dot{c}d}} = 2\delta^{\dot{a}}{}_{\dot{b}}$. It follows that $M_{\dot{a}\dot{b}}\, \delta^4(P) = 0$. ◁

By Exercise 5.1, we already know that the rotations/boosts annihilate the angle and square brackets, so now we conclude that the Poincaré generators annihilate the amplitudes; this is of course not surprising, it is the explicit realization of the statement that the amplitudes are Lorentz-invariant and conserve momentum.

For a supersymmetric theory, the Poincaré generators (5.1) are supplemented by the supersymmetry generators Q and $\tilde{Q}$; the supersymmetry generators for $\mathcal{N} = 4$ SYM were given in (4.46) in on-shell superspace. We have already discussed that the annihilation of the superamplitudes by the supersymmetry generators encodes the supersymmetry Ward identities. In particular, the supermomentum conserving Grassmann-delta function $\delta^{(8)}(\sum_i |i\rangle \eta_i)$ implies that $\tilde{Q}_A$ annihilates the superamplitude. Note that for the tree-level MHV superamplitude

$$\mathcal{A}_n^{\mathrm{MHV}} = \frac{\delta^4(P)\, \delta^{(8)}(\tilde{Q})}{\prod_{i=1}^n \langle i, i+1 \rangle}, \tag{5.5}$$

momentum and supermomentum conservation (defined in (4.48)) now appear on an equal footing. In the denominator of (5.5), we set $|n + 1\rangle = |1\rangle$.

The spacetime symmetry of $\mathcal{N} = 4$ SYM is enlarged to the superconformal group. This means that in addition to the super-Poincaré generators,

◊ four translations and six boosts and rotations in (5.1),
◊ 16 fermionic supersymmetry generators Q^{Aa} and $\tilde{Q}^{\dot{a}}_A$ in (4.46),

the superconformal algebra also has

◊ four conformal boosts $K_{a\dot{a}}$,
◊ one dilatation D,
◊ 15 $SU(4)$ R-symmetry generators $R^A{}_B$, satisfying the traceless condition $R^C{}_C = 0$,
◊ 16 fermionic conformal supersymmetry generators $\tilde{S}^A_{\dot{a}}$ and S_{aA}.

Together, these $16 + 16 = 32$ fermionic and $4 + 6 + 4 + 1 + 15 = 30$ bosonic generators form the graded Lie algebra $su(2, 2|4)$ of the superconformal group.[3] Introducing a collective index $\mathsf{A} = (a, \dot{a}, A)$ we denote the superconformal generators as $G^{\mathsf{A}}{}_{\mathsf{B}}$.

We are going to realize the generators $G^{\mathsf{A}}{}_{\mathsf{B}} \in su(2, 2|4)$ in the following form, organized here according to their mass dimensions:

$$P^{\dot{a}b} = -\sum_i |i\rangle^{\dot{a}} [i|^b ,$$

$$\tilde{Q}^{\dot{a}}_A = \sum_i |i\rangle^{\dot{a}} \eta_{iA} \qquad\qquad Q^{aA} = \sum_i [i|^a_i \partial_{\eta_{iA}}$$

$$M_{\dot{a}\dot{b}} = \sum_i \langle i|_{(\dot{a}} \partial_{|i\rangle^{\dot{b}})} \qquad D = \sum_i \left(\tfrac{1}{2} |i\rangle^{\dot{a}} \partial_{|i\rangle^{\dot{a}}} + \tfrac{1}{2} |i]_a \partial_{|i]_a} + 1 \right) \qquad M_{ab} = \sum_i |i]_{(a} \partial_{[i|^{b)}}$$

$$R_A{}^B = \sum_i \left(\eta_{iA} \partial_{\eta_{iB}} - \tfrac{1}{4} \delta_A{}^B \eta_{iC} \partial_{\eta_{iC}} \right)$$

$$\tilde{S}^A_{\dot{a}} = \sum_i \partial_{|i\rangle^{\dot{a}}} \partial_{\eta_{iA}} \qquad\qquad S_{aA} = \sum_i \partial_{[i|^a} \eta_{iA}$$

$$K_{a\dot{a}} = -\partial_{|i\rangle^{\dot{a}}} \partial_{[i|^a} .$$

$$(5.6)$$

These generators are given as a sum, $G^{\mathsf{A}}{}_{\mathsf{B}} = \sum_{i=1}^n G_i{}^{\mathsf{A}}{}_{\mathsf{B}}$, of operators $G_i{}^{\mathsf{A}}{}_{\mathsf{B}}$ that are each defined on *one* external leg i; this reflects the local nature of the symmetry. In Section 5.3, we will encounter symmetries whose generators are "non-local" in that they involve products of operators that act on different legs.

▶ **Exercise 5.2**

Show that the action of the first two terms in the dilatation operator D in (5.6) extracts the mass dimension from an expression constructed from angle and square brackets. Then show that D annihilates amplitudes when including $\delta^4(P)$.

---

[3] $U(2, 2|4)$ has 32 fermionic and 32 bosonic generators, i.e. it has two more $U(1)$s than $SU(2, 2|4)$.

We define the "helicity operator" H as

$$H = \sum_i \left[|i\rangle_{\dot{a}}\, \partial_{|i\rangle_{\dot{a}}} - |i]_a\, \partial_{|i]_a} - \eta_{iA}\, \partial_{\eta_{iA}} + 2 \right]. \tag{5.7}$$

The first three terms in H act as follows: $|i\rangle_{\dot{a}} \to |i\rangle_{\dot{a}}$, $|i]_a \to -|i]_a$, and $\eta_A \to -\eta_A$, i.e. it pulls out the correct little group weight of each on-shell variable. Hence, when H acts on a component amplitude, it extracts the sum of the helicity weights plus $2n$ from the last term in (5.7): in the $\mathrm{N}^K\mathrm{MHV}$ sector this is $\sum_i(-2h_i) + 2n = 4K + 8$. But such a component amplitude appears in the superamplitude multiplied by a factor of $4(K+2)$ Grassmann variables η_is. Therefore $H\, \mathcal{A}_n^{\mathrm{N}^K\mathrm{MHV}} = 0$.

Together with the helicity operator H, the generators in (5.6) form a closed algebra. For example, we encourage you to check that:

$$\{ S_{aA}, Q^{bB} \} = \frac{1}{2}\delta_A{}^B M_a{}^b + \delta_a{}^b R_A{}^B + \frac{1}{2}\delta_A{}^B \delta_a{}^b \left(D - \frac{1}{2}H \right). \tag{5.8}$$

Since H vanishes on the amplitude, the generators in (5.6) close into the superconformal group when acting on the on-shell amplitude.

We are now ready to study the action of superconformal symmetry on the amplitudes of $\mathcal{N} = 4$ SYM: the superamplitudes are invariant under the full superconformal symmetry group, so one should find $G^A{}_B \mathcal{A}_n = 0$.

We restrict our analysis to the MHV superamplitude (5.5). We have already discussed that this superamplitude is super-Poincaré invariant. To prove that it enjoys the full $SU(2,2|4)$ symmetry, it is sufficient to check that the amplitude vanishes under the conformal supersymmetries S_{aA} and $\tilde{S}_{\dot{a}}^A$. For example, the anticommutator of S_{aA} and $\tilde{S}_{\dot{a}}^A$ gives the conformal boost $K_{a\dot{a}}$. The anticommutator (5.8) is a another example. The following two examples show that $\mathcal{A}_n^{\mathrm{MHV}}$ is annihilated by S_{aA} and $\tilde{S}_{\dot{a}}^A$.

▷ *Example.* Let us show that $S_{aA} = \sum_i \partial_{[i|a}\, \eta_{iA}$ annihilates $\mathcal{A}_n^{\mathrm{MHV}}$. Note that

$$S_{aA}\, \delta^4(P) = -\sum_i |i\rangle^{\dot{a}}\, \eta_{iA}\, \partial_{P_{a\dot{a}}}\, \delta^4(P) = -\tilde{Q}_A^{\dot{a}}\, \partial_{P_{a\dot{a}}}\, \delta^4(P) \tag{5.9}$$

vanishes on the support of $\delta^{(8)}(\tilde{Q})$. Since $\mathcal{A}_n^{\mathrm{MHV}}$ does not have any further dependence on square spinors, we conclude $S_{aA}\, \mathcal{A}_n^{\mathrm{MHV}} = 0$. ◁

▷ *Example.* To start with, consider the action of $\tilde{S}_{\dot{a}}^A = \sum_i \partial_{|i\rangle_{\dot{a}}}\, \partial_{\eta_{iA}}$ on $\delta^{(8)}(\tilde{Q})$: it follows from direct calculation that

$$\partial_{\eta_{iA}} \delta^{(8)}(\tilde{Q}) = |i\rangle^{\dot{a}}\, \partial_{\tilde{Q}_A^{\dot{a}}} \delta^{(8)}(\tilde{Q}) \quad \text{and} \quad \tilde{S}_{\dot{a}}^A\, \delta^{(8)}(\tilde{Q}) = (n - 1 + 3)\, \partial_{\tilde{Q}_A^{\dot{a}}} \delta^{(8)}(\tilde{Q}). \tag{5.10}$$

To show the second identity, you need $\tilde{Q}_{C\dot{a}}\, \tilde{Q}_C^{\dot{b}} = \frac{1}{2}\delta_{\dot{a}}{}^{\dot{b}}\, \tilde{Q}_{C\dot{c}}\, \tilde{Q}_C^{\dot{c}}$ (no sum on C). We use this result when we write

$$\tilde{S}_{\dot{a}}^A\, \mathcal{A}_n^{\mathrm{MHV}} = \left(\tilde{S}_{\dot{a}}^A\, \delta^{(8)}(\tilde{Q}) \right) \frac{\delta^4(P)}{\prod_{i=1}^n \langle i, i+1 \rangle} + \left(\partial_{\tilde{Q}_A^{\dot{b}}} \delta^{(8)}(\tilde{Q}) \right) \left(\sum_i |i\rangle^{\dot{a}}\, \partial_{|i\rangle^{\dot{b}}}\, \frac{\delta^4(P)}{\prod_{i=1}^n \langle i, i+1 \rangle} \right). \tag{5.11}$$

We evaluate the second term using (5.2) and (5.4), and we then have

$$\tilde{S}_{\dot{a}}^{A}\, \mathcal{A}_n^{\text{MHV}} = \left[(n-1+3)\,\partial_{\tilde{Q}_A^{\dot{a}}}\delta^{(8)}(\tilde{Q}) + \left(\partial_{\tilde{Q}_A^{\dot{b}}}\delta^{(8)}(\tilde{Q})\right)(-2-n)\delta_{\dot{b}}{}^{\dot{a}}\right]\frac{\delta^4(P)}{\prod_{i=1}^n \langle i,i+1\rangle} = 0\,.$$

(5.12)

This completes the proof that $\mathcal{A}_n^{\text{MHV}}$ respects the superconformal symmetry $SU(2,2|4)$ of $\mathcal{N}=4$ SYM.[4] ◁

We end this section with a remark about the operation of **inversion**. Inversion acts on the spacetime coordinates as

$$\mathcal{I}(x^\mu) = \frac{x^\mu}{x^2}\,.$$

(5.13)

The inversion operation generates the (super)conformal symmetry algebra from the (super)Poincaré algebra, for example $K^\mu = \mathcal{I}\,P^\mu\,\mathcal{I}$.

▶ Exercise 5.3

In this exercise, we derive the form of the conformal boost generator in position space; this will be useful for us in Section 5.4. Write the momentum generator in position space as $P^{\dot{a}a} = -i\frac{\partial}{\partial x_{a\dot{a}}}$.[5] Show that $\mathcal{I}\,P^{\dot{a}a}\,\mathcal{I}$ is equivalent to

$$\mathcal{K}^{\dot{a}a} = -i\,x^{\dot{a}c}x^{\dot{c}a}\frac{\partial}{\partial x^{\dot{c}c}}$$

(5.14)

by demonstrating that $\mathcal{I}\,P^{\dot{a}a}\,\mathcal{I}$ and $\mathcal{K}^{\dot{a}a}$ give the same result when acting on $x^{\dot{b}b}$.
[Hint: you will need the same type of identity given in the hint of Exercise 5.1.]

5.2 Twistors

The representation of the superconformal generators given in (5.6) is unusual in that the generators appear with various degrees of derivatives. For example, the bosonic $SU(2,2)$ subgroup of conformal transformations has a 2-derivative generator $K_{a\dot{a}}$ as well as a multiplicative 0-derivative generator $P^{\dot{a}a}$. Since the realization of the generators depends on the choice of variables – here spinor helicity variables – we can hope to find a set of variables such that all generators are linearized as 1-derivative operators. This is actually simple and can be achieved by performing a "Fourier transformation" on the angle spinor variables:

$$\langle i|_{\dot{a}} \to i\frac{\partial}{\partial |\tilde{\mu}_i\rangle^{\dot{a}}}, \qquad \frac{\partial}{\partial |i\rangle^{\dot{a}}} \to -i\langle\tilde{\mu}_i|_{\dot{a}}\,.$$

(5.15)

---

[4] We have ignored here subtle points of non-generic momenta and anomalies.
[5] We include here the "$-i$" in the translation generator; henceforth we will often drop the is in the symmetry generators.

For example, the translation and conformal boost generators are linearized after the Fourier transform:

$$P^{\dot{a}b} \to i \sum_i [i|^b \partial_{\langle \tilde{\mu}_i|_{\dot{a}}}, \qquad K_{a\dot{a}} = i \sum_i \langle \tilde{\mu}_i|_{\dot{a}} \partial_{[i|^a}. \tag{5.16}$$

▶ **Exercise 5.4**

Show that $|i\rangle^{\dot{a}} \to -i \dfrac{\partial}{\partial \langle \tilde{\mu}_i|_{\dot{a}}}$.

▶ **Exercise 5.5**

Show that the dilatation generator then becomes

$$D = \sum_i \left(\tfrac{1}{2} |i\rangle^{\dot{a}} \partial_{|i\rangle\dot{a}} + \tfrac{1}{2} |i]_a \partial_{|i]a} + 1 \right) \;\to\; \sum_i \left(\tfrac{1}{2} |i]_a \partial_{|i]a} - \tfrac{1}{2} \langle \tilde{\mu}_i|_{\dot{a}} \partial_{\langle \tilde{\mu}_i|_{\dot{a}}} \right). \tag{5.17}$$

The new variables

$$\mathcal{W}_i^{\mathsf{A}} = \left([i|^a, \; |\tilde{\mu}_i\rangle^{\dot{a}}, \; \eta_{iA} \right), \tag{5.18}$$

with a collective index $\mathsf{A} = (\dot{a}, a, A)$, are called **supertwistors**. In terms of these variables, the generators of the superconformal algebra $su(2, 2|4)$ can be written compactly as

$$G^{\mathsf{A}}{}_{\mathsf{B}} = \sum_{i=1}^n G_i^{\mathsf{A}}{}_{\mathsf{B}} = \sum_{i=1}^n \left(\mathcal{W}_i^{\mathsf{A}} \partial_{\mathcal{W}_i^{\mathsf{B}}} - \frac{1}{4} \delta^{\mathsf{A}}{}_{\mathsf{B}} \mathcal{W}_i^{\mathsf{C}} \partial_{\mathcal{W}_i^{\mathsf{C}}} \right). \tag{5.19}$$

The $\delta^{\mathsf{A}}{}_{\mathsf{B}}$-term is necessary for the bosonic subgroups, $SU(2, 2)$ and $SU(4)$, to have traceless generators. However, the term proportional to $\delta^{\mathsf{A}}{}_{\mathsf{B}}$ simply counts the degree of $\mathcal{W}_i^{\mathsf{A}}$, so if the function one is interested in has vanishing weight in each $\mathcal{W}_i^{\mathsf{A}}$, the generators reduce to:

$$G^{\mathsf{A}}{}_{\mathsf{B}} = \sum_{i=1}^n \mathcal{W}_i^{\mathsf{A}} \partial_{\mathcal{W}_i^{\mathsf{B}}}. \tag{5.20}$$

The bosonic components, $W_i^I = ([i|^a, \; |\tilde{\mu}_i\rangle^{\dot{a}})$, of the supertwistors (5.18) are simply called **twistors**. They were first introduced by Penrose [53] in the context of describing flat Minkowski spacetime. Later they were supersymmetrized by Ferber [41] and used to form representations of the superconformal group. Under little group scaling, the supertwistors scale homogeneously, $\mathcal{W}_i \to t_i \mathcal{W}_i$, and this leaves the symmetry generators (5.19) invariant. That means that we can define the (super)twistors projectively: the bosonic twistors are points in $\mathbb{CP}^3$ while the supertwistors live in $\mathbb{CP}^{3|4}$.

Amplitudes in twistor space

Let us consider the n-gluon anti-MHV tree amplitude (2.115) in bosonic twistor space, $W_i^I = ([i|^a, \; |\tilde{\mu}_i\rangle^{\dot{a}})$. The only angle-spinor dependence is in $\delta^4(P)$, so a Fourier transformation of $|j\rangle$ gives

$$\int \left(\prod_{j=1}^n d^2|j\rangle e^{i\langle j \, \mu_j\rangle} \right) \mathcal{A}_n^{\text{anti-MHV}} = \left[\int \left(\prod_{j=1}^n d^2|j\rangle e^{i\langle j \, \mu_j\rangle} \right) \delta^4(P) \right] f(|i]). \tag{5.21}$$

Here $f(|i])$ is a function that only depends on $|i]$. To ease the integration, write the delta function itself as a Fourier integration (ignoring factors of 2π),

$$\delta^4(P) = \int d^4x\, e^{-i\, x_{a\dot{a}} \sum_j |j\rangle^a [j|^{\dot{a}}} \,. \tag{5.22}$$

The integration over $|j\rangle$ can now be carried out and we find

$$\int \left(\prod_{j=1}^n d^2|j\rangle e^{i\langle j\, \tilde{\mu}_j\rangle}\right) A_n^{\text{anti-MHV}} = \int d^4x \left(\prod_{j=1}^n \delta^2\big(\langle\tilde{\mu}_j|_{\dot{a}} + [j|^a x_{a\dot{a}}\big)\right) f(|i]) \,. \tag{5.23}$$

Hence, the twistor variables $W_i^I = ([i|^a, |\tilde{\mu}_i)^{\dot{a}})$ are localized in twistor space by delta functions that enforce, for each i, the "incidence relations"

$$\langle\tilde{\mu}_i|_{\dot{a}} + [i|^a x_{a\dot{a}} = 0 \,. \tag{5.24}$$

This equation says that, for given $x_{a\dot{a}}$, W_i^I is determined by just the input of $[i|$. Naively, this leaves two degrees of freedom, but the projective nature of W_i^I reduces this to just one degree of freedom. Thus the solution to (5.24) is parameterized by a 1-dimensional variable, say the first component of the square spinor, $[i|^1$, and that means the solution is described as a degree 1 curve, i.e. a line, in $\mathbb{CP}^3$ (it is defined as the zeroes of a degree 1 polynomial in W_i).

Since each $x_{a\dot{a}}$ defines a different line, the integration $\int d^4x$ in (5.23) can be understood as integrating over all possible lines. Thus amazingly, the n-gluon anti-MHV tree amplitude corresponds in twistor space to n twistors W_i living on a line!

A tree-level gluon amplitude with q positive helicity gluons lives on a degree $(q-1)$-curve.[6] This interesting observation was due to Witten [54], and it provided an important inspiration for the modern developments of scattering amplitudes. The geometric interpretation is given by a twistor string theory, whose tree-level amplitudes are precisely those of the 4-dimensional $\mathcal{N}=4$ SYM. More details can be found in the original paper [54]. We leave the twistor story for now, but it will sneak back into the limelight later on when supertwistors $\mathcal{W}_i^A$ make another appearance in Chapter 10.

Twistor space and spacetime

We introduced twistors as a formalism for linearizing the action of conformal symmetry. However, twistors are also naturally introduced from the point of view of conformal symmetry in spacetime. For this purpose, it is convenient to interpret the conformal group $SO(2,4)$ as the Lorentz group of a 6d space with metric signature $(-,-,+,+,+,+)$. This way conformal symmetry is realized as Lorentz symmetry if we embed the 4d spacetime into the 6d spacetime. Consider a null-subspace in 6d defined by $X \cdot X = 0$, where X^μ is a 6d vector. Since the null constraint is invariant if we rescale $X \to rX$, it is natural to identify $X \sim rX$ on the null-space. The constraint $X \cdot X = 0$ and the projective nature

---

[6] In [54], the twistor was defined with $\mathcal{W}_i^A = (|i\rangle^{\dot{a}}, [\mu_i|^a, \eta_{iA})$, where $[\mu_i|^a$ is the Fourier conjugate to $[i|^a$. Thus, instead, one had the MHV amplitude to be of degree 1 in twistor space, while N^KMHV corresponds to degree K-curves in twistor space. Note that in Section 5.4 we will introduce a variable $[\mu_i|^a$ which is *not* the same as the one in Witten's supertwistor $\mathcal{W}_i$. But it will be part of a different type of supertwistor $\mathcal{Z}_i$. Please don't be too confused.

leaves $6 - 2 = 4$ degrees of freedom, so this subspace is indeed 4-dimensional. This is the so-called "embedding formalism" that was first introduced by Dirac in 1937 [55]. (See [56] for a more recent discussion.)

We now spinorize the above discussion. Since the $SO(2, 4) \sim SU(2, 2)$, we can rewrite the 6d vector X^μ as a bi-spinor X^{IJ}. This 4×4 tensor is antisymmetric, $X^{IJ} = -X^{JI}$, and transforms in the 6-dimensional irreducible representation of $SU(2, 2)$.

The null condition now translates to:

$$X^2 = \frac{1}{2}\epsilon_{IJKL}X^{IJ}X^{KL} = 0 \,. \tag{5.25}$$

This implies that X^{IJ} has rank 2, and therefore we can write it as

$$X^{IJ} = W_i^{[I}W_j^{J]} \,, \tag{5.26}$$

where the 4-component spinors W^I are the **twistors**, now introduced in the embedding formalism. From (5.26) we see that a point X in the 4d subspace is defined in twistor space by the line formed by two twistor variables (W_i, W_j). Since the subspace of X is defined projectively, we identify $W_i \sim t W_i$ and therefore the twistor space is really $\mathbb{CP}^3$.

The $SU(2, 2)$ covariant form of the incidence relation in (5.24) is simply:

$$X^{[IJ}W_i^{K]} = 0 \,. \tag{5.27}$$

Since we have constructed our 4d spacetime with manifest conformal symmetry, we cannot properly define the length; the construction does not specify a metric in 4d. Not surprisingly, to introduce a metric, we need to break the conformal $SU(2, 2)$ symmetry. This is done by introducing a reference point I^{IJ} in the embedding space. The projectively invariant length between two points X_i and X_j in the 4d subspace can then be defined as

$$\frac{(X_i \cdot X_j)}{(I \cdot X_i)(I \cdot X_j)} \,, \tag{5.28}$$

where $(X_i \cdot X_j) \equiv \epsilon_{IJKL}X_i^{IJ}X_j^{KL}$. The choice of I corresponds to a choice of metric.

If two points X_i and X_j share a common twistor W, then it follows from (5.26) and (5.28) that they are null-separated. Since two points define a line, we conclude that a point in twistor space corresponds to a null-line in spacetime. We have then observed the general feature that points in spacetime map to lines in twistor space and that points in twistor space map to null-lines in spacetime.

5.3 Emergence of dual conformal symmetry

We have experienced the advantage of using spinor helicity formalism for scattering amplitudes, both in terms of the restrictive power of consistency conditions, such as little group scaling, and in terms of the simplicity of the final results. Many of these properties stem from two properties of the spinor helicity variables, which:

1. trivialize the on-shell condition $p^2 = 0$, and
2. transform linearly under Lorentz symmetries, so that we get manifestly Lorentz-invariant expressions for the amplitudes.

In contrast, the ordinary way of representing the amplitudes using polarization vectors realizes Lorentz invariance via gauge redundancy; but this makes the amplitudes overly complicated. Alternatively, one can work with only on-shell degrees of freedom by using light-cone (or space-cone) gauge [57]. However, as these gauges are not Lorentz invariant, the symmetry generators will act non-linearly on the kinematic variables. Thus, using the spinor helicity formalism essentially allows us to "have our cake and eat it": it allows us to work with only on-shell degrees of freedom, yet the global symmetries are linearly realized.

At this point you may have noticed that there is a glaring hole in the above story: there is an important part of the Poincaré symmetry that does not act linearly on the spinor variables: the translations. In momentum space, translation invariance corresponds to momentum conservation. This symmetry, as well as its supersymmetric partner, is respected by the scattering amplitudes in a rather *ad hoc* fashion, namely by being enforced through the presence of the delta functions:

$$\delta^4 \left(\sum_{i=1}^n p_i^{\dot{a}a} \right) \quad \text{and} \quad \delta^{(2\mathcal{N})} \left(\sum_{i=1}^n |i\rangle \eta_i^A \right). \tag{5.29}$$

Here we have indicated the $\mathcal{N}$-fold supersymmetric case, though in this section we are going to study only $\mathcal{N} = 4$ SYM. The point here is that if we follow the spirit of what the spinor helicity formalism brought us, we should try to find new variables that either simplify, or at least encode, the information of momentum and supermomentum conservation.

As an inspiration, let us visualize momentum conservation geometrically. The condition that an ordered set of n momenta p_i^μ add to zero implies that the vectors form a closed contour, e.g. for $n = 5$:

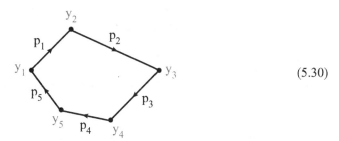

$$\tag{5.30}$$

Now there are two different ways to define the contour: it can be defined by the edges or by the cusps. The former is just the usual momentum representation. For the latter, we take the cusps to be located at points y_i^μ in a ***dual-space*** [58]. They are defined by their relation to the momentum vectors:

$$y_i^{\dot{a}a} - y_{i+1}^{\dot{a}a} = p_i^{\dot{a}a}. \tag{5.31}$$

The ***dual coordinates*** y_i (sometimes called ***zone variables*** or ***region variables***) are *not* spacetime coordinates; they are dual momentum variables defined by (5.31). In particular, they have mass-dimension 1. In dual space, n-point momentum conservation simply corresponds to the periodicity condition that $y_{n+1} = y_1$. For massless particles, the edges of the n edge polygon are light-like.

Since the ordering of the external momenta is crucial for the definition of the polygon (5.30), we need a well-defined notion of ordering. For (super) Yang–Mills theory, this is simply the color-ordering. The ordering also lets us define

$$y_{ij} \equiv y_i - y_j = p_i + p_{i+1} + \cdots + p_{j-1}. \tag{5.32}$$

The resulting variables y_{ij} precisely match the y_{ij}s introduced in (4.85) when we calculated the NMHV tree-level superamplitudes in $\mathcal{N} = 4$ SYM; this is of course no coincidence. In (4.85) we also defined fermionic variables θ_{ijA}^a; those now arise as differences $|\theta_{ij,A}\rangle \equiv |\theta_{iA}\rangle - |\theta_{jA}\rangle$ of dual-space fermionic coordinates $|\theta_{iA}\rangle$ defined as

$$|\theta_{iA}\rangle - |\theta_{i+1,A}\rangle = |i\rangle \, \eta_{iA} \,, \tag{5.33}$$

where the η_{iA}s are the on-shell superspace Grassmann variables introduced in Section 4.4 and $A = 1, 2, 3, 4$ are the $SU(4)$ R-symmetry labels.

In dual coordinates, the n-point tree-level MHV superamplitude of $\mathcal{N} = 4$ SYM is

$$\mathcal{A}_n^{\mathrm{MHV}} = \frac{\delta^4 \left(y_1 - y_{n+1}\right) \delta^{(8)} \left(\theta_1 - \theta_{n+1}\right)}{\prod_{i=1}^{n} \langle i, i+1 \rangle} \,, \tag{5.34}$$

and the tree-level NMHV superamplitude takes the form

$$\mathcal{A}_n^{\mathrm{NMHV}} = \mathcal{A}_n^{\mathrm{MHV}} \sum_{j=2}^{n-3} \sum_{k=i+2}^{n-1} R_{njk} \,, \tag{5.35}$$

where

$$R_{njk} = \frac{\langle j-1, j \rangle \langle k-1, k \rangle \ \delta^{(4)} \left(\Xi_{njk}\right)}{y_{jk}^2 \, \langle n | y_{nj} . y_{jk} | k \rangle \langle n | y_{nj} . y_{jk} | k-1 \rangle \langle n | y_{nk} . y_{kj} | j \rangle \langle n | y_{nk} . y_{kj} | j-1 \rangle} \,, \tag{5.36}$$

$$\Xi_{njk,A} = \langle n | y_{nk} . y_{kj} | \theta_{jn,A} \rangle + \langle n | y_{nj} . y_{jk} | \theta_{kn,A} \rangle \,.$$

The NMHV expression is exactly the same as the super-BCFW result in Section 4.5, where it was already given in dual coordinates. We chose here to anchor the expressions on line n.

The representations in (5.34) and (5.35) may appear somewhat disappointing since the amplitudes in dual space are basically identical to the original expressions. However, the dual space description allows us to study a new symmetry, namely (super)conformal symmetry in the dual coordinates y. This is called **dual (super)conformal symmetry**. To describe it, first note that since the defining relation (5.31) of the y_is is invariant under translations, the amplitude is guaranteed to be translational invariant in the y-space. Let the corresponding translation operator in dual y-space be $\mathcal{P}^\mu$. Next, because the conformal boost generator is $\mathcal{K}^\mu = I \mathcal{P}^\mu I$, the dual superconformal property of the amplitude can be extracted by simply studying how the amplitude transforms under *dual inversion* I:

$$I\left[y_i^\mu\right] = \frac{y_i^\mu}{y_i^2} \,, \quad I\left[|\theta_{iA}\rangle^{\dot{a}}\right] = \langle \theta_{iA}|_b \, \frac{y_i^{ba}}{y_i^2} \,, \quad I\left[[i|^a\right] = \frac{y_i^{\dot{a}b}}{y_i^2} \, |i]_b \,, \quad I\left[|i\rangle^a\right] = \langle i|_b \, \frac{y_{i+1}^{ba}}{y_{i+1}^2} \,. \tag{5.37}$$

These rules are well-defined only when we have a notion of ordering. The dual inversion rules for $|i\rangle$ and $[i|$ are defined only up to a relative scaling.

▶ **Exercise 5.6**

The inversion rules for y_i^μ and $|\theta_{iA}\rangle$ are standard. Verify the consistency of the rules for $|i\rangle$ and $[i|$ using (5.31) and $y_{i,i+2}^2 = \langle i,i+1\rangle[i,i+1]$.

[Hint: Show first that the definition (5.31) implies that $\langle i|_{\dot{a}}\, y_i^{\dot{a}a} = \langle i|_{\dot{a}}\, y_{i+1}^{\dot{a}a}$. It follows that $I[\langle i,i+1\rangle] = \langle i,i+1\rangle/y_{i+2}^2$.]

The momentum and supermomentum delta function transform under dual inversion as

$$I\big[\delta^4(y_1 - y_{n+1})\big] = y_1^8\, \delta^4(y_1 - y_{n+1})\,, \qquad I\big[\delta^{(8)}(\theta_1 - \theta_{n+1})\big] = y_1^{-8}\, \delta^{(8)}(\theta_1 - \theta_{n+1})\,. \tag{5.38}$$

Thus we see that the inversion weight of the bosonic delta function exactly cancels[7] that of the Grassmann delta function. So for the tree-level $\mathcal{N} = 4$ SYM MHV superamplitude we have

$$I\big[\mathcal{A}_n^{\text{MHV}}\big] = \left(\prod_{i=1}^n y_i^2\right) \mathcal{A}_n^{\text{MHV}}\,. \tag{5.39}$$

We conclude that under dual superconformal inversion, the tree MHV superamplitude transforms *covariantly* with equal weights on all legs.

At this point, you may wonder if this new dual conformal symmetry is secretly just another incarnation of the conventional conformal symmetry. It is straightforward to see that this is not the case. The pure Yang–Mills gluon tree amplitude is conformal invariant because it takes the same form as in $\mathcal{N} = 4$ SYM. However, under dual inversion, the split-helicity amplitude transforms as:

$$I\big[A_n[1^- 2^- 3^+ \cdots n^+]\big] = \left(\prod_{i=1}^n y_i^2\right)\big(y_1^2\big)^4 A_n[1^- 2^- 3^+ \ldots n^+]\,. \tag{5.40}$$

Clearly, this amplitude does not have homogeneous inversion properties. The situation is worse for a gluon amplitude without the split-helicity arrangement, for example $A_n[1^- 2^+ 3^- \ldots n^+]$. The result of $I\big[\langle 13\rangle\big]$ is not proportional to $\langle 13\rangle$, so $A_n[1^- 2^+ 3^- \ldots n^+]$ does not even transform covariantly under dual inversion. This shows that one can have a conformal invariant amplitude that is not dual conformal covariant. Hence the two symmetries are inequivalent. The example also shows that dual conformal invariance is a symmetry of the superamplitude, not the individual component amplitudes.

What about the tree-level NMHV superamplitude? Well, remarkably, the somewhat complicated expression in (5.36) is invariant under dual inversion, i.e. $I[R_{njk}] = R_{njk}$.

---

[7] Note that in $D = 4$ this cancellation only happens for $\mathcal{N} = 4$. In $D = 3$, a similar cancellation happens for $\mathcal{N} = 6$; indeed a theory with $\mathcal{N} = 6$ supersymmetry exists, namely ABJM theory, and its amplitudes also respect dual superconformal symmetry. We discuss this theory in more detail in Section 11.3.6. A similar result in $D = 6$ would require a supersymmetric theory with 24 supercharges. It is not clear if an interacting theory exists that can realize this symmetry in $D = 6$.

Thus $\mathcal{A}_n^{\mathrm{NMHV}}$ in (5.35) has the same homogeneous dual inversion weight as the tree-level MHV superamplitude. In fact, using super-BCFW recursion relations it can be shown [46, 49] that *all* tree superamplitudes of $\mathcal{N} = 4$ SYM transform covariantly under dual inversion,

$$I[\mathcal{A}_n^{\mathrm{tree}}] = \left(\prod_{i=1}^n y_i^2\right) \mathcal{A}_n^{\mathrm{tree}}. \tag{5.41}$$

We prove this statement using recursion relations in Section 7.3.

▶ **Exercise 5.7**

Use the inversion rules (5.37) to show that R_{njk} is invariant. Note that it is crucial that the spinor-products in R_{njk} can be arranged to involve adjacent lines.
[Hint: The identity $(y_{nj}y_{jk} + y_{nk}y_{kj})_a{}^b + y_{jk}^2 \delta_a{}^b = 0$ is useful for calculating $I[\Xi_{njk}]$.]

Due to the non-trivial weights in (5.41), the dual conformal boost generator does not vanish on the amplitudes. Rather, it generates an "anomaly" term,

$$\mathcal{K}^\mu \mathcal{A}_n^{\mathrm{tree}} = \left(-\sum_{i=1}^n y_i^\mu\right) \mathcal{A}_n^{\mathrm{tree}}. \tag{5.42}$$

If we bring this term to the LHS and redefine $\tilde{\mathcal{K}}^\mu \equiv \mathcal{K}^\mu + \sum_{i=1}^n y_i^\mu$, then the new generator $\tilde{\mathcal{K}}^\mu$ annihilates the amplitudes.

The dual conformal symmetry can be enlarged into an $SU(2,2|4)$ dual superconformal symmetry. Recall that $\mathcal{N} = 4$ SYM is also superconformal invariant, with the same $SU(2,2|4)$ group. If we combine the two sets of generators, we obtain an infinite dimensional algebra called a ***Yangian*** [59]. The generators of this algebra are organized by levels. For the $SU(2,2|4)$ Yangian, level 0 consists of the ordinary superconformal generators $G^{\mathsf{A}}{}_{\mathsf{B}} = \sum_{i=1}^n G^{\mathsf{A}}_{i\,\mathsf{B}}$, where $\mathsf{A} = (I, A)$ with $I = (\dot{a}, a)$ the index of $SU(2,2)$ and A the $SU(4)$ R-symmetry index. At level 1, the generators are bi-local in their particle index:

$$\text{level 0: } \sum_{i=1}^n G^{\mathsf{A}}_{i\,\mathsf{B}}$$

$$\text{level 1: } \sum_{i<j}^n (-1)^{|\mathsf{C}|} [G^{\mathsf{A}}_{i\,\mathsf{C}}\, G^{\mathsf{C}}_{j\,\mathsf{B}} - (i \leftrightarrow j)]$$

$$\vdots \tag{5.43}$$

Here, $|\mathsf{C}|$ is 0 for $\mathsf{C} = I$ and 1 for $\mathsf{C} = A$. It turns out that the shifted dual conformal boost generator $\tilde{\mathcal{K}}^\mu$ (not the unshifted one, $\mathcal{K}^\mu$) belongs to level 1. So the "anomaly" in (5.42) was not a nuisance, but rather it was needed in order for the tree-level superamplitude of $\mathcal{N} = 4$ SYM to be Yangian invariant! Beyond level 1, the new generators can be obtained simply by repeated (anti)-commutation of level 1 and level 0 generators. For further information about Yangian symmetry we refer to the original work [59]. The message here is that the tree-level superamplitudes of $\mathcal{N} = 4$ SYM are Yangian invariant.

5.4 Momentum twistors

Now that we know the $\mathcal{N} = 4$ SYM tree superamplitudes to have dual superconformal symmetry (in fact even Yangian symmetry), let us again set out on a path to find new variables that transform covariantly under the new symmetry. This is especially justified given that R_{njk} is fairly unwieldy in its current form (5.36): we would like to write it as an expression that is manifestly invariant under dual superconformal symmetry. Also, the presence of both helicity spinors and the vectors y_i^μ in R_{njk} is a further redundancy of variables that we would like to eliminate.

As a first step, we redefine the dual-space coordinate y_i^μ in terms of spinor variables. Recall that we introduced the y_i^μ coordinates by their relation to the momenta: $p_i = y_i^\mu - y_{i+1}^\mu$. This relation implies that $\langle i|_{\dot{a}}\, y_i^{\dot{a}a} = \langle i|_{\dot{a}}\, y_{i+1}^{\dot{a}a}$. Instead of referring to the momentum, we can take this relation to be the defining relation for the dual coordinates y_i^μ: these are called the ***incidence relations*** and take the form

$$[\mu_i|^a \equiv \langle i|_{\dot{a}}\, y_i^{\dot{a}a} = \langle i|_{\dot{a}}\, y_{i+1}^{\dot{a}a}\,. \tag{5.44}$$

The incidence relations define the new variables $[\mu_i|^a$. The statement of the incidence relations is that for a given $Z_i^I = (|i\rangle, [\mu_i|)$, with $I = (\dot{a}, a)$ an $SU(2,2)$ index, any two points, y_i^μ and y_{i+1}^μ, in y-space that satisfies (5.44) must be null-separated by the vector $y_i^\mu - y_{i+1}^\mu = p_i^{\dot{a}a}$. Thus a point $Z_i^I = (|i\rangle, [\mu_i|)$ in Z-space corresponds to the null-line in y-space determined by the two points, y_i^μ and y_{i+1}^μ.

On the other hand, any point in y-space is determined by a line in Z-space. To see this, note that the point $y_i^{\dot{a}a}$ is involved in two incidence relations: $[\mu_i| = \langle i|y_i$ and $[\mu_{i-1}| = \langle i-1|y_i$. Combining these leads to

$$|i\rangle^b [\mu_{i-1}|^a - |i-1\rangle^b [\mu_i|^a = \left(|i\rangle^b \langle i-1|_{\dot{a}} - |i-1\rangle^b \langle i|_{\dot{a}}\right) y_i^{\dot{a}a} = \langle i-1, i\rangle\, y_i^{ba}\,, \tag{5.45}$$

so that

$$y_i^{\dot{a}a} = \frac{|i\rangle^{\dot{a}} [\mu_{i-1}|^a - |i-1\rangle^{\dot{a}} [\mu_i|^a}{\langle i-1, i\rangle}\,. \tag{5.46}$$

This means that y_i is determined by Z_{i-1}^I and Z_i^I: these two points define a line in Z-space. The relationship between y-space and Z-space is illustrated in Figure 5.1.

▶ **Exercise 5.8**

Use identities from the Appendix to show that $|\mu_i]_a = -(y_i)_{a\dot{b}}|i\rangle^{\dot{b}}$.

We have translated the dual coordinates y_i to $Z_i^I \equiv (|i\rangle^{\dot{a}}, [\mu_i|^a)$. The new four-component spinor variables Z_i^I are called ***momentum twistors*** [24]. The name stems from the analogy with spacetime twistors introduced in Section 5.2: a point in position space maps to a line in twistor space, and vice versa. Writing the first equality in the incidence relation (5.44) in

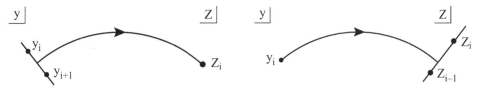

Fig. 5.1   The map between dual space y^μ and momentum twistor space $Z_i^I = (|i\rangle, [\mu_i|)$. The left-hand figure illustrates the incidence relations (5.44): a null-line in dual space, defined by the two points y_i and y_{i+1}, corresponds to a point $Z_i^I = (|i\rangle, [\mu_i|)$ in momentum twistor space. The right-hand figure shows how a point y_i in dual space maps to a line in momentum twistor space via the relation (5.46).

the form $([\mu_i|^a - \langle i|_{\dot{a}}\, y_i^{\dot{a}a}) = 0$ should remind you of the twistor space incidence relation (5.24). The second equality in (5.44) geometrizes the Weyl equation $p_i|i\rangle = 0$.

The defining incidence relations (5.44) imply that $|\mu_i] \to t_i|\mu_i]$ under little group scaling (2.95), so $Z_i^I \to t_i Z_i^I$. Hence the momentum twistors are defined projectively.

So what have we achieved by going from y_i to $Z_i^I \equiv (|i\rangle^{\dot{a}}, [\mu_i|^a)$? Well, the new variables transform linearly under dual conformal transformations. The generators $\mathcal{G}^I{}_J$ of the dual conformal group can now be written compactly together with the group algebra as[8]

$$\mathcal{G}^I{}_J \equiv \sum_i Z_i^I \frac{\partial}{\partial Z_i^J}, \qquad \left[\mathcal{G}^I{}_J, \mathcal{G}^K{}_L\right] = \delta_J{}^K \mathcal{G}^I{}_L - \delta^I{}_L \mathcal{G}^K{}_J, \qquad (5.47)$$

with $I, J, \ldots = (\dot{a}, a)$. The operator $\mathcal{G}^I{}_J$ can be thought of as a 4×4 matrix with a block diagonal 2×2 structure. To make the generators more concrete, consider the 2×2 block with $I = a$ and $J = \dot{a}$: it is $\mathcal{G}^a{}_{\dot{a}} = \sum_i [\mu_i|^a \frac{\partial}{\partial|i\rangle^{\dot{a}}}$. Its index structure, and the fact that it has mass-dimension 1, indicate that this should be the dual conformal boost $\mathcal{K}^a{}_{\dot{a}}$. Analogously with the regular conformal boost, given in Exercise 5.3, the dual conformal boost generator can be written in dual y-space as $\mathcal{K}^a{}_{\dot{a}} = -\sum_i \epsilon_{\dot{a}\dot{c}}\, y_i^{\dot{b}a}\, y_i^{\dot{c}b} \frac{\partial}{\partial y_i^{\dot{b}b}}$. Comparing this expression to $\mathcal{G}^a{}_{\dot{a}}$, it is not obvious that $\mathcal{G}^a{}_{\dot{a}} = \mathcal{K}^a{}_{\dot{a}}$, but the following exercises show you how it works.

▶ **Exercise 5.9**

Show that $\mathcal{K}^a{}_{\dot{a}}$ and $\mathcal{G}^a{}_{\dot{a}}$ are equivalent by demonstrating that they give the same result when acting on $y_i^{\dot{c}c}$ in (5.46).

Since the y_is and the momenta $p_i = -|i\rangle[i|$ are related, the change of variables from $(|i\rangle, y_i)$ to $Z_i = (|i\rangle, [\mu_i|)$ should make it possible to express $[i|$ in terms of $|i\rangle$ and $[\mu_i|$. Indeed, one finds

$$[i| = \frac{\langle i+1, i\rangle[\mu_{i-1}| + \langle i, i-1\rangle[\mu_{i+1}| + \langle i-1, i+1\rangle[\mu_i|}{\langle i-1, i\rangle\langle i, i+1\rangle}. \qquad (5.48)$$

▶ **Exercise 5.10**

Derive (5.48) using the incidence relations (5.44) and Schouten identities.

---

[8] For simplicity we consider $U(2, 2)$. The tracelessness condition can be implemented as in (5.19).

Since the momentum twistors Z^I carry the dual conformal $SU(2,2)$ index I, we can form a dual conformal invariant by contracting four Z^Is with the Levi-Civita symbol of $SU(2,2)$: we use a **4-bracket** for this invariant (the minus sign is chosen for later convenience),

$$\langle i,j,k,l \rangle \equiv -\epsilon_{IJKL} Z_i^I Z_j^J Z_k^K Z_l^L = \langle ij \rangle [\mu_k \mu_l] + \langle ik \rangle [\mu_l \mu_j] + \langle il \rangle [\mu_j \mu_k]$$
$$+ \langle kl \rangle [\mu_i \mu_j] + \langle lj \rangle [\mu_i \mu_k] + \langle jk \rangle [\mu_i \mu_l]. \tag{5.49}$$

We have expanded out the product in terms of $SL(2, \mathbb{C})$ invariants $[\mu_i \mu_j] \equiv [\mu_i|^a |\mu_j]_a$.

We can gain intuition for the 4-brackets by evaluating them in special cases. For example

$$\langle k, j-1, j, r \rangle = \langle j-1, j \rangle \langle k | y_{kj} y_{jr} | r \rangle. \tag{5.50}$$

▶ **Exercise 5.11**

Prove (5.50) by first using (5.49) to rewrite the LHS as a sum of $\langle ij \rangle [\mu_k \mu_l]$s. Then apply (5.44) and the Schouten identity to pull out an overall factor $\langle j-1, j \rangle$.

A special case of (5.50) is

$$\langle j-1, j, k-1, k \rangle = \langle j-1, j \rangle \langle k-1 | y_{k-1, j} \, y_{jk} | k \rangle = \langle j-1, j \rangle \langle k-1, k \rangle \, y_{jk}^2, \tag{5.51}$$

i.e.

$$y_{jk}^2 = \frac{\langle j-1, j, k-1, k \rangle}{\langle j-1, j \rangle \langle k-1, k \rangle}. \tag{5.52}$$

Since $1/y_{ij}^2$ are propagators, the relation (5.52) will appear repeatedly in our discussions.

Looking at (5.50) and (5.52) makes us realize that these are exactly the type of objects that appear in the denominators of the R-invariants (5.36) of the NMHV tree amplitudes. Indeed

R_{njk}
$$= \frac{\langle j-1, j \rangle^4 \langle k-1, k \rangle^4 \, \delta^{(4)}(\Xi_{njk})}{\langle n, j-1, j, k-1 \rangle \langle j-1, j, k-1, k \rangle \langle j, k-1, k, n \rangle \langle k-1, k, n, j-1 \rangle \langle k, n, j-1, j \rangle}. \tag{5.53}$$

Note that it was possible to arrange the five denominator factors such that the input of the 4-brackets go cyclically through the set of five labels $(n, j-1, j, k, k-1)$. Now the denominator is manifestly dual conformal invariant since it is composed entirely of $SU(2,2)$-invariant 4-brackets. However, the 4-brackets transform under little group scaling with weight 1 for each line in the argument. Thus the denominator is not really an invariant since the Z^Is are defined only projectively. So let us take a closer look at the numerator; for this purpose we need Grassmann-companions for the Z^Is.

Similarly to the bosonic incidence relations (5.44), we use the spinor-fermionic coordinate $|\theta_{iA}\rangle^{\dot{a}}$ to introduce a Grassmann-odd (spacetime-)scalar coordinate χ_{iA}:

$$\chi_i^A = \langle i \, \theta_{iA} \rangle = \langle i \, \theta_{i+1,A} \rangle \,. \tag{5.54}$$

With these new fermionic twistor variables, we have extended the $SU(2,2)$ momentum twistors Z^I to $SU(2,2|4)$ **momentum supertwistors**

$$\mathcal{Z}_i^{\mathsf{A}} \equiv \left(|i\rangle^{\dot{a}}, [\mu_i|^a \mid \chi_{iA} \right), \quad \text{where } \mathsf{A} = (\dot{a}, a, A). \tag{5.55}$$

Under little group scaling, $\mathcal{Z}_i$ scales uniformly, $\mathcal{Z}_i^{\mathsf{A}} \to t_i \mathcal{Z}_i^{\mathsf{A}}$, so the momentum supertwistors are defined projectively and are elements in $\mathbb{CP}^{3|4}$.

▶ **Exercise 5.12**

Derive the fermionic versions of (5.46) and (5.48).

In the R-invariant (5.53), the fermionic delta function argument is $\Xi_{njk,A} = \langle n|y_{nk}.y_{kj}|\theta_{jn,A}\rangle + \langle n|y_{nj}.y_{jk}|\theta_{kn,A}\rangle$. To rewrite it in terms of momentum supertwistors, we need the identity

$$\langle k|y_{kr}y_{rj}|\theta_{jA}\rangle = -\frac{\langle k,r-1,r,j-1\rangle \chi_{jA} - \langle k,r-1,r,j\rangle \chi_{j-1,A}}{\langle r-1,r\rangle\langle j-1,j\rangle} \,. \tag{5.56}$$

▶ **Exercise 5.13**

Derive (5.56) by manipulating the RHS using (5.50) and employing the Schouten identity.

▶ **Exercise 5.14**

Use (5.56) to show that

$$\Xi_{njk,A} = -\frac{\left[\langle j-1,j,k-1,k\rangle \chi_{nA} + \text{cyclic} \right]}{\langle j-1,j\rangle\langle k-1,k\rangle} \,. \tag{5.57}$$

[Hint: use the hint in Exercise 5.7.]

Plugging (5.57) into (5.53), we then have

$$R_{njk}$$
$$= \frac{\delta^{(4)}\left(\langle j-1,j,k-1,k\rangle \chi_n + \text{cyclic}\right)}{\langle n,j-1,j,k-1\rangle\langle j-1,j,k-1,k\rangle\langle j,k-1,k,n\rangle\langle k-1,k,n,j-1\rangle\langle k,n,j-1,j\rangle} \,. \tag{5.58}$$

The "$+$ cyclic" is the instruction to sum cyclically over the labels $(n, j-1, j, k, k-1)$, similarly to the product structure in the denominator. Now R_{njk} is manifestly invariant under both the little group scaling and dual $SU(2,2)$. Together with the overall factor of $\mathcal{A}_n^{\text{MHV}}$, we then have the building blocks of the NMHV superamplitude in a form that makes it manifestly dual superconformal invariant.

The expression (5.58) for R_{njk} is cyclic in the five labels $(n, j-1, j, k-1, k)$. This motivates us to introduce a **5-bracket** defined as

$$[i, j-1, j, k-1, k] \equiv R_{ijk}. \tag{5.59}$$

The 5-bracket is cyclic in its five arguments, so for example $[6, 1, 2, 3, 4] = [1, 2, 3, 4, 6]$.

In terms of the 5-bracket, the n-point NMHV tree superamplitude is simply given by

$$\mathcal{A}_n^{\text{NMHV}} = \mathcal{A}_n^{\text{MHV}} \sum_{j=2}^{n-3} \sum_{k=j+2}^{n-1} [n, j-1, j, k-1, k]. \tag{5.60}$$

Let us review what we have accomplished so far. Starting with the simple observation that momentum conservation is imposed in a rather ad hoc fashion, we introduced the auxiliary variables y_i such that momentum conservation is encoded in a geometric fashion. This led us to the realization of a new symmetry of the tree amplitude for $\mathcal{N} = 4$ SYM, namely superconformal symmetry in the dual space y_i. The new symmetry set us on a journey to search for new variables, the momentum (super)twistors, that linearize the transformation rules. This culminated in the simple symmetric form of the n-point NMHV tree superamplitude in (5.60). For N^KMHV, equation (5.60) generalizes to a sum involving products of K 5-brackets (see Section 7.3).

The 5-bracket in (5.60) corresponds to the terms in the super-BCFW expansion of the superamplitude; specifically we have seen in Section 4.5.2 how each R_{n2k} arises from an MHV×MHV BCFW diagram while the remaining R_{njk}s with $j > 2$ appear via recursion from the BCFW diagram with NMHV×anti-MHV subamplitudes. As discussed, this means that the representation (5.60) is not unique, since there are many equivalent BCFW expansions for a given amplitude, depending on the choice of lines in the BCFW shift. This implies that the dual conformal invariants (5.59) must have some linear dependencies. For example, compare for $n = 6$ the result of the recursions relations based on the BCFW supershifts $[6, 1\rangle$ and $[1, 2\rangle$: they have to give the same result, so

$$[6, 1, 2, 3, 4] + [6, 1, 2, 4, 5] + [6, 2, 3, 4, 5] = [1, 2, 3, 4, 5] + [1, 2, 3, 5, 6]$$
$$+ [1, 3, 4, 5, 6]. \tag{5.61}$$

Using the cyclic property of the 5-bracket, we can write this

$$[2, 3, 4, 6, 1] + [2, 3, 4, 5, 6] + [2, 4, 5, 6, 1] = [3, 4, 5, 6, 1] + [3, 5, 6, 1, 2]$$
$$+ [3, 4, 5, 1, 2]. \tag{5.62}$$

Now you see that the LHS looks like the result of a $[2, 3\rangle$ supershift, while the RHS comes from a $[3, 4\rangle$ supershift. Actually, the LHS and, independently, the RHS are invariant under $i \to i + 2$. We can also reverse the labels in the 5-brackets at no cost to get

$$[2, 3, 4, 6, 1] + [2, 3, 4, 5, 6] + [2, 4, 5, 6, 1] = [3, 1, 6, 5, 4] + [3, 2, 1, 6, 5]$$
$$+ [3, 2, 1, 5, 4]. \tag{5.63}$$

This states that the "parity conjugate" supershifts $[2, 3\rangle$ and $[3, 2\rangle$ give identical results. In fact, we can now conclude that any adjacent supershifts are equivalent.

The presence of these equivalence-relations between the dual conformal invariants may strike you as rather peculiar and you may wonder if it has a deeper meaning. Furthermore, while the expressions in (5.58) and (5.60) are extremely simple, they lack one key aspect when compared to the Parke–Taylor superamplitude (5.5): cyclic invariance. The presence of dual conformal symmetry relies heavily on the cyclic ordering of the amplitude, and hence it is somewhat surprising that the manifestly dual conformal invariant form of the superamplitude (5.60) breaks manifest cyclic invariance. One may say that we are asking too much of the amplitude, but considering the payoff we have reaped from the innocent chase of manifest momentum conservation, we will boldly push ahead with our pursuit of "having cakes and eating them" in Chapters 9 and 10.

▶ **Exercise 5.15**

Use cyclicity of the 5-brackets to show that the tree-level 6-point NMHV superamplitude can be written in the manifestly cyclic invariant form

$$A_6^{\text{NMHV}} = A_6^{\text{MHV}} \times \frac{1}{2}\Big(R_{146} + \text{cyclic}\Big), \tag{5.64}$$

where "+cyclic" means the sum over advancing the labels cyclically, i.e. $R_{146} + R_{251} +$ four more terms.

Momentum twistors

For our further studies, it is worth making a few observations about the momentum twistors. We introduced the dual y_is in order to make momentum conservation manifest; but the y_is could not be chosen freely since they are subject to the constraint $(y_i - y_{i+1})^2 = 0$ that ensures the corresponding momenta p_i to be on-shell. On the other hand, the momentum twistors Z_i are free variables: they are subject to the scaling equivalence $Z_i \sim t Z_i$, so they live in projective space $\mathbb{CP}^3$. (The momentum supertwistors $\mathcal{Z}_i$ are elements of $\mathbb{CP}^{3|4}$.) We can choose n points Z_i in $\mathbb{CP}^3$, subject to no constraints, then study the n lines defined by consecutive points (Z_i, Z_{i+1}), with the understanding that the nth line is (Z_n, Z_1). Equation (5.46) maps each line (Z_i, Z_{i+1}) to y_i and the incidence relation (5.44) guarantees that the points y_i and y_{i+1} are null-separated; thus the corresponding momenta $p_i = y_i - y_{i+1}$ are on-shell. The collection of lines (Z_i, Z_{i+1}) gives, per definition, a closed contour and that ensures momentum conservation, $y_{n+1} = y_1$. So all in all, the map to momentum twistors geometrizes the kinematic constraints of momentum conservation and on-shellness by simply stating these requirements as the intersection of n lines $(i, i+1) \equiv (Z_i, Z_{i+1})$ at the points $(i) \equiv Z_i$ in momentum twistor space $\mathbb{CP}^3$. The momentum supertwistors similarly make conservation of supermomentum automatic.

Pursuing the geometric picture a little further, we consider intersections of lines and planes in momentum twistor space. The momentum twistors are points $(i) \equiv Z_i$ in $\mathbb{CP}^3$. Two points in $\mathbb{CP}^3$ define a line (i, j) that can be parameterized as $Z_i + \tau Z_j$ for $\tau \in \mathbb{C}$. Now any two distinct points on that line can be used to define it, so a reparameterization-invariant definition of a line in $\mathbb{CP}^3$ is given by the antisymmetric product $Z_i^{[I} Z_j^{J]}$. This applies similarly for planes, which are defined in terms of the antisymmetrization of three points. So we have:

$$
\begin{aligned}
\text{line:} \quad & (i, j) = Z_i^{[I} Z_j^{J]}, \\
\text{plane:} \quad & (i, j, k) = Z_i^{[I} Z_j^{J} Z_k^{K]}.
\end{aligned}
\tag{5.65}
$$

In $\mathbb{CP}^3$, the point $(p) \equiv Z_p$ that corresponds to the intersection of line $(i, j) \equiv (Z_i, Z_j)$ with a plane $(k, l, m) \equiv (Z_k, Z_l, Z_m)$ is given by

$$
(p) = (i, j) \bigcap (k, l, m) = Z_i \langle j, k, l, m \rangle - Z_j \langle i, k, l, m \rangle .
\tag{5.66}
$$

The symbol $\bigcap$ indicates the intersection of the two objects. Similarly, the line (p, q) that corresponds to the intersection of two planes (Z_i, Z_j, Z_k) and (Z_l, Z_m, Z_n) is given by

$$
(p, q) = (i, j, k) \bigcap (l, m, n) = (i, j) \langle k, l, m, n \rangle + (j, k) \langle i, l, m, n \rangle + (k, i) \langle j, l, m, n \rangle .
\tag{5.67}
$$

A more detailed discussion of twistor geometry can be found in [60].

Propagators $1/y_{ij}^2$ are expressed in terms of momentum twistors via (5.52). Hence, the on-shell condition $y_{ij}^2 = 0$ becomes the requirement $\langle i - 1, i, j - 1, j \rangle = 0$. This is the statement that the four momentum twistors labeled by $i - 1, i, j - 1, j$ are linearly dependent. Geometrically, that means they lie in the same plane in $\mathbb{CP}^3$. In Section 7.3 we will see that in momentum twistor space a pole $y_{ij}^2 = 0$ in the BCFW-shifted amplitude is characterized as the intersection between the line $(i - 1, i)$ and the plane $(i - 1, j - 1, j)$; this motivates our interest in formulas such as (5.66).

It is now time to venture beyond tree-level and wrestle loops: in the second part of the book, we discuss various approaches to loop-amplitudes.

LOOPS

In Part I of the book, we focused exclusively on tree-level amplitudes, the leading contributions in the perturbative expansion of a scattering process. The loop-corrections are of course highly relevant, both in particle physics applications and for our understanding of the mathematical structure of the S-matrix. Here, in Part II of the book, we explore the structure of loop amplitudes.

Feynman rules require momentum conservation at each vertex of a Feynman diagram. At tree-level, this fixes all momenta of the internal lines in terms of the external momenta. At loop-level, momentum conservation leaves one momentum undetermined per loop and one must integrate over all such unfixed momenta. Thus in D-dimensions, one has a D-dimensional loop-integral for each loop. For example, in ϕ^3 theory one of the diagrams that contributes to the 1-loop 4-point amplitude is

$$\begin{array}{c} p_2 \\ p_1 \quad \ell \quad p_4 \end{array}^{p_3} = \int \frac{d^D\ell}{(2\pi)^D} \frac{1}{\ell^2 (\ell - p_1)^2 (\ell - p_1 - p_2)^2 (\ell + p_4)^2} . \tag{6.1}$$

As another example, consider a diagram of the 1-loop correction to the 4-point all-scalar amplitude in scalar QED:

$$\begin{array}{c} p_1 \qquad p_2 \\ p_4 \qquad p_3 \end{array} = e^4 \int \frac{d^D\ell}{(2\pi)^D} \frac{[(\ell - 2p_1) \cdot (\ell + 2p_2)][(\ell - p_1 + p_4) \cdot (\ell + p_2 - p_3)]}{\ell^2 (\ell - p_1)^2 (\ell - p_1 - p_2)^2 (\ell + p_4)^2} ,$$

$$\tag{6.2}$$

where e is the electric charge of the scalar. Note that the integral (6.2) has loop-momentum dependence in the numerator.

In general, an L-loop amplitude can be written schematically as

$$\mathcal{A}_n^{L\text{-loop}} = \sum_j \int \left(\prod_{l=1}^{L} \frac{d^D\ell_l}{(2\pi)^D} \right) \frac{1}{S_j} \frac{n_j c_j}{\prod_{\alpha_j} p_{\alpha_j}^2} , \tag{6.3}$$

where the sum runs over all possible L-loop Feynman diagrams j. For each diagram, the ℓ_l denote the L loop-momenta, α_j label the propagators, and S_j is the symmetry factor associated with the diagram. The kinematic numerator factors n_j are polynomials of Lorentz-invariant contractions of external- and loop-momenta and polarization vectors (or other external wavefunctions). The constants c_j capture the information about couplings and gauge group factors.

The color-structure we have worked with for super Yang–Mills amplitudes is rather simple: each color-ordered amplitude is associated with a single trace of gauge group generators. At loop-level, more intricate trace-structures arise. In this book, however, we will be concerned with the *planar limit* of Yang–Mills amplitudes. In this context, the planar limit means that the amplitudes have only one trace-structure, just as the tree amplitudes. They are called *planar amplitudes*. The color-ordered Feynman rules of Section 2.5 apply and the diagrams that contribute to an amplitude are precisely those in which no lines cross.

At loop-level, it is useful to distinguish the following three objects:

1. The *loop-integrand* is the rational function inside the loop-momentum integration.
2. The *loop-integral* is the combination of the integrand and the loop-momentum integration measure: this is a formal object, since we have not specified the integration region of the loop-momentum or addressed divergences.
3. The *loop amplitude* is the result of carrying out the loop-integrations in the loop-integral. If we integrate over physical momentum space $\mathbb{R}^{1,3}$, the integral may have infrared (IR) and/or ultraviolet (UV) divergences. We need to regulate such divergences in order to make the integrated result, the loop amplitude, well-defined.

Our focus will mainly be on the loop-integrand and the loop-integral. However, it is useful to understand the analytic structure of the loop amplitude, since this reveals which constraints the loop-integrand, and hence integral, must satisfy.

It is not surprising that integrals of loop amplitudes can be divergent. They integrate a rational function – the loop-integrand – over L copies of $\mathbb{R}^{1,3}$ and this includes regions where the loop-momentum is infinite $|\ell| \to \infty$ as well as regions where integrand-denominator may vanish; both cases give potential divergences. We discuss the UV and IR divergences in Section 6.1. After regulating the amplitude and separating out the divergences, the finite pieces of the amplitude are typically not just rational functions. For massless lower-loop amplitudes, one sometimes encounters functions that can be expressed in the form of iterative integrals or products of iterative integrals. For example,

$$\log(x) = \int_0^x \frac{dt}{t}, \quad \log(x)\log(y) = \left(\int_0^x \frac{dt}{t}\right)\left(\int_0^y \frac{ds}{s}\right), \quad \mathrm{Li}_2(x) = \int_0^x \frac{dt}{t} \int_0^t \frac{du}{1-u}. \tag{6.4}$$

The last function, Li_2, is a dilogarithm. It is the $q = 2$ special case of the polylogarithm Li_q. Starting with the familiar logarithm $\mathrm{Li}_1(x) = -\log(1-x)$, the polylogarithms are defined iteratively as

$$\mathrm{Li}_q(x) = \int_0^x dt \, \frac{\mathrm{Li}_{q-1}(t)}{t}. \tag{6.5}$$

Such functions are usually referred to as **_transcendental functions_**. They are associated with a **_degree of transcendentality_**, given by the dimension of the integral involved. The three functions in (6.4) are of transcendentality 1, 2, and 2, while it is r for the polylog Li_r. Note that in this categorization, π has transcendentality 1, since $i\pi = \log(-1)$.

The analytic structure of loop amplitudes is clearly more complicated than for tree amplitudes: where tree amplitudes are simple rational functions with only simple poles, the loop-integrations typically give rise to generalized logarithms, as in (6.4), and other special functions. These have branch cuts in addition to poles. The well-understood pole structure of tree amplitudes was instrumental for developing the on-shell recursion relations (Chapter 3), so at first sight it looks challenging to develop a similar "on-shell" approach for loop amplitudes. Nonetheless, the analytic structure of the loop-integrands can be exploited to reconstruct the amplitude from lower-order on-shell data. The purpose of this and the following two chapters is to show how to do this.

In the present chapter, we begin with a brief description of the physics of loop-divergences. We then focus on the widely applicable and very successful unitarity method. Chapter 7 studies BCFW recursion for the loop-integrands and in Chapter 8 we discuss Leading Singularities and on-shell diagrams.

6.1 UV and IR divergences

If a loop-integral diverges as the loop-momentum becomes large, the loop-diagram is said to have an **_ultraviolet (UV) divergence_**. At large loop-momentum, the denominators of the propagators behave as ℓ^2, so a given integrand behaves as

$$\int d^D\ell \, \frac{\ell^N}{(\ell^2)^P} \sim \int d|\ell| \, \frac{|\ell|^{N+D-1}}{|\ell|^{2P}}, \tag{6.6}$$

where D is the number of spacetime dimensions, P is the number of propagators, and N counts the powers of loop-momenta in the numerator. With $D + N - 2P = 0, 1, 2, \ldots$, the integral is logarithmically, linearly, quadratically divergent, etc. To regulate the UV behavior of the integral, one can introduce an explicit UV cutoff Λ, but it is often more convenient to analytically continue the spacetime to $4 - 2\epsilon$ dimensions, and use ϵ as a regularization parameter. This is **_dimensional regularization_**, which is widely used in quantum field theory due to the appealing feature that it preserves gauge invariance.

If the amplitude has UV divergences, this is telling us that, at short distances, the theory needs to be modified either by changing the constants in the original action or by adding new local operators to the action. The way in which the theory is modified depends crucially on the form of the divergence. Since the loop-integral has a definite mass dimension, the divergence must be multiplied by a kinematic invariant with the corresponding dimension. If this invariant is the tree amplitude, then we can simply modify the constants, such as overall coupling factors, of the tree amplitude to absorb this divergence. This modification can be translated into the modification of the coupling constants and masses in the Lagrangian description of the theory. If, on the other hand, the kinematic invariant in front of the loop-divergence is not proportional to the tree amplitude, then this means that one has to modify the theory by adding new operators to the action such that the divergences can be absorbed.

The procedure of removing UV-divergences from loop amplitudes is called **_renormalization_**. If the UV-divergences of all loop amplitudes for a given theory can be removed by adding a *finite* number of terms to the action (including zero), then the theory is said to be **_renormalizable_**. Renormalization is an extremely important subject in quantum field theory. It is crucial for making physical predictions from gauge theories and the techniques of renormalization are well-tested in the remarkable and successful comparisons of theoretical predictions with experimental results.

Another type of singularity is **_infrared (IR) divergences_**. They can come from two sources: soft and collinear divergences. To facilitate the discussion, let us consider an integral with two adjacent massless legs:

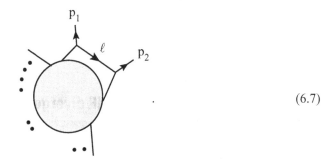

$$(6.7)$$

The relevant part of the integrand is

$$\int \frac{d^D \ell}{(\ell + p_1)^2 \ell^2 (\ell - p_2)^2} . \tag{6.8}$$

Let us consider light-cone coordinates

$$p^+ \equiv p^0 + p^1, \quad p^- \equiv p^0 - p^1, \quad \vec{p}_\perp \equiv (p^2, p^3, \dots, p^{D-1}). \tag{6.9}$$

Then $p^2 = -p^+ p^- + p_\perp^2$. We can always choose the coordinates in such a way that p_1 is in the p^- direction. To simplify the discussion, we choose p_2 to be in the p^+ direction; this is not general but the result of the discussion will be. The loop-integral is now (dropping overall factors)

$$\int \frac{d\ell^+ \, d\ell^- \, d^{D-2}\ell_\perp}{(\ell^2 - 2\ell^+ p_1^-)\, \ell^2 \,(\ell^2 - 2\ell^- p_2^+)} . \tag{6.10}$$

Let us consider the integration region where the loop-momentum becomes soft, $\ell^\mu \to 0$. Practically, we scale the integral to this region by changing integration variable from ℓ^+ to t as

$$\ell^+ \to t, \quad \ell^- \to t\ell^-, \quad \vec{\ell}_\perp \to t\vec{\ell}_\perp . \tag{6.11}$$

All terms in ℓ^2 scale homogeneously as t^2 which is subleading to $\ell.p_i$ for small t. Thus, for t near zero, the integral (6.10) gives

$$\int \frac{t^{D-1}\, dt}{t^4} = \int \frac{dt}{t^{5-D}} \,. \tag{6.12}$$

In $D = 4$, this integral is logarithmic divergent near $t = 0$.[1] We can regulate it by formally continuing the spacetime dimensions to $4 - 2\epsilon$, and one now has:

$$\int \frac{t^{3-2\epsilon}\, dt}{t^4} = \int \frac{dt}{t^{1+2\epsilon}} = \frac{1}{2\epsilon} + O(\epsilon^0) \,. \tag{6.13}$$

This divergence is called a ***soft divergence***, since it occurs when the loop-momentum is "softer" than the external legs.

Next, consider a limit in which the loop-momentum becomes collinear with p_1: we scale the components to align ℓ with the p^- direction,

$$\ell^+ \to t^2 \,, \qquad \ell^- \to \ell^- \,, \qquad \vec{\ell}_\perp \to t\,\vec{\ell}_\perp \,. \tag{6.14}$$

Again, ℓ^2 scales as t^2, so for small t we have that $(\ell^2 - 2\ell^+ p_1^-)$ scales as t^2 and that $(\ell^2 - 2\ell^- p_2^+)$ goes as t^0 at leading order. Thus the overall behavior of the integral at small t is again (6.13), so it goes as $1/\epsilon$ in $4-2\epsilon$ dimensions. This is a ***collinear divergence***, it corresponds to the limit where the loop-momentum becomes collinear with one of the external momenta.

The soft divergences can occur simultaneously with collinear divergences, so in dimensional regularization $D = 4-2\epsilon$, the ℓ-integration of the diagram (6.7) behaves at leading order as $1/\epsilon^2$. For an L-loop amplitude, the leading IR divergences are usually of order $1/\epsilon^{2L}$. Soft and collinear limits for amplitudes in massless gauge theories have been studied since the late 1970s and this is an entire subject on its own; see for example the brief outline in [61] and references therein.

The appearance of IR divergences does not correspond to a need to modify our original theory. Rather, it is a consequence of a subtlety in defining scattering amplitudes for massless particles. To define a scattering amplitude, one must have a well-defined notion of asymptotic states, i.e. what exactly are the particles that we are scattering? If identical massless particles become collinear, it is difficult to distinguish them from a single massless particle: a single particle with definite light-like momentum is indistinguishable from hundreds of soft particles whose momenta are collinear and near zero. This ambiguity for asymptotic massless states is reflected in the IR divergences of loop amplitudes as well as in the singularities of the soft and collinear limits of the tree-level amplitude (Section 2.8).

The tree and loop soft and collinear singularities are intimately related, as can be seen from the computation of the scattering cross-section. The cross-section of a 4-point process for four gluons in pure Yang–Mills theory involves the norm-squared product of gluon

---

[1] It is also divergent for $t \to \infty$. But in that limit, our approximation where the loop-momentum is small compared to the external momenta is no longer valid.

amplitudes, appropriately summed over states. Organizing the perturbative expansion in powers of the gauge coupling g, we find

$$\frac{d\sigma}{d\Omega} \propto \sum_i \langle g_1 g_2 | S | i \rangle \langle i | S | g_3 g_4 \rangle = \underbrace{\begin{array}{c}\text{tree} \ \vdots \ \text{tree}\end{array}}_{g^4} + \underbrace{\begin{array}{c}\text{1-L} \ \vdots \ \text{tree}\end{array}}_{g^6} + \underbrace{\begin{array}{c}\text{tree} \ \vdots \ \text{tree}\end{array}}_{g^6} + O(g^8).$$

(6.15)

The sum $\sum_i$ represents the sum over all on-shell multi-particle states and S is the S-matrix. The dashed line in each diagram represents a sum over the particle species of the on-shell internal leg. The 4-gluon tree and 1-loop amplitudes are order g^2 and g^4, respectively, while the 5-gluon tree-amplitude is order g^3. Hence the product of the 4-gluon tree and 1-loop amplitudes contributes at the same order, namely g^6, as the product of two 5-gluon tree amplitudes. Now diagrammatically, one can see that the collinear divergence of the 1-loop 4-gluon amplitude in one diagram can indeed be regarded as the collinear divergence of the 5-point tree amplitude looked at in another way:

$$\begin{array}{c}\text{(diagram)}\end{array} \qquad \begin{array}{c}\text{(diagram)}\end{array}.$$

(6.16)

The collinear divergences of the 5-point tree amplitudes "conspire" to cancel the IR divergence of the 4-point 1-loop amplitudes to yield a finite *cross-section*. A simple toy model example of such cancellations can be found in [62, 63].

The preceding discussion illustrates that the IR divergences of loop amplitudes are intimately tied to the soft and collinear behavior of tree amplitudes. Historically, this has served as an inspirational starting point for the development of tree-level recursion relations [13] discussed in Section 3. Note that if a scattering process is absent at tree-level and is non-vanishing only at loop-level, then the leading-order result cannot have IR singularities; it would have no tree-level counterpart to cancel its IR divergences in the cross-section. For example, the all-minus or single-plus gluon amplitudes vanish at tree-level, as we have seen in (2.112), but in pure Yang–Mills theory they are non-vanishing at 1-loop order. It can be shown by explicit calculation that the results for these gluon amplitudes are free of IR divergences, and in fact are simple rational functions e.g. [64, 65]

$$\text{pure Yang–Mills:} \quad A_n^{\text{1-loop}}\left[1^- 2^- \dots n^-\right] = \frac{\sum [i_1 i_2]\langle i_2 i_3\rangle[i_3 i_4]\langle i_4 i_1\rangle}{[12][23]\cdots[n1]}, \quad (6.17)$$

where the sum in the numerator is over $1 \leq i_1 < i_2 < i_3 < i_4 \leq n$ and we have dropped overall normalization factors.

It is useful for the following discussion to consider a set of basic *scalar integrals* and their integrated results in dimensional regularization. The 1-loop scalar integrals $I_m^{(i)}$ are the m-gon Feynman diagrams that appear in ϕ^n-theory:

$$(6.18)$$

In the scalar integrals, the dependence on loop- and external momenta is contained solely in the propagators. We categorize 1-loop scalar integrals as **bubbles**, **triangles**, **boxes**, etc. if they have $2, 3, 4$, etc. propagators. These integrals are denoted by $I_{2;n}(K_i, K_j)$, $I_{3;n}(K_i, K_j, K_k)$, and $I_{4;n}(K_i, K_j, K_k, K_l)$, with the subscript indicating the number of corners of the diagrams as well as the number of external legs involved. K_i is the sum of external momenta at each corner. The 4-point box, triangle, and bubble scalar integrals are

$$
\begin{aligned}
&= I_{4;4}(p_1, p_2, p_3, p_4) = \int \frac{d^D \ell}{(2\pi)^D} \frac{1}{\ell^2 (\ell - p_1)^2 (\ell - p_1 - p_2)^2 (\ell + p_4)^2}, \\
&= I_{3;4}(p_1, p_2, p_3 + p_4) = \int \frac{d^D \ell}{(2\pi)^4} \frac{1}{\ell^2 (\ell - p_1)^2 (\ell - p_1 - p_2)^2}, \\
&= I_{2;4}(p_1 + p_2, p_3 + p_4) = \int \frac{d^D \ell}{(2\pi)^4} \frac{1}{\ell^2 (\ell - p_1 - p_2)^2}.
\end{aligned}
\tag{6.19}
$$

We met the scalar box integral in (6.1). In $D = 4$, power counting shows that the box and triangle integrals $I_{4;n}$ and $I_{3;n}$ cannot have UV divergences, only IR divergences due to the massless corners. The exceptions are the cases in which all corners are massive, i.e. all $K_i^2 \neq 0$; such massive triangle and box integrals are IR finite. The bubble integrals $I_{2;n}$ have UV divergences, but no IR divergences because both corners are massive, $K_i^2 \neq 0$. The scalar bubble integral with a massless corner vanishes in dimensional regularization. This is because the integral has mass-dimension 2, yet there is no dimensionful quantity to give it dimension since the only invariant entering the integral is $K_i^2 = 0$.

Setting $D = 4 - 2\epsilon$, the integrated results of the 4-point box, triangle, and bubble integrals are

$$
\begin{aligned}
I_{4;4}(p_1, p_2, p_3, p_4) &= \frac{\gamma_\Gamma}{su} \left(\frac{2}{\epsilon^2} \left[(-\mu^{-2} s)^{-\epsilon} + (-\mu^{-2} u)^{-\epsilon} \right] - \ln^2 \left(\frac{s}{u} \right) - \pi^2 \right) + O(\epsilon), \\
I_{3;4}(p_1, p_2, p_3 + p_4) &= \frac{\gamma_\Gamma}{\epsilon^2} (-\mu^{-2} s)^{-1-\epsilon} + O(\epsilon), \\
I_{2;4}(p_1 + p_2, p_3 + p_4) &= \gamma_\Gamma \left(\frac{1}{\epsilon} - \ln(-\mu^{-2} s) + 2 \right) + O(\epsilon).
\end{aligned}
\tag{6.20}
$$

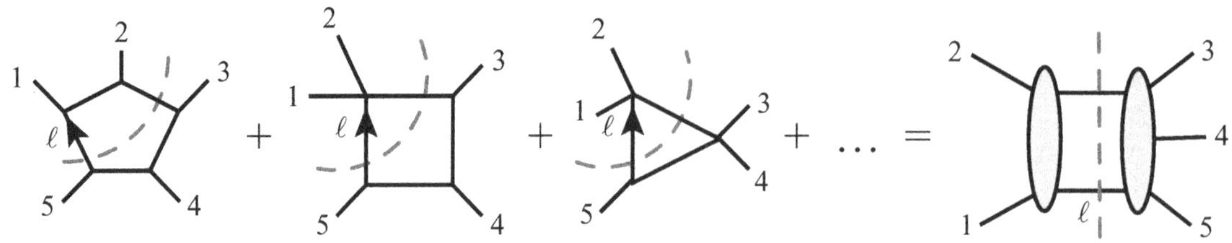

Fig. 6.1 The sum of residues from all Feynman diagrams with propagators ℓ^2 and $(\ell - p_1 - p_2)^2$ on-shell must give the product of two tree amplitudes.

Here, $\gamma_\Gamma = \Gamma(1 + \epsilon)\Gamma^2(1 - \epsilon)/\Gamma(1 - 2\epsilon)$ and μ is the regularization scale. These results and their n-point generalizations can be found in [66]. The 1-loop box and triangle integrals are $1/\epsilon^2$, as expected from the combination of the soft and collinear divergences. The $1/\epsilon$ divergence in the bubble integral is a UV divergence.

The scalar box, triangle, and bubble integrals provide a basis for 1-loop amplitudes in $D=4$ dimensions. We will see how this works next when we study the unitarity method.

6.2 Unitarity method

We begin with an example. Consider the color-ordered planar n-point 1-loop gluon amplitude in pure Yang–Mills theory. Suppose we identify[2] the loop-momentum ℓ such that in each Feynman diagram, ℓ is the momentum that flows between legs 1 and n, as indicated for $n = 5$ in Figure 6.1. Then we can unambiguously collect all the distinct Feynman diagrams under one integral,

$$\int d^D \ell \sum_j J_j \,. \tag{6.21}$$

The integrands J_j take the form indicated in (6.3). To compute the full amplitude we need to integrate ℓ over $\mathbb{R}^4$ (after Wick rotation from $\mathbb{R}^{1,3}$), but let us focus on the subplane where the loop-momentum satisfies the two *cut conditions*

$$\ell^2 = (\ell - p_1 - p_2)^2 = 0 \,. \tag{6.22}$$

On this subplane, integrands of the form

$$J_j = \frac{1}{S_j} \frac{c_j n_j}{\dots (\ell^2) \dots (\ell - p_1 - p_2)^2 \dots} \tag{6.23}$$

become singular. The singularity corresponds to a kinematic configuration where the two propagators, $1/\ell^2$ and $1/(\ell - p_1 - p_2)^2$, go on-shell. The sum of the residues from all such integrands must be equivalent to the product of two *on-shell* tree amplitudes, as shown schematically in Figure 6.1.

---

[2] More about this choice in Section 7.1.

This observation is very useful for constraining the loop-integrand: if the enemy gives us a rational function and claims that it is the 1-loop integrand of some (unitary) theory, we can apply cuts, such as (6.22), and test the claim by checking if the function factorizes correctly into products of tree amplitudes. This way, our knowledge of tree amplitudes can be recycled into information about the loop-integrand!

The operation of taking loop propagators on-shell is called a ***unitarity cut***. It originates from the unitary constraint of the S-matrix. To see how, recall that unitarity requires $S^\dagger S = 1$. Writing $S = 1 + iT$, where T represents the interacting part of the S-matrix, unitarity requires $-i(T - T^\dagger) = T^\dagger T$. If we examine this constraint order by order in perturbation theory, it tells us that the imaginary part of the T-matrix at a given order is related to the product of lower-order results. In particular, the imaginary part of the 1-loop amplitude is given by a product of two tree amplitudes. This is illustrated by the diagram

$$\tag{6.24}$$

The product of two tree amplitudes on the RHS involves a sum over all possible on-shell states that can "cross" the cut. Only states from the physical spectrum of the theory are included in this sum. In gauge theory, Feynman diagram calculations of loop amplitudes require ghosts in the loops: the purpose of the ghosts is to cancel unphysical modes in the loops. In contrast, in a unitarity cut (6.24) we restrict the loop-momenta to be on-shell and only physical modes are included in the two on-shell amplitudes on the RHS of (6.24).

The cutting rules also include integrals of any remaining freedom in the loop-momentum after imposing the cut constraints, such as (6.22), and momentum conservation. The integral over all allowed kinematic configurations, with respect to the diagram (6.24), can be written as

$$\int d^D \ell \, \delta_+\big(\ell^2\big) \, \delta_+\big((\ell - p_1 - p_2)^2\big) . \tag{6.25}$$

The subscript $_+$ means that we are choosing the solution to the on-shell condition that has a positive time component, $\ell^0 > 0$, i.e. it is associated with a particle (as opposed to an anti-particle) crossing the cut. Note that (6.25) just replaces the two cut propagators with their on-shell conditions, exactly as we did in Figure 6.1.

The imaginary part of the amplitude probes its branch cut structure, hence the unitarity cut allows us to relate the "pole structure" of the integrand with the "branch cut structure" of the loop-integral. One can reconstruct the integrand by analyzing different sets of unitarity cuts. The unitarity cuts can also involve more than two "cut" lines, i.e. several internal lines taken on-shell, such that the loop amplitude factorizes into multiple lower-loop on-shell amplitudes. An N-line cut simply means that N internal lines are taken on-shell. Reconstruction of the full loop amplitude from systematic application of unitarity cuts is called the ***generalized unitarity method*** [67]. It has been applied to a wide range of scattering problems, from next-to-leading order precision QCD predictions to the ultraviolet behavior of perturbative supergravity theories. The generalized unitarity method [67] deserves much more attention than we are able to offer here. Our introduction to the unitarity method covers just the minimum needed for you to see the idea and appreciate its power. The method has been reviewed extensively and you can learn more about it and its applications to supersymmetric as well as non-supersymmetric theories in the reviews [68–71]. Next, we discuss its implementation at 1-loop level.

6.3 1-loop amplitudes from unitarity

The information from unitarity cuts can be utilized most efficiently if we know, a priori, a complete basis of integrals that can appear in the scattering amplitudes. It can be shown [72–76] that in D-dimensions all 1-loop amplitudes can be rewritten as a sum of m-gon 1-loop scalar integrals I_m for $m = 2, 3, \ldots, D$:

$$A^{\text{1-loop}} = \sum_i C_D^{(i)} I_D^{(i)} + \sum_j C_{D-1}^{(j)} I_{D-1}^{(j)} + \cdots + \sum_k C_2^{(k)} I_2^{(k)} + \text{rational terms}. \quad (6.26)$$

Here, $C_m^{(i)}$ are kinematics-dependent coefficients for the m-gon scalar integrals $I_m^{(i)}$ that were introduced towards the end of Section 6.1. The 4-point box, triangle, and bubble scalar integrals (6.19) were evaluated in (6.20). The n-point box integral $I_4^{(i)}$ takes the form (henceforth dropping the subscript n)

$$I_4^{(i)} = \int \frac{d^D \ell}{(2\pi)^D} \frac{1}{\ell^2 (\ell - K_1^{(i)})^2 (\ell - K_1^{(i)} - K_2^{(i)})^2 (\ell + K_4^{(i)})^2},$$

$$(6.27)$$

where $(K_1^{(i)}, K_2^{(i)}, K_3^{(i)}, K_4^{(i)})$ are sums of the external momenta at each corner. The label i indicates a particular choice of distributing the external lines on the four subamplitudes, i.e. different 4-line cuts.

It may seem surprising that all 1-loop integrals can be written in terms of a basis of scalar 1-loop integrals as in (6.26), since integrals obtained from Feynman diagrams generally involve numerators that depend on loop-momentum. The simplification is that one can use the external momenta to form a basis for any vectors in the integrals. For example, a 4-point 1-loop "vector integral" can be expanded as

$$\int d^D \ell \frac{\ell^\mu}{\ell^2 (\ell - p_1)^2 (\ell - p_1 - p_2)^2 (\ell + p_4)^2} = a_1 p_1^\mu + a_2 p_2^\mu + a_3 p_3^\mu. \quad (6.28)$$

Here we are simply using the fact that the integral must be a Lorentz vector, and since by momentum conservation there are only three linearly independent momentum vectors in the 4-particle input, the answer must be a linear combination thereof. The coefficients in (6.28) are solved using the ***Passarino–Veltman reduction*** procedure [76]. The idea is simply to dot in various combinations of external momenta and simplify the result. For example, dotting p_1 into (6.28), the numerator of the integral becomes

$$(\ell \cdot p_1) = \frac{1}{2}\left[\left(\ell + p_1\right)^2 - \ell^2\right].\tag{6.29}$$

The two terms of the RHS each cancel a loop-propagator, leaving behind two scalar integrals. Then (6.28) gives

$$\int d^D \ell \left[\frac{1}{2\ell^2\left(\ell + p_1 + p_2\right)^2\left(\ell - p_4\right)^2} - \frac{1}{2\left(\ell + p_1\right)^2\left(\ell + p_1 + p_2\right)^2\left(\ell - p_4\right)^2}\right]$$
$$= -\frac{1}{2}\left(a_2 s + a_3 t\right).\tag{6.30}$$

Repeating the above procedure, one can obtain three linear equations for a_1, a_2, and a_3. Solving this linear system is straightforward and gives the vector integral (6.28) as a linear combination of scalar integrals.

▶ **Exercise 6.1**

Only scalar triangle integrals appear on the LHS of (6.30). Show that scalar box integrals appear if we contract (6.28) with p_2.

In general, since there are only D independent vectors in D-dimensions, the set of needed scalar integrals can be reduced to the set of scalar integrals shown in (6.26). Tensor-integrals can be simplified similarly. A more detailed discussion of integral reductions can be found in Section 4.2 of [77], also see [3].

The basis expansion (6.26) makes the task of computing the 1-loop amplitude a matter of determining the coefficients $C_m^{(i)}$. Since the scalar integrals have distinct propagator structures, only a subset contributes to a given unitarity cut. By applying multiple unitarity cuts, one obtains a set of linear equations that relate the $C_m^{(i)}$'s to the results of the cuts, each computed as a product of tree amplitudes. Solving these linear equations gives us the coefficients $C_m^{(i)}$'s as a combination of products of tree amplitudes. By (6.26), this determines the 1-loop amplitude, up to the possibility of rational terms that we discuss below.

The integral coefficients $C_m^{(i)}$ can be found systematically by organizing the calculation according to the number of propagators present in the scalar integrals $I_m^{(i)}$. In D-dimensions, the loop-momentum ℓ has D components, so one can find isolated solutions for ℓ such that all propagators in the scalar integral $I_D^{(i)}$ are on-shell: this corresponds to a D-line cut and this is the maximally possible cut in D dimensions. It is called a ***maximal cut*** and is discussed in much more detail in Chapter 8.

Since the cut constraints are quadratic in loop-momentum, there are two isolated solutions, denoted $\ell^{*(1)}$ and $\ell^{*(2)}$, to a maximal cut. The corresponding coefficient $C_D^{(i)}$ is

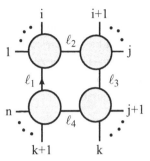

Fig. 6.2 1-loop box diagram with $K_1^{(i)} = p_1 + \cdots + p_i$, $K_2^{(i)} = p_{i+1} + \cdots + p_j$, $K_3^{(i)} = p_{j+1} + \cdots + p_k$ and $K_4^{(i)} = p_{k+1} + \cdots + p_n$. The corresponding box coefficient $C_4^{(i)}$ in (6.32) is the product of the four tree amplitudes at each corner.

completely determined by the product of D tree amplitudes as

$$C_D^{(i)} = \frac{1}{2} \sum_{\ell = \ell^{*(1)}, \ell^{*(2)}} A_{n_1}^{\text{tree}} \cdots A_{n_D}^{\text{tree}}. \tag{6.31}$$

Note that one averages over the two solutions, $\ell^{*(1)}$ and $\ell^{*(2)}$. At 1-loop, the relative weight between the two solutions can be determined by considering special integrands that integrate to zero. The associated maximal cut must also vanish and this fixes the above prescription. See [78] for a concise discussion.

▷ *Example.* Let us make (6.31) concrete. For an n-point 1-loop amplitude in $D = 4$, the coefficient of the box integral shown in Figure 6.2 is given by

$$C_4^{(\text{Fig. 6.2})} = \frac{1}{2} \sum_{\ell = \ell^{*(1)}, \ell^{*(2)}} \Bigg[\sum_{\text{states}} A_{n_1}\big[-\ell_1, 1, \ldots, i, \ell_2\big] \times A_{n_2}\big[-\ell_2, i+1, \ldots, j, \ell_3\big]$$
$$\times A_{n_3}\big[-\ell_3, j+1, \ldots, k, \ell_4\big] \times A_{n_4}\big[-\ell_4, k+1, \ldots, n, \ell_1\big] \Bigg], \tag{6.32}$$

where $\sum_{\text{states}}$ indicates a state sum for each internal line $\ell_1 = \ell$, $\ell_2 = \ell - (p_1 + \cdots + p_i)$, $\ell_3 = \ell - (p_1 + \cdots + p_j)$, and $\ell_4 = \ell + (p_{k+1} + \cdots + p_n)$. The vectors $\ell^{*(1)}$ and $\ell^{*(2)}$ solve the on-shell conditions $\ell_1^2 = \ell_2^2 = \ell_3^2 = \ell_4^2 = 0$ of the 4-line cut. ◁

Coefficients $C_m^{(i)}$ with $m < D$ are not quite as simple to calculate, but they can be obtained systematically. After determining all D-gon coefficients, we treat $(D-1)$-cuts. Both $I_D^{(i)}$ and $I_{D-1}^{(j)}$ integrals can potentially contribute to $(D-1)$-cuts, but since we have already determined all the $C_D^{(i)}$s, we can unambiguously determine all $C_{D-1}^{(j)}$s. Similarly, all integral coefficients can be determined iteratively. This way, the generalized unitarity method offers a systematic way to determine the 1-loop amplitude in terms of tree amplitudes. Detailed discussions of extracting 1-loop integral coefficients in $D = 4$ can be found in [47, 79].

▷ *Example.* As a concrete example, the integral coefficients for the fermion-loop correction of the 1-loop 4-gluon amplitude $A_4[1^+, 2^-, 3^+, 4^-]$ are

$$C_4(p_1, p_2, p_3, p_4) = A_4^{\text{tree}} \frac{-u^2 s^2 (u^2 + s^2)}{t^4},$$

$$C_3(p_1, p_2 + p_3, p_4) = A_4^{\text{tree}} \frac{us(u^3 + us^2)}{t^4}, \tag{6.33}$$

$$C_2(p_1 + p_4, p_2 + p_3) = A_4^{\text{tree}} \frac{2u(5s^2 + stu + 2u^2)}{6t^3},$$

where the integral coefficients are labeled in the same way as the scalar integrals in (6.19). All other integral coefficients are related to the above by cyclic permutation. ◁

▶ Exercise 6.2

Use the integrated results in (6.20) for the 4-point box, triangle, and bubble integrals, and the above integral coefficients to compute the final result for the fermion loop-correction to the 1-loop 4-gluon amplitude $A_4[1^+, 2^-, 3^+, 4^-]$. Show that there are only UV and no IR divergences. Explain why. What does the UV divergence indicate about renormalization of the theory at 1-loop level?

Of course, there is a big elephant in the room – you already met it in (6.26): it is the **rational terms**. Rational terms are rational functions that do not possess branch cuts, so they are undetectable by unitarity cuts. The rational terms arise from the need to regularize the loop integrals. In dimensional regularization, the loop-momentum ℓ is really $(D - 2\epsilon)$-dimensional. If we separate the loop-momentum into a D-dimensional part, $\ell^{(D)}$, and a (-2ϵ)-dimensional part, $\mu^{-2\epsilon}$, there can also be contributions from the μ-integrals. An example of an integrand that gives a branch-cut-free contribution is the $(D/2 + 1)$-gon scalar integral with μ^2 numerator: it integrates to a finite value

$$\int \frac{d\ell^{(D)} d\mu^{-2\epsilon}}{(2\pi)^{D-2\epsilon}} \, I_{D/2+1}[\mu^2] = -\frac{1}{(4\pi)^{D/2}} \frac{1}{(D/2)!} + O(\epsilon). \tag{6.34}$$

One cannot capture this from the ordinary unitarity cut since it is just a rational function (here, a constant). The unitarity cut forces the loop-momentum to be on-shell in D-dimensions and this implies $\mu^2 = 0$, so the contribution from the above integrand vanishes. On the other hand, if one considers unitarity cuts where the internal lines become massless in $D - 2\epsilon$ dimensions $(\ell^{(D)})^2 + \mu^2 = 0$, or equivalently *massive* in D-dimensions with mass $m^2 = \mu^2$, such terms *are* detectable. Thus rational terms can be reconstructed from unitarity cuts if we allow the states crossing the cut to be massive.

Rational terms are *absent* for supersymmetric Yang–Mills theories because supersymmetry cancellations ensure that the powers of loop-momentum in the integrals do not lead to rational terms after integral reduction. For non-supersymmetric theories – for example $\lambda\phi^4$, QED, QCD – rational terms are present and they are the most time-consuming ones to compute. We will not discuss this important issue in further detail, but simply refer you to

[80, 81] and references therein. Instead, we illustrate the unitarity method by working out an explicit example.

▷ *Example.* In this example, we calculate the **4-point 1-loop superamplitude in** $\mathcal{N} = 4$ **SYM** using the generalized unitarity method. We do so by first considering a 2-line cut and then infer from it which terms contribute in the integral basis expansion (6.26).

The s-channel unitarity cut is

$$\text{Cut}_s = \sum_{\text{states}} \mathcal{A}_4[-\ell_1, 1, 2, \ell_2] \times \mathcal{A}_4[-\ell_2, 3, 4, \ell_1]. \tag{6.35}$$

Using the analytic continuation (3.17), the tree-level MHV superamplitudes are

$$\mathcal{A}_4[-\ell_1, 1, 2, \ell_2] = \frac{\delta^{(8)}(L)}{\langle \ell_1 1 \rangle \langle 12 \rangle \langle 2\ell_2 \rangle \langle \ell_2 \ell_1 \rangle}, \tag{6.36}$$

$$\mathcal{A}_4[-\ell_2, 3, 4, \ell_1] = \frac{\delta^{(8)}(R)}{\langle \ell_2 3 \rangle \langle 34 \rangle \langle 4\ell_1 \rangle \langle \ell_1 \ell_2 \rangle}.$$

The arguments of the Grassmann delta functions are $L = -|\ell_1\rangle \eta_{\ell_1} + |1\rangle \eta_1 + |2\rangle \eta_2 + |\ell_2\rangle \eta_{\ell_2}$, and $R = -|\ell_2\rangle \eta_{\ell_2} + |3\rangle \eta_3 + |4\rangle \eta_4 + |\ell_1\rangle \eta_{\ell_1}$. As in the case of tree-level recursion (see the discussion around (4.55)), the intermediate state sum is performed as an integration of the on-shell Grassmann variables η_{ℓ_i} associated with each internal line [36]. These integrals are easy to perform when we use $\delta^{(8)}(L)\delta^{(8)}(R) = \delta^{(8)}(L + R)\delta^{(8)}(R) = \delta^{(8)}(\tilde{Q})\delta^{(8)}(R)$. We find

$$\text{Cut}_s = \frac{\delta^{(8)}(\tilde{Q})}{\langle 12 \rangle \langle 34 \rangle} \int d^4\eta_{\ell_1} d^4\eta_{\ell_2} \frac{\delta^{(8)}(R)}{\langle \ell_1 1 \rangle \langle 2\ell_2 \rangle \langle \ell_2 \ell_1 \rangle \langle \ell_2 3 \rangle \langle 4\ell_1 \rangle \langle \ell_1 \ell_2 \rangle}$$

$$= -\frac{\delta^{(8)}(\tilde{Q})}{\langle 12 \rangle \langle 34 \rangle} \frac{\langle \ell_1 \ell_2 \rangle^2}{\langle \ell_1 1 \rangle \langle 2\ell_2 \rangle \langle \ell_2 3 \rangle \langle 4\ell_1 \rangle}. \tag{6.37}$$

On the unitarity cut, one can convert the loop-momentum part of the denominator in the above expression into propagators:

$$\text{Cut}_s = \mathcal{A}_4^{\text{tree}}[1234] \times \frac{-su}{(\ell_2 + p_2)^2 (\ell_1 + p_4)^2}\Big|_{\ell_1^2 = \ell_2^2 = 0}. \tag{6.38}$$

▶ **Exercise 6.3**

Show that (6.38) follows from (6.37).

Let us now consider the possible integrals from (6.26) that contribute to the s-channel cut: the integrals that contain the propagators with $\ell_1^2 = \ell^2$ and $\ell_2^2 = (\ell - p_1 - p_2)^2$ are the box-integral $I_4(p_1, p_2, p_3, p_4)$, the triangle-integrals $I_3(p_1, p_2, p_3 + p_4)$, and $I_3(p_3, p_4, p_1 + p_2)$, and the bubble-integrals $I_2(p_1 + p_2, p_3 + p_4)$ and $I_2(p_2 + p_3, p_4 + p_1)$. In each case, we have indicated the distribution of the external lines. The result (6.38) for Cut$_s$ shows that there are two uncut propagators left after cutting ℓ_1 and ℓ_2, so this excludes the triangle- and bubble-integrals. Thus we conclude that only the box-integral is present, i.e. $A_4^{\text{1-loop}}[1234] = C_4 \, I_4(p_1, p_2, p_3, p_4)$. The box-coefficient C_4 is readily determined from (6.38), giving[3]

$$A_4^{\text{1-loop}}[1234] = su \, A_4^{\text{tree}}[1234] \, I_4(p_1, p_2, p_3, p_4). \tag{6.39}$$

To make sure that (6.39) is the correct result for the amplitude, we examine other distinct cuts to see if there could be terms that vanish in the s-channel cut and were therefore not captured in our analysis. The only other available cut is the u-channel cut. (Color-ordering excludes the t-channel cut.) But since the RHS of (6.39) is invariant under cyclic permutations, it is guaranteed to produce the correct u-channel cut. Hence (6.39) is indeed the correct 1-loop 4-point amplitude for $\mathcal{N} = 4$ SYM. We discuss the evaluation of I_4 in dimensional regularization in Section 6.4. ◁

Working through the details in the example above, you will notice that the unitarity method does take some work and you may wonder how it compares with a brute-force 1-loop Feynman diagram calculation. The answer is that the unitarity method is superior, since it heavily reduces the number of diagrams needed and it avoids gauge obscurities. For the unitarity method, the input is gauge-invariant on-shell amplitudes. You might find it curious that the first computation of the 1-loop 4-gluon amplitude $A_4^{\text{1-loop}}$ in $\mathcal{N} = 4$ SYM was not done in QFT, but in string theory: in 1982, Green, Schwarz, and Brink [82] obtained $A_4^{\text{1-loop}}$ as the low-energy limit of the superstring scattering amplitude with four external gluon states.

We close this section with some general comments on the 1-loop integral expansion (6.26). The representation of 1-loop amplitudes in terms of scalar integrals provides an interesting categorization scheme in terms of whether or not particular classes of integrals – in 4d: boxes, triangles, bubbles, and rationals – appear or not. For example, we mentioned earlier that rational terms are absent in $\mathcal{N} > 0$ super Yang–Mills theory. One can ask which theories involve only box-integrals, i.e. no triangle- or bubble-integrals and no rational terms. In 4d, such so-called "no-triangle" theories include $\mathcal{N} = 4$ SYM [67], $\mathcal{N} = 8$ supergravity [83, 84], and $\mathcal{N} = 2$ SYM coupled to specific tensor matter fields [85]. For pure $\mathcal{N} = 6$ supergravity, only box- and triangle-integrals appear [86], while for pure $\mathcal{N} \leq 4$ supergravity all integrals in (6.26), including rational terms, appear.

One relevant aspect of the above analysis is that in 4d only the bubble integrals $I_2^{(i)}$ contain ultraviolet divergences. In dimensional regularization, all bubble integrals have a common leading $1/\epsilon$-term, and hence contribute $\frac{1}{\epsilon} \sum_i C_2^{(i)}$ to the amplitude. As a result, the beta function for a given theory vanishes at 1-loop order precisely when $\sum_i C_2^{(i)} = 0$. In fact, in a renormalizable theory, one must have $\sum_i C_2^{(i)} \sim A_n^{\text{tree}}$ and the proportionality constant is related to the 1-loop beta function [47, 85, 87, 88]. Note that even though bubble coefficients are non-trivial for pure $\mathcal{N} \leq 4$ supergravity theories, their sum $\sum_i C_2^{(i)}$ must vanish since the theory is known to be free of ultraviolet divergences at 1-loop order.

Next, we offer a quick survey of results for loop amplitudes in planar $\mathcal{N} = 4$ SYM.

---

[3] The minus sign compared with (6.38) comes from the $(-i)^2$ in the cut propagators.

6.4 1-loop amplitudes in planar $\mathcal{N} = 4$ SYM

When we introduced the $\mathcal{N} = 4$ SYM theory in Section 4.4, we mentioned that this is a conformal theory, there is no running of the coupling. This means that there are no UV divergences in the on-shell scattering amplitudes at any loop-order. At 1-loop order, this can be seen from the fact that in $\mathcal{N} = 4$ SYM only the box-integrals contribute in the scalar integral basis expansion (6.26), i.e.

$$\mathcal{N} = 4 \text{ SYM in 4d:} \qquad A_n^{\text{1-loop}} = \sum_i C_{\text{box}}^{(i)} I_{\text{box}}^{(i)} \,. \tag{6.40}$$

Only bubble-integrals give UV divergences at 1-loop in 4d, so it is clear that the 1-loop amplitudes in $\mathcal{N} = 4$ SYM are UV finite. Loop amplitudes in $\mathcal{N} = 4$ SYM theory do have IR divergences and we will see more of them in the following.

We used the unitarity method in Section 6.3 to construct the 1-loop 4-point superamplitude in planar $\mathcal{N} = 4$ SYM. We found that the answer (6.39) could be written in terms of a single scalar box-integral:

$$\mathcal{A}_4^{\text{1-loop}}[1234] = su\, \mathcal{A}_4^{\text{tree}}[1234]\, I_4(p_1, p_2, p_3, p_4) \,. \tag{6.41}$$

The scalar box integral I_4, evaluated in dimensional regularization $D = 4 - 2\epsilon$, was given in (6.20). It follows that

$$\mathcal{A}_4^{\text{1-loop}}[1234] = \mathcal{A}_4^{\text{tree}}[1234] \left\{ \frac{2}{\epsilon^2} \left[\left(-\mu^{-2} y_{13}^2 \right)^{-\epsilon} + \left(-\mu^{-2} y_{24}^2 \right)^{-\epsilon} \right] \right.$$
$$\left. - \ln^2 \left(\frac{y_{13}^2}{y_{24}^2} \right) - \pi^2 + O(\epsilon) \right\} \,, \tag{6.42}$$

where μ is the regularization scale and $y_{ij} = y_i - y_j$ are the zone-variables defined in (5.31). In terms of Mandelstam variables, we have $s = -y_{13}^2$ and $u = -y_{24}^2$.

Let us now use the notation

$$\mathcal{A}_{n;L}^{N^K \text{MHV}} = n\text{-point } L\text{-loop } N^K \text{MHV superamplitude of planar } \mathcal{N} = 4 \text{ SYM} \,, \tag{6.43}$$

with the color-ordering $12 \ldots n$ of external particles implicit. Since supersymmetry and $SU(4)$ R-symmetry[4] Ward identities hold at each loop-order, this decomposition of the loop amplitude is sensible and $\mathcal{A}_{n;L}^{N^K \text{MHV}}$ has Grassmann degree $4(K + 2)$.

It is convenient to factor out an MHV tree-level superamplitude and write the loop-expansion as

$$\mathcal{A}_{n;L}^{N^K \text{MHV}}(\epsilon) = \mathcal{A}_{n;0}^{\text{MHV}} \left(\mathcal{P}_{n;0}^{N^K \text{MHV}} + \lambda\, \mathcal{P}_{n;1}^{N^K \text{MHV}}(\epsilon) + \cdots \right) \,, \tag{6.44}$$

---

[4] The $SU(4)$ R-symmetry is non-anomalous [89, 90].

where $\lambda \sim g^2 N$ is the t'Hooft coupling written in terms of the gauge coupling g and the rank of the gauge group $SU(N)$. We include the dependence on the ϵ-regulator explicitly in the loop amplitudes. At tree-level, we have

$$\mathcal{P}_{n;0}^{\text{MHV}} = 1 \quad \text{and} \quad \mathcal{P}_{n;0}^{\text{NMHV}} = \sum_{j=3}^{n-2} \sum_{k=j+2}^{n} R_{1jk}. \tag{6.45}$$

The NMHV result is given in terms of the dual superconformal invariants R_{1jk} defined in (4.83) and discussed further in Section 5.4.

▶ Exercise 6.4

Why is it possible to factor out $\mathcal{A}_{n;0}^{\text{MHV}}$ even at loop-level?

The ϵ-regulator explicitly breaks the conformal and dual conformal symmetry. You can see that in the expression (6.42) for the 1-loop 4-point superamplitude: not even the finite part $O(\epsilon^0)$ respects dual conformal inversion (5.37):

$$I(y_{ij}^2) = \frac{y_{ij}^2}{y_i^2 y_j^2}. \tag{6.46}$$

So the raw output of the loop amplitudes does not entertain the ordinary or dual conformal symmetries of the $\mathcal{N} = 4$ SYM theory. However, the IR divergences take a universal form that facilitates construction of IR-finite quantities that turn out to respect the symmetries.

At 1-loop order, the IR divergent part of $\mathcal{A}_{n;1}^{\text{N}^K\text{MHV}}$ is captured entirely by the MHV superamplitude in the sense that

$$\mathcal{A}_{n;1}^{\text{N}^K\text{MHV}}(\epsilon) = \mathcal{A}_{n;0}^{\text{N}^K\text{MHV}} \times \text{IRdiv}\left[\mathcal{P}_{n;1}^{\text{MHV}}(\epsilon)\right] + O(\epsilon^0), \tag{6.47}$$

where

$$\text{IRdiv}\left[\mathcal{P}_{n;1}^{\text{MHV}}(\epsilon)\right] = \frac{1}{\epsilon^2} \sum_{i=1}^{n} (-\mu^{-2} y_{i,i+2}^2)^{-\epsilon}. \tag{6.48}$$

Note that for $n = 4$, this reproduces the IR divergent terms in (6.42). The *finite* part of the 4-point MHV superamplitude is

$$\mathcal{F}_{4;1}^{\text{MHV}}(\epsilon) \equiv \mathcal{P}_{4;1}^{\text{MHV}}(\epsilon) - \text{IRdiv}\left[\mathcal{P}_{4;1}^{\text{MHV}}(\epsilon)\right] = -\ln^2\left(\frac{y_{13}^2}{y_{24}^2}\right) - \pi^2 + O(\epsilon). \tag{6.49}$$

The universal form (6.47) of the 1-loop IR divergences implies that the *ratio functions*[5]

$$\mathcal{R}_{n;1}^{\text{N}^K\text{MHV}}(\epsilon) \equiv \mathcal{P}_{n;1}^{\text{N}^K\text{MHV}}(\epsilon) - \mathcal{P}_{n;0}^{\text{N}^K\text{MHV}} \mathcal{P}_{n;1}^{\text{MHV}}(\epsilon) \tag{6.50}$$

---

[5] The RHS of (6.50) can be viewed as the $O(\lambda)$ term in the small λ expansion of the ratio $\mathcal{A}_n^{\text{N}^K\text{MHV}}/\mathcal{A}_n^{\text{MHV}}$.

are IR finite. Moreover, it has been proposed [50, 91] that $\mathcal{R}_{n;1}^{N^K\text{MHV}}(0)$s are actually dual conformal invariant. This was shown at NMHV level for $n \leq 9$ in [50, 91] and for general n in [92, 93] using generalized unitarity. To give you a sense of the expressions, we present the result [91] for the ratio function for the 6-point 1-loop NMHV superamplitude. It is

$$\mathcal{R}_{6;1}^{\text{NMHV}}(0) = \frac{1}{2}\left(R_{146}\, V_{146} + \text{cyclic}\right), \tag{6.51}$$

where

$$V_{146} = -\log u_1 \log u_2 + \frac{1}{2}\sum_{k=1}^{3}\left[\log u_k \log u_{k+1} + \text{Li}_2(1-u_k)\right] - \frac{\pi^2}{6}. \tag{6.52}$$

The dilogarithm Li_2 was defined in (6.4). The u_is are dual conformal cross-ratios,

$$u_1 = \frac{y_{13}^2 y_{46}^2}{y_{14}^2 y_{36}^2}, \qquad u_2 = \frac{y_{24}^2 y_{51}^2}{y_{25}^2 y_{41}^2}, \qquad u_3 = \frac{y_{35}^2 y_{62}^2}{y_{36}^2 y_{52}^2}, \tag{6.53}$$

so each V_{ijk} is a dual conformal invariant, as you can see by applying dual inversion (6.46). The "+ cyclic" in (6.51) is the instruction to sum over the cyclic sum of the external state labels; note that V_{251} is just V_{146} with $u_1 \to u_2 \to u_3 \to u_1$.

Recalling that the BCFW recursion relations for the tree-level 6-point NMHV superamplitude only have three terms, you might be surprised to see six terms in the 1-loop result (6.51). However, in Exercise 5.15 we used the cyclically invariant 5-brackets $[i, j-1, j, k-1, k] = R_{ijk}$ to rewrite the tree-level superamplitude as $\mathcal{A}_{6;0}^{\text{NMHV}}/\mathcal{A}_{6;0}^{\text{MHV}} = \mathcal{P}_{6;0}^{\text{NMHV}} = \frac{1}{2}(R_{146} + \text{cyclic})$. This was done in anticipation of the 1-loop ratio function (6.51), and now you see that (6.51) is just like the tree-level result but with each R_{ijk} dressed with a dual conformal invariant V_{ijk}. Adding loop-orders 0 and 1, we can therefore write 6-point ratio function

$$\mathcal{R}_{6}^{\text{NMHV}}(0) = \frac{1}{2}\left(R_{146}\left(1 + \lambda\, V_{146}\right) + \text{cyclic}\right) + O(\lambda^2). \tag{6.54}$$

There are two properties worth noting about the 1-loop ratio function $\mathcal{R}_{6;1}^{\text{NMHV}}(0)$:

- It is dual conformal invariant, but not dual *super*conformal invariant. For a discussion of this, see [94].
- V_{146} – and hence $\mathcal{R}_{6;1}^{\text{NMHV}}(0)$ – has uniform transcendentality 2. This can be extended to the ϵ-dependent terms if ϵ is assigned transcendentality -1.

Both of these properties carry over to all $\mathcal{R}_{n;1}^{\text{NMHV}}(0)$. At higher-loop order in *planar* $\mathcal{N} = 4$ SYM, the degree of transcendentality is expected to be uniformly $2L$.

At higher-point, there are more dual conformal invariant cross-ratios available than just the three u_is for $n = 6$. Consequently, the NMHV 1-loop ratio functions $\mathcal{R}_{n;1}^{\text{NMHV}}(0)$ are more involved; however, they are all known explicitly and they take a similar form to (6.51). You can find the results for $\mathcal{R}_{n;1}^{\text{NMHV}}(0)$ in [93].

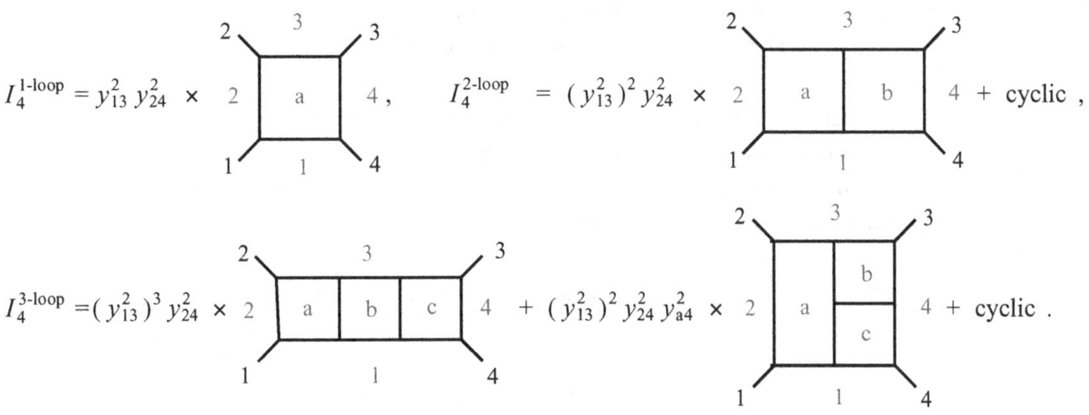

$$I_4^{\text{1-loop}} = y_{13}^2\, y_{24}^2 \times \quad \text{[diagram]} \quad ,$$

$$I_4^{\text{2-loop}} = (y_{13}^2)^2\, y_{24}^2 \times \quad \text{[diagram]} \quad + \text{cyclic} ,$$

$$I_4^{\text{3-loop}} = (y_{13}^2)^3\, y_{24}^2 \times \quad \text{[diagram]} \quad + (y_{13}^2)^2\, y_{24}^2\, y_{a4}^2 \times \quad \text{[diagram]} \quad + \text{cyclic} .$$

The integrands of $\mathcal{N} = 4$ SYM 4-point amplitude to 3-loop order. These are the unique scalar integrands that are dual conformal invariant.

Fig. 6.3

6.5 Higher-loop amplitudes in planar $\mathcal{N} = 4$ SYM

The generalized unitarity method can be applied successfully to higher-loop amplitudes, both at the planar and non-planar level; for a recent review see [68]. The application of unitarity is most efficient when a complete integral basis is available; for example, for 1-loop amplitudes in 4d, the basis consists of the scalar box-, triangle-, and bubble-integrals in (6.26).

Beyond 1-loop, there is not a complete understanding of the basis integrals for amplitudes in generic quantum field theories, although partial results have been achieved at 2-loops in the planar limit, see [78, 95–99]. One thing worth noting is that the integral basis is finite [100].

Without a given basis of integrals, one strategy is to construct the most general integral Ansatz that satisfies certain criteria, such as dimension-counting, and then use various integral identities to recast the Ansatz into a basis of independent integrals. Further symmetries, such as dual conformal invariance in planar $\mathcal{N} = 4$ SYM, can be a strong handle on finding a complete integral basis. As an example, the diagrams in Figure 6.3 correspond to the only dual conformal invariant scalar integrals for the 4-point 1-, 2- and 3-loop integrands of planar $\mathcal{N} = 4$ SYM. The coefficients of each integral are fixed by applying unitarity cuts [101, 102], so that the LHSs of the equations in Figure 6.3 are the full integrands for the 4-point 1-, 2- and 3-loop amplitudes in planar $\mathcal{N} = 4$ SYM. The evaluation of these integrals leads to interesting results that we discuss next.

The analytical result [103] for the **2-loop 4-point amplitude** in planar $\mathcal{N} = 4$ SYM was shown by Anastasiou, Bern, Dixon, and Kosower (ABDK) [101] to be expressible in terms of the 1-loop amplitude as

$$\mathcal{P}_{4;2}^{\text{MHV}}(\epsilon) = \frac{1}{2}\left[\mathcal{P}_{4;1}^{\text{MHV}}(\epsilon)\right]^2 + \mathcal{P}_{4;1}^{\text{MHV}}(2\epsilon)\, f^{(2)}(\epsilon) + C^{(2)} + O(\epsilon), \qquad (6.55)$$

where the MHV tree factor is stripped off as in (6.44), $f^{(2)}(\epsilon) = -\zeta_2 - \zeta_3 \epsilon - \zeta_4 \epsilon^2$ and $C^{(2)} = -\zeta_2^2/2$. Here $\zeta_s = \sum_{k=1}^{\infty} k^{-s}$ is the Riemann zeta function; note $\zeta_2 = \frac{\pi^2}{6}, \zeta_3 \approx 1.202$, and $\zeta_4 = \frac{\pi^4}{90}$.

It is interesting that the 2-loop 4-point amplitude in planar $\mathcal{N} = 4$ SYM can be written in terms of the 1-loop result. But the plot thickens! By explicit calculation of the **3-loop 4-point amplitude** in planar $\mathcal{N} = 4$ SYM, Bern, Dixon, and Smirnov (BDS) [102] found that the iterative structure continues:

$$\mathcal{P}_{4;3}^{\text{MHV}}(\epsilon) = -\frac{1}{3}\left[\mathcal{P}_{4;1}^{\text{MHV}}(\epsilon)\right]^3 + \mathcal{P}_{4;1}^{\text{MHV}}(\epsilon)\,\mathcal{P}_{4;2}^{\text{MHV}}(\epsilon) + f^{(3)}(\epsilon)\,\mathcal{P}_{4;1}^{\text{MHV}}(3\epsilon) + C^{(3)} + O(\epsilon).$$

$$(6.56)$$

Here $f^{(3)}(\epsilon) = \frac{11}{2}\zeta_4 + O(\epsilon)$ and $C^{(3)}$ is a constant.

The 2- and 3-loop results indicate an exponentiation structure. This motivates the **ABDK/BDS Ansatz for the full MHV superamplitude in $\mathcal{N} = 4$ SYM**:

$$\mathcal{P}_n^{\text{MHV(BDS)}}(\epsilon) = \exp\left[\sum_{L=1}^{\infty} \lambda^L \left(f^{(L)}(\epsilon)\,\mathcal{P}_{n;1}^{\text{MHV}}(L\epsilon) + C^{(L)} + O(\epsilon)\right)\right]. \qquad (6.57)$$

This Ansatz is almost correct: keep reading! In the ABDK/BDS Ansatz, the functions $f^{(L)}$ are of the form $f^{(L)}(\epsilon) = f_0^{(L)} + \epsilon f_1^{(L)} + \epsilon^2 f_2^{(L)}$, and the constants $C^{(L)}$ and $f_{0,1,2}^{(L)}(\epsilon)$ are independent of the number of external legs n. In particular, at 1-loop order $f^{(1)}(\epsilon) = 1$ and $C^{(1)} = 0$, and at 2-loops the results for $f^{(2)}(\epsilon)$ and $C^{(2)}$ were given below (6.55).

▶ **Exercise 6.5**

Show that (6.57) reproduces the 4-point 2- and 3-loop expressions (6.55) and (6.56).

Of course, the way one would go about testing the ABDK/BDS exponentiation Ansatz (6.57) is by direct calculation of the n-point L-loop amplitudes at $L = 2, 3, \ldots$. But how many 2-loop amplitudes have you ever calculated? Yeah, it is not an easy task, nonetheless progress has been made. It has been shown numerically in [104, 105] that the exponentiation Ansatz correctly produces the **5-point 2-loop amplitude**. It is very interesting that something new happens at 6- and higher-point: while *the ABDK/BDS Ansatz matches the IR divergent structure, it does not fully produce the correct finite part. The ABDK/BDS Ansatz determines the finite part of the amplitude only up to a function of dual conformal cross-ratios of the external momenta.* This function is called the **remainder function** and it is defined as

$$\mathbf{r}_{n;L}(\epsilon) \equiv \mathcal{P}_{n;L}^{\text{MHV}}(\epsilon) - \mathcal{P}_{n;L}^{\text{MHV(BDS)}}(\epsilon), \qquad (6.58)$$

where $\mathcal{P}_{n;L}^{\text{MHV}}(\epsilon)$ is the actual MHV L-loop amplitude and $\mathcal{P}_{n;L}^{\text{MHV(BDS)}}(\epsilon)$ is the $O(\lambda^L)$ terms in the expansion of the exponential Ansatz (6.57). The remainder function does not show up for $n = 4, 5$ because in those cases there are no available conformal cross-ratios.

The first indication of the remainder function came from a strong coupling calculation by Alday and Maldacena [106] who proposed [107] to use the AdS/CFT correspondence to calculate $\mathcal{P}_n^{\mathrm{MHV}}$. Subsequently, it was verified numerically that a remainder function is needed for the parity-even part of the **6-point 2-loop MHV amplitude** [108], whereas ABDK/BDS successfully determines the parity-odd part [109]. The analytic form of the remainder function $\mathbf{r}_{6;2}$ for the 6-point 2-loop MHV amplitude was calculated (as a hexagonal Wilson-loop) by Del Duca, Duhr, and Smirnov [110, 111]. The result, written in terms of the three dual conformal cross-ratios $u_{1,2,3}$ in (6.53), is a respect-inducing 17-page long sum of generalized polylogarithms; all terms have transcendentality 4. In an impressive application of a mathematical tool known as *the Symbol*, Goncharov, Spradlin, Vergu, and Volovich [112] managed to simplify this complicated result for $\mathbf{r}_{6;2}$ to an expression that involves only regular polylogs – Li_s and log – and fits in just a few lines of LaTeX.

The simple answer [112] for $\mathbf{r}_{6;2}$ is an important step towards a better understanding of loop amplitudes in planar $\mathcal{N} = 4$ SYM. The Symbol is now being used to understand higher-loop amplitudes, however, there will be amplitudes in planar $\mathcal{N} = 4$ SYM involving integrals that the Symbol does not help with. Thus techniques are eventually needed beyond the Symbol.

We have reviewed the unitarity method and shown you how it allows us to construct L-loop amplitudes from *on-shell* lower-loop input. The approach explores the analytic structure of the loop-integrands in a somewhat different way from the recursive techniques you know from tree-level amplitudes. BCFW is available at the level of loop-integrands, and that is the subject of the next chapter.

7 BCFW recursion for loops

The unitarity cuts discussed in Chapter 6 involve cuts of *at least two* propagators. In contrast, the tree-level recursion relation of Chapter 3 looks more like a single-line cut construction, since it uses the factorization-structure of a single propagator going on-shell. It is tempting to ask if loop amplitudes can similarly be reconstructed from the singularities associated with taking a single propagator on-shell. In this chapter, we study a recursive approach that generalizes the tree-level BCFW construction to loop-integrands.

We begin in Section 7.1 with a discussion of what is meant by *the* loop-integrand of an amplitude and how and when it is well-defined. The rest of the chapter focuses on super-BCFW recursion relations for loop-integrands of planar superamplitudes in $\mathcal{N} = 4$ SYM. In this setting it is quite natural to work with the momentum twistor variables introduced in Section 5.4; we will in particular make use of the description given there of points, lines, and planes in momentum twistor space. It is interesting how the description of the BCFW shift becomes geometric in the momentum twistor description, for example, the on-shell conditions for the internal line in a diagram become a statement of intersection of lines and planes. We demonstrate the geometrization in Section 7.2 and apply it in Section 7.3 to derive the super-BCFW result for the N^kMHV tree-level superamplitude in $\mathcal{N} = 4$ SYM. Thence, suitably warmed up and prepared, we introduce loop-recursion in Section 7.4. In Section 7.5, we give a fully worked-through example of its application to the 4-point 1-loop superamplitude of $\mathcal{N} = 4$ SYM. We end with brief comments about higher loop amplitudes in Section 7.6.

7.1 Loop-integrands

As seen in Chapter 6, loop amplitudes have a complicated analytic structure, so we focus on the *loop-integrand* which is just a rational function with poles at the location of the propagators, much similar to the tree amplitudes. Suppose we do a BCFW shift on the external legs, for example $p_1^\mu \to p_1^\mu + z\, q^\mu$ and $p_n^\mu \to p_n^\mu - z\, q^\mu$, with $q^2 = 0$ as usually. We can deduce from the Feynman diagrams that the shifted loop-integrands possess two types of poles in z:

1. poles in loop-independent propagators;
2. poles in propagators involving loop-momentum.

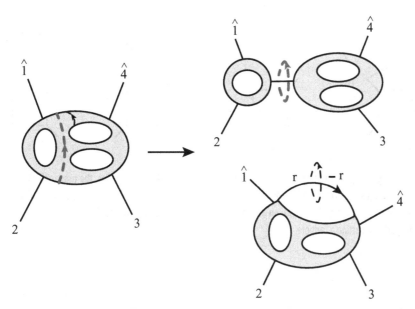

Representation of the two types of poles occurring in a BCFW shift of a 4-point 3-loop integrand. Type 1 are poles associated with loop-momentum *independent* propagators and type 2 poles are loop-momentum *dependent* propagators. The former factorizes the 3-loop integrand into a product of a 2-loop and a 1-loop integrand while the latter gives a forward limit of a 2-loop amplitude with two extra legs. **Fig. 7.1**

The residue of a type 1 pole corresponds to factorization of the integrand into a product of two lower-loop integrands. The residue of a type 2 pole in an L-loop n-point integrand is an $(n+2)$-point $(L-1)$-integrand with two adjacent legs evaluated in the ***forward limit***

$$p_i^\mu \to r^\mu\,, \qquad p_{i+1}^\mu \to -r^\mu\,, \quad \text{with} \ \ r^2 = 0\,. \tag{7.1}$$

This is illustrated for the example of a 4-point 3-loop amplitude in Figure 7.1. The poles of type 2 are what we would call single-line cuts in the unitarity method [113, 114].

Thus – provided that the large-z behavior is well-understood – it appears that one can straightforwardly set up a recursion relation for loop-integrands. However, there are subtleties we have to resolve:

- The first issue has to do with the identification of the loop-momenta in the loop-integrand. In the amplitude, we have to integrate the loop-momenta, so the ℓ_i are just dummy variables that can be redefined while still giving the same integrated answer. But the integrand itself can have different pole structures depending on how the ℓ_i are identified. As an example, consider the 1-loop 4-point box-integral (6.1) and compare the equivalent parameterizations $I_4^{(a)}$ and $I_4^{(b)} = I_4^{(a)}(\ell \to \ell + p_1)$. BCFW shifting legs 1 and 2 yields

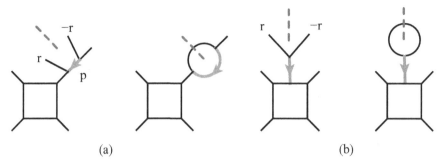

(a)   (b)

Fig. 7.2   Examples of diagrams problematic for the forward limit. In the lefthand diagram of (a), the propagator between the forward legs diverges due to momentum conservation. The righthand part of diagram (a) illustrates that such a diagram corresponds to the single-cut of a bubble on an external leg. Similar remarks apply to the diagrams in (b), where the limit corresponds to cutting a tadpole.

two distinct analytic functions in z:

$$I_4^{(a)}(\hat{1}, \hat{2}, 3, 4) = \frac{1}{\ell^2 (\ell - p_1 - zr)^2 (\ell - p_1 - p_2)^2 (\ell + p_4)^2},$$

$$I_4^{(b)}(\hat{1}, \hat{2}, 3, 4) = \frac{1}{(\ell + p_1 + zr)^2 \ell^2 (\ell - p_2 + zr)^2 (\ell - p_2 - p_3 + zr)^2}.$$

(7.2)

In general there is no canonical way to identify how the loop-momentum is parameterized, so that is the first subtlety that needs to be resolved. It basically comes down to the definition of what we mean by *the* "un-integrated integrand."

- The second subtlety has to do with the forward limit (7.1). When loop-momentum-dependent propagators go on-shell, there is a residue corresponding to a lower-loop $(n+2)$-point integrand in the forward limit, but that limit suffers from singularities. For example, from the explicit Feynman diagrams one sees that if the forward legs are attached to the same external line, then due to momentum conservation there is a $1/p^2$ singularity as $p^2 \to 0$. Such diagrams can be identified with cuts of bubbles on external legs or tadpole diagrams. This is illustrated in Figure 7.2. In massless theories, these integrate to zero in dimensional regularization and do not contribute to the loop amplitude. However, prior to integration, they are part of the integrand and will contribute to the single cuts. Therefore an important, but difficult, task is to identify these contributions in the forward limit such that one can consistently remove them.

Resolution of the above subtleties has been partially achieved for non-supersymmetric theories [113] and *completely resolved in supersymmetric theories in the planar limit* [114, 115]. In particular, it was shown that for supersymmetric theories the problematic terms associated with the tadpole as well as for external bubbles cancel in the state sum over the supermultiplet, and thus one has a perfectly well-defined residue. Furthermore, in the planar limit, the loop-momenta in the integrand can be defined unambiguously. This is done by defining the ℓ_is with a specific relation to the ordering of the external momenta. For example at 1-loop, one choice is to declare that ℓ is the momentum associated with the internal line immediately before line 1. For the 4-point 1-loop box integral, this selects integrand $I_4^{(a)}$ in (7.2).

The identification of the loop-momentum is naturally done in dual space y, that we defined in (5.31) in order to make momentum conservation manifest. We noted there that the y_is are also sometimes called zone variables; that is because we can think of them as labeling the "zones," or regions, that the external lines of the amplitude separate the plane into. This assumes a well-defined ordering of the external lines based on the color-ordering – and at loop-level it further requires the graphs to be planar. Let us illustrate this for a 6-point tree-graph and the 4-point 1-loop box diagram:

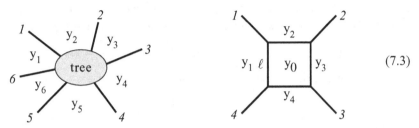

(7.3)

At loop-level there are "internal" zones, one for each loop. This offers the opportunity to switch the integration variable from ℓ_i to the new internal zone variables. For example in the 4-point 1-loop box graph, we can use y_0 as the loop-parameter instead of ℓ. They are related by $\ell = y_1 - y_0$, similarly to the relationship $p_i = y_i - y_{i+1}$. The dual variables therefore give an unambiguous definition of the loop-momentum and this facilitates the loop-level recursion relations for planar integrands.

As a useful example, let us express the 4-point scalar box-integral from (7.3) in dual variables. Following the rules for identifying the momentum on each internal line in terms of the zone-variables of the two adjacent zones, we have $\ell_i^2 = (y_0 - y_i)^2 = y_{0i}^2$. The box-integral is therefore simply

$$
I_4(p_1, p_2, p_3, p_4) = \int \frac{d^4\ell}{(2\pi)^4} \frac{1}{\ell^2 (\ell - p_1)^2 (\ell - p_1 - p_2)^2 (\ell + p_4)^2}
$$
$$
\rightarrow \int \frac{d^4 y_0}{y_{01}^2\, y_{02}^2\, y_{03}^2\, y_{04}^2}. \tag{7.4}
$$

To reiterate, for *planar amplitudes in supersymmetric theories, all subtleties in defining the loop-integrand have been overcome.* In the following, we introduce a BCFW recursion relation that generates the planar loop-integrands for $\mathcal{N} = 4$ SYM; it was developed in [115] and also considered in [116].

7.2 BCFW shift in momentum twistor space

The planar integrand is well-defined in the dual coordinates y_i, so it would seem natural to formulate the BCFW shifts in the dual representation. However, it is actually more interesting to use the momentum supertwistors $\mathcal{Z}_i^A$ that we introduced in Section 5.4. The $\mathcal{Z}_i^A$s can be chosen freely in $\mathbb{CP}^{3|4}$ and make momentum conservation, supermomentum conservation, and on-shell conditions automatic. We can therefore set up the BCFW shift without worrying about these constraints. The simplest possibility is to shift just a single momentum twistor, say $\mathcal{Z}_i$:

$$\hat{\mathcal{Z}}_i = \mathcal{Z}_i + w\mathcal{Z}_{i+1}. \tag{7.5}$$

All other $\mathcal{Z}_j$s are unshifted. The shift parameter w is a complex variable, $w \in \mathbb{C}$. Geometrically, (7.5) is the statement that the point $\hat{\mathcal{Z}}_i$ lies on the line $(i, i+1) = (\mathcal{Z}_i, \mathcal{Z}_{i+1})$.[1]

Let us translate (7.5) back to the spinor helicity formalism. In components (7.5) says

$$|\hat{i}\rangle = |i\rangle + w|i+1\rangle, \quad |\hat{\mu}_i] = |\mu_i] + w|\mu_{i+1}], \quad \hat{\chi}_{iA} = \chi_{iA} + w\chi_{i+1,A}. \tag{7.6}$$

Using (5.46) and the incidence relations (5.44), one finds that

$$\hat{y}_i = y_i + z|i-1\rangle[i|, \tag{7.7}$$

where

$$z = \frac{w\langle i, i+1\rangle}{\langle i-1, i\rangle + w\langle i-1, i+1\rangle}. \tag{7.8}$$

All other y_js are unshifted.

▶ **Exercise 7.1**

Use (5.46) and (5.44) to show that y_{i+1} and y_{i-1} are unshifted. Then derive (7.7).

The shift in y-space makes sense geometrically, because by Figure 5.1 the point $\hat{y}_{i+1}$ is determined by the line $(\hat{i}, i+1)$ which is equivalent to the line $(i, i+1)$ since (7.5) exactly tells us that the point $\hat{\mathcal{Z}}_i$ lies on $(i, i+1)$. So we conclude $\hat{y}_{i+1} = y_{i+1}$. On the other hand, $\hat{y}_i$ is determined by the line $(i-1, \hat{i})$ which is different from $(i-1, i)$, so $\hat{y}_i \neq y_i$ for $w \neq 0$.

Translating from dual y-space to momentum space, we have

$$\hat{p}_i = \hat{y}_i - y_{i+1} = -(|i\rangle - z|i-1\rangle)[i|, \tag{7.9}$$

$$\hat{p}_{i-1} = y_{i-1} - \hat{y}_i = -|i-1\rangle([i-1| + z[i|). \tag{7.10}$$

No other momenta shift. We immediately read off that this is a $[i-1, i\rangle$ BCFW shift,

$$|\hat{i}\rangle = |i\rangle - z|i-1\rangle, \quad |\widehat{i-1}] = |i-1] + z|i]. \tag{7.11}$$

Since there is also a shift of the Grassmann-components of the momentum supertwistors, (7.5) actually induces a BCFW $[i-1, i\rangle$-supershift.

▶ **Exercise 7.2**

Show that the Grassmann-part of the shift in (7.5) is equivalent to $\hat{\eta}_{i-1} = \eta_{i-1} + z\,\eta_i$.

---

[1] As in Section 5.4, we denote the line in momentum twistor space defined by two points (Z_j, Z_k) as (j, k), and the plane defined by three points (Z_j, Z_k, Z_l) as (j, k, l).

▶ **Exercise 7.3**

It may seem surprising that the shift $\hat{\mathcal{Z}}_i = \mathcal{Z}_i + w \mathcal{Z}_{i+1}$ is equivalent to a $[i - 1, i\rangle$-supershift; one might have expected a shift involving lines i and $i + 1$ instead. Actually, the shift (7.5) is also equivalent to a $[i + 1, i\rangle$ shift: this is because the momentum twistors are defined projectively, so we could supplement (7.5) with an overall scaling. For example, one finds that the angle spinor shift in (7.11) is equivalent to

$$|\hat{i}\rangle = \frac{\langle i - 1, i \rangle}{\langle i - 1, i \rangle + w \langle i - 1, i + 1 \rangle} \left(|i\rangle + w|i + 1\rangle \right). \tag{7.12}$$

Manipulate $|\hat{i}\rangle$ in (7.11) to find (7.12).

Our next task is to describe the kinematics associated with the internal lines in the BCFW diagrams – it turns out to have a nice geometric description in momentum twistor space. Consider a typical BCFW diagram associated with a factorization channel P_I:

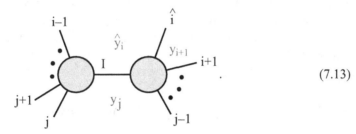

$$(7.13)$$

For simplicity, let us for now suppose that there are no loop-momenta in P_I. The shifted momentum on the internal line is

$$\hat{P}_I^2 = (\hat{p}_i + p_{i+1} + \cdots + p_{j-1})^2 = (\hat{y}_i - y_j)^2 = \hat{y}_{ij}^2 = \frac{\langle i - 1, \hat{i}, j - 1, j \rangle}{\langle i - 1, i \rangle \langle j - 1, j \rangle}. \tag{7.14}$$

In the last step, we have written the distance in dual y-space in terms of the 4-bracket, as in (5.52), and we have also used that $\langle i - 1, \hat{i} \rangle = \langle i - 1, i \rangle$. The shift $\hat{Z}_i = Z_i + w\, Z_{i+1}$ says that the point $\hat{Z}_i$ lies on the line $(i, i + 1)$ and its position on that line is parameterized by w. The condition $\hat{P}_I^2 = 0$ is the statement that w is chosen such that $\langle i - 1, \hat{i}, j - 1, j \rangle = 0$, so this value w_* is such that the point $\hat{Z}_i$ lies in the plane $(i - 1, j - 1, j)$. In other words, $\hat{Z}_i$ can be characterized as the point of intersection between the line $(i, i + 1)$ and the plane $(i - 1, j - 1, j)$, viz.

$$(\hat{i}) = (i, i + 1) \bigcap (i - 1, j - 1, j). \tag{7.15}$$

The intersection formula was given in (5.66) in terms of the 4-brackets. The geometry is illustrated in Figure 7.3(a).

We now determine the momentum twistor Z_I associated with the internal line $\hat{P}_I$. Take a look at the BCFW diagram in (7.13). The point y_j in dual space can be determined by the line $(j - 1, j)$ in momentum twistor space. But by inspection of (7.13), y_j can also be determined by the line (I, J). This means that the three points Z_I, Z_{j-1}, and Z_j lie on the same line. Similarly, the point $\hat{y}_i$ can be determined by the line $(i - 1, \hat{i})$ or by the line

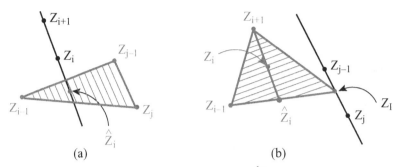

Fig. 7.3 The geometry of (7.15) and (7.16): (a) shows that the on-shell condition $\hat{P}_I^2 = 0$ fixes the shifted momentum twistor $\hat{Z}_i$ to be at the intersection of the line $(i, i + 1)$ with the plane $(i - 1, j, j - 1)$. In (b) the momentum twistor Z_I is located at the intersection of line $(j, j - 1)$ and the plane $(i - 1, i, i + 1)$.

$(I, \hat{i})$, so Z_I, Z_{i-1}, and $\hat{Z}_i$ lie on the same line. Since Z_I lies on both lines, we conclude that Z_I can be characterized as the intersection point of the lines $(i - 1, \hat{i})$ and $(j - 1, j)$. We previously learned that $\hat{Z}_i$ lies in the plane $(i - 1, j - 1, j)$, and therefore the plane contains the line $(i - 1, \hat{i})$. Thus we conclude that Z_I is the point where the line $(j - 1, j)$ intersects the plane $(i - 1, i, i + 1)$:

$$(I) = (j, j - 1) \bigcap (i - 1, i, i + 1). \qquad (7.16)$$

The geometry is shown in Figure 7.3(b). These results will be useful in the following.

Now we are ready to study the BCFW recursion relations in momentum twistor space. Consider the BCFW shift (7.5) of an n-point L-loop integrand $\mathcal{I}_n^L$ (for tree-level you can translate "integrand" to "superamplitude" in your head). The recursion relations are based on the contour argument for $\int \frac{dw}{w} \hat{\mathcal{I}}_n^L(w)$, just as in Section 3.1. Poles at finite values of w are equivalent, via (7.8), to poles at finite z: they arise from propagators with momentum $\hat{y}_{ij}$ going on-shell and the corresponding BCFW diagrams are those in (7.13). However, in completing the contour integral argument we also need to consider the large-w limit. It is clear from the relation (7.8) that z goes to a finite value z_* as $w \to \infty$. Specifically,

$$z \xrightarrow{w \to \infty} z_* \equiv \frac{\langle i, i + 1 \rangle}{\langle i - 1, i + 1 \rangle}. \qquad (7.17)$$

Thus the pole at infinity in the w-plane maps to a finite point in the z-plane and we will have to consider this contribution too. At $w = \infty$, the shifted angle spinors are

$$|\hat{i}\rangle \Big|_{w \to \infty} = \frac{\langle i - 1, i \rangle}{\langle i - 1, i + 1 \rangle} |i + 1\rangle. \qquad (7.18)$$

Thus, in the limit $w \to \infty$, the spinors $|\hat{i}\rangle$ and $|i + 1\rangle$ become proportional, and that implies that $\hat{P}_{i,i+1} = \hat{p}_i + p_{i+1}$ is on-shell: $\hat{P}_{i,i+1}^2 = \langle \hat{i}, i + 1 \rangle [i, i + 1] \to 0$ for $w \to \infty$. Or equivalently, we may note that z_* is exactly the solution to $\langle \hat{i}, i + 1 \rangle = 0$. Hence the pole at $w = \infty$ corresponds to a factorization channel of an integrand into a 3-point anti-MHV part – the only possibility that can support the special kinematics $|\hat{i}\rangle \propto |i+1\rangle \propto |\hat{P}_{i,i+1}\rangle$ – and the remainder L-loop integrand:

(7.19)

Note that the lines in this figure are labeled by momentum supertwistors, so we place a hat just on line i since that is the only shifted momentum twistor.

▶ **Exercise 7.4**

Show that for $w \to \infty$, the kinematics give

$$|\widehat{i-1}\rangle = |i-1\rangle, \qquad |\widehat{i-1}] = |i-1] - \frac{\langle i, i+1 \rangle}{\langle i+1, i-1 \rangle}|i],$$

$$|\hat{P}_I\rangle = |i+1\rangle, \qquad |\hat{P}_I] = -\left(\frac{\langle i, i-1 \rangle}{\langle i+1, i-1 \rangle}|i] + |i+1] \right). \qquad (7.20)$$

To summarize, BCFW recursion relations in momentum twistor space express the integrand (superamplitude) as a boundary contribution $\mathcal{B}_\infty$ from $w = \infty$ plus a sum of residues at finite w. Schematically, we have

$$\mathcal{I}_n^L = \hat{\mathcal{I}}_n^L(w=0) = \mathcal{B}_\infty^L - \sum_{w_* \neq 0} \left(\text{Residues of } \frac{\hat{\mathcal{I}}_n^L(w)}{w} \text{ at finite } w_* \right). \qquad (7.21)$$

The boundary term $\mathcal{B}_\infty$ is computable and is given by the diagram (7.19). The rest of the residues come from diagrams such as (7.13).

To become familiar with how this works in practice, we first apply the recursion relations to tree-level superamplitudes before moving on to loop-integrands in Section 7.4.

7.3 Momentum twistor BCFW at tree-level

In Section 4.5.2, we used the super-BCFW recursion relations to show that the NMHV tree-level superamplitude of $\mathcal{N} = 4$ SYM can be written as the MHV superamplitude times a sum of the dual superconformal invariants R_{ijk}; see (4.88). We then rewrote the NMHV formula in terms of momentum twistors in Section 5.4 and found (5.60),

$$\mathcal{A}_n^{\text{NMHV}} = \mathcal{A}_n^{\text{MHV}} \sum_{j=2}^{n-3} \sum_{k=j+2}^{n-1} [n, j-1, j, k-1, k], \qquad (7.22)$$

where the 5-brackets $[n, j - 1, j, k - 1, k] \equiv R_{njk}$ are invariant under cyclic permutations of the five labels. It was claimed then that the N^KMHV tree superamplitudes took a similar form but with the sum involving products of K 5-brackets. We now prove this statement using the momentum twistor formulation of super-BCFW. This also serves to prove that the tree superamplitudes of $\mathcal{N} = 4$ SYM are dual superconformal covariant.

Adapted to tree-level, the $\hat{Z}_i$-shift BCFW relation (7.21) reads

$$\mathcal{A}_n = \mathcal{B}_\infty - \sum_{j=i-3}^{i+2} \left(\text{Residues of } \frac{\hat{\mathcal{A}}_n(w)}{w} \text{ at } \hat{y}_{ij}^2 = 0 \right). \tag{7.23}$$

We begin with a detailed evaluation of the boundary term.

The boundary term $\mathcal{B}_\infty$. Per definition, the boundary contribution is the residue of the pole at infinity,[2]

$$\mathcal{B}_\infty = - \oint_{\mathcal{C}_\infty} \frac{dw}{w} \hat{\mathcal{A}}_n(w). \tag{7.24}$$

Here $\mathcal{C}_\infty$ is a contour that surrounds $w = \infty$ counterclockwise. Since we are more familiar with the shift in momentum space, let us change variables from w to z. With the help of (7.8) we find

$$\mathcal{B}_\infty = \oint_{\mathcal{C}_{z_*}} dz \, \frac{z_*}{z(z - z_*)} \hat{\mathcal{A}}_n(z), \tag{7.25}$$

where $z_* = \frac{\langle i, i+1 \rangle}{\langle i-1, i+1 \rangle}$ is the value of z at $w = \infty$. Now we need to find out how the shifted n-point amplitude behaves for z near z_*. We have already established in (7.19) that the N^KMHV superamplitude factorizes as N^KMHV$_{n-1}\times$anti-MHV$_3$ at $z = z_*$. Let us focus on the **MHV case** ($K = 0$) first to see explicitly how this comes about. Under the $[i-1, i\rangle$-supershift, the Grassmann delta function in the MHV superamplitude is inert, and the only part of the amplitude affected by the shift is the denominator factor $\langle \hat{i}, i + 1 \rangle$. This exactly is the factorization pole for $z \to z_*$. Therefore, near z_* we can write

$$\hat{\mathcal{A}}_n^{\text{MHV}}(z) \xrightarrow{z \to z_*} \hat{\mathcal{A}}_{n-1}^{\text{MHV}}(z_*) \frac{1}{\hat{P}_I^2} \hat{\mathcal{A}}_3^{\text{anti-MHV}}(z_*) = \frac{P_I^2}{\hat{P}_I^2} \left[\hat{\mathcal{A}}_{n-1}^{\text{MHV}}(z_*) \frac{1}{P_I^2} \hat{\mathcal{A}}_3^{\text{anti-MHV}}(z_*) \right]. \tag{7.26}$$

We know from super-BCFWing the MHV superamplitude in Section 4.5.1 that the factor $[\dots]$ in (7.26) equals $\mathcal{A}_n^{\text{MHV}}$ (remember, for MHV only one diagram contributed in the recursion relations based on a BCFW shift of adjacent lines). The prefactor is $\frac{P_I^2}{\hat{P}_I^2} = -z_*/(z - z_*)$. Thus

$$\mathcal{B}_\infty^{\text{MHV}} = -\mathcal{A}_n^{\text{MHV}} \oint_{\mathcal{C}_{z_*}} dz \, \frac{z_*^2}{z(z - z_*)^2} = \mathcal{A}_n^{\text{MHV}}. \tag{7.27}$$

---

[2] We ignore the $2\pi i$ of the Cauchy theorem since all such factors drop out at the end.

In the second equality, we evaluated the double pole integral using

$$\oint dz \, \frac{f(z)}{(z - z_*)^2} = \frac{d}{dz_*} \oint dz \, \frac{f(z)}{(z - z_*)} = f'(z_*) \tag{7.28}$$

for the case $f(z) = 1/z$.

What we have achieved for the MHV case here is a verification of the simple statement that the MHV tree-level superamplitude satisfies the super-BCFW recursion relations, which for MHV include only one term, namely the $\text{MHV}_{n-1} \times \text{anti-MHV}_3$ diagram. We knew that already more than 50 pages ago (Section 4.5.1), but the point is that here we have set up the calculation in a way that facilitates the generalization to $N^K \text{MHV}$ level. And that is what we do next.

$N^K \text{MHV}$ case. Assume inductively that the $(n - 1)$-point tree-level $N^K \text{MHV}$ superamplitude can be written as an MHV prefactor times a function we will call $Y_{n-1}^{(K)}$. We know that $Y_n^{(0)} = 1$ and that $Y_n^{(1)}$ is the sum of 5-brackets given in (7.22). In particular, $Y_n^{(0)}$ and $Y_n^{(1)}$ are dual superconformal invariants. The aim here will be to use BCFW to derive a recursive expression for $Y_n^{(K)}$ in terms of Ys with smaller K and n. This will prove that $Y_n^{(K)}$ is a dual superconformal invariant for all n and k.

The calculation of the contribution from $w = \infty$ follows the same steps as the MHV case, except that the factorization (7.26) is now replaced by

$$\hat{\mathcal{A}}_n^{\text{MHV}}(z) \xrightarrow{z \to z_*} \hat{\mathcal{A}}_{n-1}^{\text{MHV}}(z_*) \, \widehat{Y}_{n-1}^{(K)}(z_*) \, \frac{1}{\hat{P}_I^2} \, \hat{\mathcal{A}}_3^{\text{anti-MHV}}(z_*) \,. \tag{7.29}$$

Let us take a closer look at the Y-factor. It is naturally a function of momentum supertwistors

$$\widehat{Y}_{n-1}^{(K)}(z_*) = \widehat{Y}_{n-1}^{(K)}(\ldots, \mathcal{Z}_{i-1}, \mathcal{Z}_I, \mathcal{Z}_{i+2}, \ldots) \,. \tag{7.30}$$

Now in our analysis of the kinematics, we learned that the momentum twistor $\mathcal{Z}_I$ is characterized as the intersection (7.16) between the line $(j - 1, j)$ and the plane $(i - 1, i, i + 1)$. In our case here, we have $j = i + 2$, so (7.16) says that $\mathcal{Z}_I$ is the point of intersection between the line $(i + 1, i + 2)$ and the plane $(i - 1, i, i + 1)$. Obviously this intersection point is $\mathcal{Z}_{i+1}$, i.e.

$$\text{For } w = \infty \text{ case:} \quad (I) = (i + 1, i + 2) \bigcap (i - 1, i, i + 1) = (i + 1) \,. \tag{7.31}$$

So we can freely substitute $\mathcal{Z}_I \to \mathcal{Z}_{i+1}$ to find

$$\widehat{Y}_{n-1}^{(K)}(z_*) = Y_{n-1}^{(K)}(\ldots, \mathcal{Z}_{i-1}, \mathcal{Z}_{i+1}, \mathcal{Z}_{i+2}, \ldots) \,. \tag{7.32}$$

This factor is independent of z and we can therefore repeat our argument from the MHV case to find

$$\mathcal{B}_\infty^{N^K \text{MHV}} = \oint_{C_\infty} \frac{dw}{w} \hat{\mathcal{A}}_n^{N^K \text{MHV}}(w) = \mathcal{A}_n^{\text{MHV}} \, Y_{n-1}^{(K)}(\mathcal{Z}_1, \ldots, \mathcal{Z}_{i-1}, \mathcal{Z}_{i+1}, \ldots \mathcal{Z}_n) \,. \tag{7.33}$$

This completes the calculation of the boundary term.

Residues at finite w. Now we extract the residues of the finite poles in the w-plane. They arise from propagators $1/\hat{y}_{ij}^2$ going on-shell. Writing the shifted propagator in terms of momentum twistor, we find

$$
\frac{1}{\hat{y}_{ij}^2} = \frac{\langle \hat{i}, i-1 \rangle \langle j, j-1 \rangle}{\langle \hat{i}, i-1, j, j-1 \rangle} = \frac{\langle i, i-1 \rangle \langle j, j-1 \rangle}{\langle i, i-1, j, j-1 \rangle + w \langle i+1, i-1, j, j-1 \rangle}
$$
$$
= \frac{w_*}{y_{ij}^2} \frac{1}{w - w_*}, \tag{7.34}
$$

where $w_* = -\frac{\langle i, i-1, j, j-1 \rangle}{\langle i+1, i-1, j, j-1 \rangle}$. This means that

$$
-\int_{\mathcal{C}(w_*)} \frac{dw}{w} \frac{1}{\hat{y}_{ij}^2} f(w) = -\int_{\mathcal{C}(w_*)} \frac{dw}{w} \frac{w_*}{y_{ij}^2} \frac{1}{w - w_*} f(w) = \frac{1}{y_{ij}^2} f(w_*). \tag{7.35}
$$

So the contribution from the shifted propagator is simply the unshifted propagator, exactly the same as the usual BCFW rules. In a given factorization channel, we can write the left and the right subamplitudes as:

$$
\begin{aligned}
\mathcal{A}_{n_L} &= \mathcal{A}_{n_L}^{\mathrm{MHV}} Y_{n_L}^{(K_L)}(\mathcal{Z}_I, \mathcal{Z}_{j-1}, \dots, \mathcal{Z}_{i+1}, \mathcal{Z}_i), \\
\mathcal{A}_{n_R} &= \mathcal{A}_{n_R}^{\mathrm{MHV}} Y_{n_R}^{(K_R)}(\mathcal{Z}_I, \mathcal{Z}_{i-1}, \dots, \mathcal{Z}_{j+1}, \mathcal{Z}_j).
\end{aligned} \tag{7.36}
$$

Here the Grassmann degrees obey $K_R + K_L + 1 = K$; in particular this is why there were no such diagrams for the MHV case. The contribution of the BCFW channel is simply

$$
\left(\sum_{\mathrm{states}} \frac{\hat{\mathcal{A}}_{n_L}^{\mathrm{MHV}} \hat{\mathcal{A}}_{n_R}^{\mathrm{MHV}}}{P_I^2} \right) \widehat{Y}_{n_L}^{(K_L)}(\mathcal{Z}_I, \mathcal{Z}_{j-1}, \dots, \mathcal{Z}_{i+1}, \hat{\mathcal{Z}}_i) \, \widehat{Y}_{n_R}^{(K_R)}(\mathcal{Z}_I, \mathcal{Z}_{i-1}, \dots, \mathcal{Z}_{j+1}, \mathcal{Z}_j). \tag{7.37}
$$

The $\mathcal{Z}_I$ appearing in the dual superconformal invariants $Y_{\mathrm{L,R}}$ can be written in terms of the external line supermomentum twistors using the characterization of $\mathcal{Z}_I$ as an intersection point (7.16) and the formula (5.66), similarly for $\hat{\mathcal{Z}}_i$, via the intersection rule (7.15).

As a consequence, the state sum – which is an integration over the η_I variables – acts solely on the MHV prefactors. Furthermore, the factor in parenthesis is in (7.37) simply the BCFW term of the NMHV amplitude arising from the P_I factorization channel. We have calculated this in Section 4.5.2, and later learned that in the momentum twistor language the answer is written in terms of the 5-bracket:

$$
\left(\sum_{\mathrm{states}} \frac{\hat{\mathcal{A}}_{n_L}^{\mathrm{MHV}} \hat{\mathcal{A}}_{n_R}^{\mathrm{MHV}}}{P_I^2} \right) = \mathcal{A}_n^{\mathrm{MHV}} [i-1, i, i+1, j-1, j]. \tag{7.38}
$$

Thus we have finally arrived at the BCFW recursion relation for tree-level amplitudes in $\mathcal{N} = 4$ SYM, written in momentum twistor space:

$$
\mathcal{A}_n^{\mathrm{N}^K \mathrm{MHV}} = \mathcal{A}_n^{\mathrm{MHV}} \left\{ Y_{n-1}^{(K)}(\dots, \mathcal{Z}_{i-1}, \mathcal{Z}_{i+1}, \mathcal{Z}_{i+2}, \dots) + \sum_{j=i+3}^{i-2} [i-1, i, i+1, j-1, j] \right.
$$
$$
\left. \times \widehat{Y}_{n_L}^{(K_L)}(\mathcal{Z}_I, \mathcal{Z}_j, \mathcal{Z}_{j+1}, \dots, \mathcal{Z}_{i-1}) \times \widehat{Y}_{n_R}^{(K_R)}(\mathcal{Z}_I, \hat{\mathcal{Z}}_i, \mathcal{Z}_{i+1}, \dots, \mathcal{Z}_{j-1}) \right\}. \tag{7.39}
$$

The above relation corresponds to the shift defined in (7.5), and the momentum twistors $\hat{\mathcal{Z}}_i$ and $\mathcal{Z}_I$ are given by (7.15) and (7.16) respectively. Also, there are implicit sums over $K_R + K_L + 1 = K$ and $n_L + n_R = n + 2$.

▶ **Exercise 7.5**

Show that (7.39) gives the known NMHV expression (7.22) for $K = 1$.

The result (7.39) verifies the claim that all tree-level amplitudes of $\mathcal{N} = 4$ SYM can be written as an MHV prefactor times polynomials of 5-brackets: given that this is true for the NMHV amplitudes, (7.39) ensures that the 5-brackets are recycled into the higher-K results. Since the 5-brackets are manifestly dual superconformal invariant, so are all tree-level superamplitudes of $\mathcal{N} = 4$ SYM.

7.4 Momentum twistor BCFW for planar loop-integrands

To initiate the discussion of BCFW recursion for planar loop-integrands of $\mathcal{N} = 4$ SYM, let us examine a specific example to get some intuition for the good looks and behaviors of integrands. In other words, we start with the answer and let that guide our discussion.

In Section 6.2, we used the generalized unitarity method to construct the 1-loop $\mathcal{N} = 4$ SYM superamplitude. We found

$$\mathcal{A}_4^{\text{1-loop}}[1234] = su\, \mathcal{A}_4^{\text{tree}}[1234]\, I_4(p_1, p_2, p_3, p_4)\,, \tag{7.40}$$

where I_4 is the 1-loop box-integral which we wrote in dual y-space in (7.4) as

$$I_4(p_1, p_2, p_3, p_4) = \int \frac{d^4 y_0}{y_{01}^2\, y_{02}^2\, y_{03}^2\, y_{04}^2}\,. \tag{7.41}$$

Here the propagator-terms $y_{0i}^2 = (y_0 - y_i)^2$ involve the zone-variable y_0 associated with the loop-momentum, as indicated in (7.3).

The expressions (7.40), (7.41) determine the loop-integrand for the 4-point 1-loop $\mathcal{N} = 4$ SYM superamplitude to be (using $-s = y_{13}^2$ and $-u = y_{24}^2$)

$$\mathcal{I}_4^{\text{1-loop}}[1234] = \mathcal{A}_4^{\text{tree}}[1234]\, \frac{y_{13}^2\, y_{24}^2}{y_{01}^2\, y_{02}^2\, y_{03}^2\, y_{04}^2}\,. \tag{7.42}$$

Now, to translate this to momentum twistor space, recall that a point y in dual space maps to a line in momentum twistor space. So let us take y_0 to be mapped to some line (A, B) determined by two points $\mathcal{Z}_A$ and $\mathcal{Z}_B$; then the loop-integral $\int d^4 y_0$ maps to an integral over all inequivalent lines (A, B). There is a story here of how to define the integration measure appropriately – we postpone this until later in this section in order to first discuss the structure of the loop-integrands.

Using (5.52) to rewrite all the dual variables y in the integrand in terms of 4-brackets, in particular $y_{0i}^2 = \frac{\langle A,B,i-1,i\rangle}{\langle AB\rangle\langle i-1,i\rangle}$, we arrive at the expression

$$\mathcal{I}_4^{\text{1-loop}}[1234] = -\mathcal{A}_4^{\text{tree}}[1234]\,\frac{\langle 1234\rangle^2\langle AB\rangle^4}{\langle AB12\rangle\langle AB23\rangle\langle AB34\rangle\langle AB41\rangle}\,. \tag{7.43}$$

Note that all $\langle i-1,i\rangle$s dropped out. The factor $\langle AB\rangle^4$ will eventually be absorbed in the integration measure and what remains is manifestly dual conformal invariant.

The expression (7.43) is an example of what a 1-loop integrand looks like in momentum twistor space. Under a super-BCFW shift $\hat{\mathcal{Z}}_4 = \mathcal{Z}_4 + w\mathcal{Z}_3$, the integrand has a pole that involves the loop-momentum: it comes from $\langle AB\hat{4}1\rangle = 0$. The residue of such a pole is the new input we need for the loop-level recursion relations.

▶ **Exercise 7.6**

What type of super-BCFW shift is induced in momentum space by the momentum twistor shift $\hat{\mathcal{Z}}_4 = \mathcal{Z}_4 + w\mathcal{Z}_3$?

Next, we outline the form of the recursion relations for general L-loop integrands.

Structure of the BCFW recursion for loop-integrands

Without loss of generality, we consider the recursion relations derived from a shift of the nth momentum twistor,

$$\hat{\mathcal{Z}}_n = \mathcal{Z}_n + w\mathcal{Z}_{n-1}\,. \tag{7.44}$$

For an n-point L-loop integrand, there will be three distinct types of contributions to the recursion relations:

1. The boundary contribution from $w \to \infty$. This contribution is calculated just as in the tree-level case of the previous section, so we simply state the result (suppressing the K of the N^KMHV classification)

$$\text{term at } w \to \infty: \quad \mathcal{A}_{n,\text{MHV}}^{\text{tree}}\,Y_{n-1}^L\big(\mathcal{Z}_1,\ldots,\mathcal{Z}_{n-1}\big)\,. \tag{7.45}$$

   Here Y_{n-1}^L is, by the inductive assumption, an L-loop dual superconformal invariant.

2. Residues of factorization channels from propagators that do *not* involve loop-momenta correspond to poles in $\hat{y}_{1j}^2 \propto \langle \hat{n}, 1, j-1, j\rangle = 0$. The results for these also follow the same steps as the tree-level case, and one finds

$$\mathcal{A}_{n,\text{MHV}}^{\text{tree}}\sum_{j=3}^{n-2}\big[j-1,j,n-1,n,1\big]\,Y_{\text{L}}^{L_1}\big(\mathcal{Z}_{I_j},\mathcal{Z}_j,\mathcal{Z}_{j+1},\ldots,\hat{\mathcal{Z}}_{n_j}\big)\,Y_{\text{R}}^{L_2}$$
$$\times\big(\mathcal{Z}_{I_j},\mathcal{Z}_1,\mathcal{Z}_2,\ldots,\mathcal{Z}_{j-1}\big)\,, \tag{7.46}$$

   where $\hat{\mathcal{Z}}_{n_j} = (n-1,n)\bigcap(1,j-1,j)$ and $\mathcal{Z}_{I_j} = (j,j-1)\bigcap(n-1,n,1)$. This includes an implicit sum over loop-orders L_1 and L_2 in the sub-integrands such that $L_1 + L_2 = L$. Also, the N^KMHV level was suppressed, and one must sum over the Grassmann degrees associated with the sub-integrands such that $K_{\text{L}} + K_{\text{R}} = K - 1$.

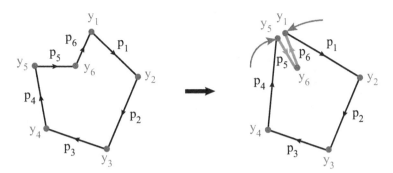

The forward limit illustrated in dual coordinates. The limit corresponds to y_5 and y_1 approaching a point while satisfying $y_{14}^2 = 0$.

Fig. 7.4

3. Residues of factorization channels from propagators that *do* involve loop-momenta; they correspond to

$$\langle A\,B\,\hat{n}\,1 \rangle = 0 \,. \tag{7.47}$$

These are the new forward limit contributions at loop-level, so we will take a closer look at them now.

Forward limit contributions

For an L-loop n-point integrand, the residue of the pole (7.47) is an $(L-1)$-loop $(n+2)$-point integrand whose two extra legs are evaluated in the forward limit (7.1), as shown in Figure 7.1. The following example of $n = 4$ will illustrate the idea of how to do this.

▷ *Example.* Start with a 6-point $(L - 1)$-loop integrand. Translated to dual coordinates, the forward limit of p_5 and p_6 approaching $p_5 = -p_6 = r$ is the limit of taking $y_1 \to y_5$ while y_6 remains fixed. This is illustrated in Figure 7.4. In momentum twistor space, the point y_1 is determined by the line $(1, 6)$ and y_5 by $(4, 5)$, so y_1 and y_5 can be identified only when (Z_1, Z_6, Z_5, Z_4) lie on the same line. It is easy to achieve this configuration if the line $(1, 4)$ intersects line $(6, 5)$, because then we can send Z_5 and Z_6 to the intersection point $(6, 5) \cap (1, 4)$. Note that this does not change y_6, but the result is $y_1 \to y_5$.

However, momentum twistors live in $\mathbb{CP}^3$ where two lines generically do not intersect. So we cannot take the limit as naively as above. Instead, we modify the momentum twistor $Z_4 \to \hat{Z}_4 = Z_4 + w Z_3$, and tune w such that the new line $(1, \hat{4})$ intersects $(5, 6)$: let $Z_{\hat{B}}$ be the point of intersection. Since $\hat{Z}_4$ per construction lies on the line $(3, 4)$, we can characterize $Z_{\hat{B}}$ as the intersection point between the line $(5, 6)$ and the plane $(3, 4, 1)$ (see Figure 7.5):

$$Z_{\hat{B}} = (5, 6) \cap (3, 4, 1) \,. \tag{7.48}$$

Because the lines $(1, \hat{4})$ and $(5, 6)$ are arranged to intersect, it follows that $\hat{Z}_4$ lies in the plane $(5, 6, 1)$, see Figure 7.5. But $\hat{Z}_4$ is also on the line $(3, 4)$, so the shifted

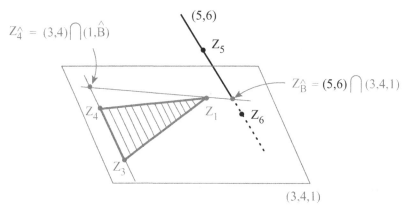

Fig. 7.5 The geometry of the forward limit Figure 7.4 illustrated here in momentum twistor space. The point $Z_{\hat{B}}$ is defined as the intersection of line (Z_5, Z_6) with plane (Z_1, Z_3, Z_4). The BCFW-deformed $\hat{Z}_4$ is fixed at the interaction of line (Z_3, Z_4) and $(Z_{\hat{B}}, Z_1)$. The gray points lie in the same plane.

momentum twistor can be identified in terms of the unshifted lines as

$$\hat{Z}_4 = (3, 4) \bigcap (5, 6, 1). \tag{7.49}$$

The setup with (7.48) and (7.49) allows us to take the forward limit directly by sending Z_5, Z_6 to the intersection point $Z_{\hat{B}}$. We can summarize the deformation and forward limit $p_5 = -p_6 = r$ as

$$(Z_1, Z_2, Z_3, Z_4, Z_5, Z_6) \to (Z_1, Z_2, Z_3, \hat{Z}_4, Z_5, Z_6)\Big|_{Z_5, Z_6 \to Z_{\hat{B}}}. \tag{7.50}$$

It is important to note that $\hat{Z}_4$ satisfies

$$\langle 5, 6, \hat{4}, 1 \rangle = 0. \tag{7.51}$$

Comparing (7.51) with (7.47), we recognize the single-cut condition (or equivalently, momentum dependent pole in BCFW) provided that Z_5 and Z_6 are identified as the loop-momentum twistors Z_A and Z_B. This is also the statement that y_6 has been identified as the Lth loop integration region, as Figure 7.4 indicates that it should be. ◁

Let us return to the general case. The forward limit is taken for $(L - 1)$-loop $(n + 2)$-point integrands by sending the momentum twistors $(\hat{Z}_n, Z_A, Z_B)$ to the forward configuration discussed in the example. One must multiply by an overall MHV factor as well as the result for the cut propagator. The result (which we discuss further below) is

$$\mathcal{A}_{n,\text{MHV}}^{\text{tree}} \times f(A, B, n-1, n, 1) \times \left(Y_{n+2}^{L-1}[\mathcal{Z}_1, \mathcal{Z}_2, \dots, \hat{\mathcal{Z}}_{n_{AB}}, \mathcal{Z}_A, \mathcal{Z}_B]\Big|_{A,B \to \hat{B}} \right), \tag{7.52}$$

where adapting (7.48) and (7.49) to the n-point case with $5 \to A$ and $6 \to B$ identifies

$$\hat{\mathcal{Z}}_{n_{AB}} = (n-1, n) \bigcap (A, B, 1) \quad \text{and} \quad \hat{B} = (A, B) \bigcap (n-1, n, 1). \tag{7.53}$$

Here, $f(A, B, n-1, n, 1)$ represents the kinematic function which includes the cut propagator as well as possible Jacobian factors that arise from solving the single cut constraint, $\langle A, B, \hat{n}, 1 \rangle = 0$. We will determine this function shortly, but first we address one important missing piece: what to do about the Z_A and Z_B momentum twistors and how the extra loop-momenta integral emerges from the forward limit. That is the next step.

The integration measure

The forward limit is taken of a higher point amplitude/integrand, so we need to devise a way to remove the information of the two extra external legs. The most naive proposal is to apply the following integration:

$$\int d^{4|4} Z_A d^{4|4} Z_B .$$ (7.54)

Surprisingly, this is essentially the correct answer! The reason it is correct is rather nontrivial. Let us first consider the bosonic part of the integration. The integration over Z_A and Z_B can be decomposed into two pieces, one is the integration over all possible lines (A, B), and the other is the movement of Z_A and Z_B along a particular line (A, B). To aid this separation, consider the following $GL(2)$ transformation on (Z_A, Z_B),

$$\begin{pmatrix} Z_{A'} \\ Z_{B'} \end{pmatrix} = \begin{pmatrix} c_{A'}{}^A & c_{A'}{}^B \\ c_{B'}{}^A & c_{B'}{}^B \end{pmatrix} \begin{pmatrix} Z_A \\ Z_B \end{pmatrix} .$$ (7.55)

The above 2×2 matrix exactly parameterizes the movement along the line (A, B), because the new pair (A', B') defines the same line as (A, B). In the forward limit, we are sending Z_A and Z_B on a given line to the intersection of (A, B) with the plane $(n-1, n, 1)$, so this limit corresponds to a particular solution for the $GL(2)$ matrix. In light of this, it will be convenient to separate the bosonic integral as

$$\int d^4 Z_A d^4 Z_B = \int \frac{d^4 Z_A d^4 Z_B}{\text{Vol}[GL(2)]} \int_{GL(2)} .$$ (7.56)

The $\text{Vol}[GL(2)]$ in the denominator indicates that as one integrates over the 4×2 dimensional space of Z_A and Z_B, one needs to mod out the 2×2 c-matrix in (7.55) that parameterizes an arbitrary $GL(2)$ transformation. The explicit integration measure for $\int_{GL(2)}$ can be fixed by requiring it to be $SL(2)$ invariant and having $GL(1)$ weight 4 in both A and B.[3] This fixes the form to be

$$\int_{GL(2)} = \int \langle c_{A'} d c_{A'} \rangle \langle c_{B'} d c_{B'} \rangle \langle c_{A'} c_{B'} \rangle^2 ,$$ (7.57)

where $\langle c_{A'} c_{B'} \rangle = c_{A'}{}^A c_{B'}{}^B - c_{A'}{}^B c_{B'}{}^A$. After one has separated out the $GL(2)$ integral, the remaining integration measure can be naturally related to the y_0 measure. To see this, note that after stripping off the $GL(2)$ part, the remaining measure is purely integrating over all distinct lines (A, B). Recall that distinct lines in twistor space define distinct points in dual space, so this tells us that this measure is precisely proportional to $\int d^4 y_0$. The precise

---

[3] This follows from the fact that there are four-components in Z_A and Z_B.

momentum twistor integral that is equivalent to the loop-integral over y_0 is

$$\int d^4 y_0 = \int \frac{d^4 Z_A d^4 Z_B}{\text{Vol}[GL(2)]\langle AB \rangle^4} , \qquad (7.58)$$

where the four extra factors of $\langle AB \rangle$ in the denominator are necessary for the measure to be projective. The angle bracket $\langle AB \rangle$ breaks the $SL(4)$ invariance because it picks only the angle spinor piece of the momentum twistors. This breaks dual conformal invariance – but that is expected because the $d^4 y_0$ inverts non-trivially under dual conformal inversion. From our example (7.43), we see that the $\langle AB \rangle^4$-factor in the measure is exactly canceled by the $\langle AB \rangle^4$-factor that appeared when we rewrote the box-integral in momentum twistor space. This is a general feature which follows from (or, if you prefer, is necessary for) the dual conformal invariance of the loop-integrand. Henceforth, we simply implicitly assume the cancellation of the $\langle AB \rangle^4$s. Let us for later reference write what the 1-loop 4-point superamplitude looks like when dressed in full momentum twistor regalia:

$$\mathcal{A}_4^{\text{1-loop}}[1234] = -\mathcal{A}_4^{\text{tree}}[1234] \int \frac{d^4 Z_A d^4 Z_B}{\text{Vol}[GL(2)]} \frac{\langle 1234 \rangle^2}{\langle AB12 \rangle \langle AB23 \rangle \langle AB34 \rangle \langle AB41 \rangle} . \qquad (7.59)$$

Back to the forward-limit discussion. To integrate over all possible configurations of the forward limit, one should only integrate over all distinct lines (A, B). This requires us to remove the $GL(2)$-part of the integration in (7.54). Thus we have two problems to solve, how to put the higher-point amplitude on the forward limit and how to remove the $GL(2)$ redundancy. Fortunately, we can scare two birds with one stone![4] We begin by simply presenting the resolution: the correct prescription for the computation of the forward limit is

$$\mathcal{A}_{n,\text{MHV}}^{\text{tree}} \int \frac{d^{4|4} \mathcal{Z}_A d^{4|4} \mathcal{Z}_B}{\text{Vol}[GL(2)]} \int_{\text{GL}(2)} [A, B, n-1, n, 1] \times Y_{n+2}^{L-1}[\mathcal{Z}_1, \mathcal{Z}_2, \ldots, \hat{\mathcal{Z}}_{n_{AB}}, \mathcal{Z}_A, \mathcal{Z}_{\hat{B}}],$$
$$(7.60)$$

where $\mathcal{Z}_B$ is sent to the intersection $\hat{B} = (A, B) \bigcap (n-1, n, 1)$. Note the appearance of the factor $[A, B, n-1, n, 1]$. The role this factor plays is two-fold:

- It contains the invariants $\langle A, n-1, n, 1 \rangle$ and $\langle B, n-1, n, 1 \rangle$ in the denominator. The vanishing of these invariants is precisely the forward limit, and therefore these poles can be used to localize the $GL(2)$ integral on to the forward limit. (The two birds fly away.)
- It also contains the factor $1/\langle A, B, n, 1 \rangle$, which is precisely the single-cut propagator.

Hence the $GL(2)$ integration is understood to encircle poles that correspond to the forward limit. One may ask if $[A, B, n-1, n, 1]$ is the unique function that satisfies the above two points. The answer is no, however, it can be easily justified by dual conformal invariance. The recursion better preserves this symmetry. With Y_{n+2}^{L-1} already an invariant, $[A, B, n-1, n, 1]$ is the unique invariant that satisfies the above two properties. Thus, using

---

[4] No need for excessive aggression.

symmetry arguments we did not need to know a priori what the function $f(A, B, n - 1, n, 1)$ in (7.52) should be; it is whatever $[A, B, n - 1, n, 1]$ evaluates to once the $GL(2)$ integral is localized. This is admittedly rather abstract, but we are going to realize the contents of the discussion here explicitly when we compute the 4-point 1-loop amplitude in Section 7.5.

Finally, we need to sum over all $\mathcal{N} = 4$ SYM states that can run in the forward limit loop. In (7.60) this is naturally achieved in a way that preserves the dual superconformal symmetry by simply extending the bosonic momentum twistor integration to include the Grassmann-components, χ_A and χ_B. We are now ready to put everything together.

Result of the BCFW recursion for L-loop integrands

Summarizing the preceding discussion, the loop-level BCFW recursion relation is given by

$$
\begin{aligned}
&A_n^{L\text{-loop}} \\
&= A_{n,\text{MHV}}^{\text{tree}} \bigg\{ Y_{n-1}^L (\mathcal{Z}_1, \ldots, \mathcal{Z}_{n-1}) \\
&\quad + \sum_{j=3}^{n-2} [j-1, j, n-1, n, 1] Y_{\text{L}}^{L_1} (\mathcal{Z}_{I_j}, \mathcal{Z}_j, \mathcal{Z}_{j+1}, \ldots, \hat{\mathcal{Z}}_{n_j}) Y_{\text{R}}^{L_2} (\mathcal{Z}_{I_j}, \mathcal{Z}_1, \mathcal{Z}_2, \ldots, \mathcal{Z}_{j-1}) \\
&\quad + \int \frac{d^{4|4} \mathcal{Z}_A d^{4|4} \mathcal{Z}_B}{\text{Vol}[GL(2)]} \int_{GL(2)} [A, B, n-1, n, 1] \, Y_{n+2}^{L-1} [\mathcal{Z}_1, \mathcal{Z}_2, \ldots, \hat{\mathcal{Z}}_{n_{AB}}, \mathcal{Z}_A, \mathcal{Z}_{\hat{B}}] \bigg\}.
\end{aligned}
\tag{7.61}
$$

In the second line, L_1 and L_2 are summed over subject to $L_1 + L_2 = L$, as are the Grassmann degrees $K_1 + K_2 = K - 1$, and we have

$$
\hat{\mathcal{Z}}_{n_j} = (n-1, n) \bigcap (1, j-1, j), \quad \mathcal{Z}_{I_j} = (j, j-1) \bigcap (n-1, n, 1),
$$
$$
\hat{\mathcal{Z}}_{n_{AB}} = (n-1, n) \bigcap (A, B, 1).
\tag{7.62}
$$

In the last line of (7.61), the Grassmann degree of Y_{n+2}^{L-1} is $4(K+1)$, i.e. one level higher than the amplitude on the LHS of the equation.

Before moving on to an explicit application of the loop-integrand recursion relations, it is important to note that (7.61) provides us with the tool to prove dual conformal properties of loop-integrals, just as we did for the tree-level recursion. Assuming the n-point $L' < L$-loop as well as the $(n+2)$-point $(L-1)$-loop integral is given by an MHV tree amplitude times a dual conformal invariant function, then – through (7.61) – the n-point L-loop amplitude will have the same property.

We will now apply the recursion relations developed in this section to show how the 4-point 1-loop integrand (7.43), can be derived recursively from the recursion relation with the input of just a tree amplitude. Sharpen your pencils and keep your eraser close at hand.

7.5 Example: 4-point 1-loop amplitude from recursion

The 4-point 1-loop amplitude is the simplest example that can illustrate all the novelties of the loop-recursion. Let us examine the potential terms in the recursion formula (7.61):

The *first term* with $Y_3^{L=1}$ is absent. This is because there are no 3-point 1-loop amplitudes. Another way to understand this is that this contribution came from the pole at $w \to \infty$. If we sneak-peak at the answer for the 4-point 1-loop amplitude (7.43), we realize that while the shifted MHV prefactor does have a $w \to \infty$ pole (as we saw and used in Section 7.3), its residue is actually zero for the 1-loop integrand because the $1/\langle A B \hat{4} 1\rangle \to 0$ as $w \to \infty$. In other words, we have a consistent picture for the absence of the first term $Y_3^{L=1}$ in (7.61).

The *second term* in (7.61) is absent, again because $Y_3^{L=1} = 0$. This is consistent with (7.43) not having any momentum-independent poles at finite w.

The *third term* in (7.61) is

$$\mathcal{A}_4^{\text{1-loop}} = \mathcal{A}_{4,\text{MHV}}^{\text{tree}} \int \frac{d^{4|4} \mathcal{Z}_A d^{4|4} \mathcal{Z}_B}{\text{Vol}[GL(2)]} \int_{GL(2)} [A, B, 3, 4, 1] \times Y_6[\mathcal{Z}_1, \mathcal{Z}_2, \mathcal{Z}_3, \hat{\mathcal{Z}}_{4_{AB}}, \mathcal{Z}_A, \mathcal{Z}_{\hat{B}}].$$

$$(7.63)$$

Unfortunately, we now have to evaluate this thing!!

Y_6 is the tree-level NMHV dual conformal invariant for $n = 6$, discussed previously in (5.60):

$$Y_6[\mathcal{Z}_1, \mathcal{Z}_2, \mathcal{Z}_3, \hat{\mathcal{Z}}_{4_{AB}}, \mathcal{Z}_A, \mathcal{Z}_{\hat{B}}] = [\hat{B}, 1, 2, 3, \hat{4}] + [\hat{B}, 1, 2, \hat{4}, A] + [\hat{B}, 2, 3, \hat{4}, A].$$

$$(7.64)$$

The hatted momentum twistors can be found explicitly using the intersection formulas (5.66). Since the twistors are defined projectively, one can freely include a scaling-factor:

$$(\hat{4}) = (3, 4) \bigcap (A, B, 1) \implies \mathcal{Z}_{\hat{4}} = \frac{1}{\langle 3AB1\rangle} \left(\mathcal{Z}_4 \langle 3AB1\rangle - \mathcal{Z}_3 \langle 4AB1\rangle \right), \quad (7.65)$$

$$(\hat{B}) = (A, B) \bigcap (3, 4, 1) \implies \mathcal{Z}_{\hat{B}} = \frac{1}{\langle A341\rangle} \left(-\mathcal{Z}_A \langle B341\rangle + \mathcal{Z}_B \langle A341\rangle \right). \quad (7.66)$$

For convenience, we pick overall factors such that the "hatted" twistors have the same projective weights as the un-hatted ones. Note that some 4-brackets remain unshifted: $\langle 3, \hat{4}, ., .\rangle = \langle 3, 4, ., .\rangle$ and $\langle A, \hat{B}, ., .\rangle = \langle A, B, ., .\rangle$.

Let us first do the fermionic integrals in $d^{4|4} \mathcal{Z}_A d^{4|4} \mathcal{Z}_B$, i.e. $d^4 \chi_A d^4 \chi_B$. We have to saturate the Grassmann integrals, so it is only relevant to look at the χ_A- and χ_B-terms in the Grassmann delta functions. Begin with the 5-bracket $[A, B, 3, 4, 1]$ that multiplies each of the three terms in Y_6. Its Grassmann delta function involves

$$[A, B, 3, 4, 1] \propto \delta^{(4)}\left(\chi_A \langle B341\rangle - \chi_B \langle A341\rangle + \cdots \right). \quad (7.67)$$

It follows from (7.66) that $\chi_{\hat{B}} \propto \chi_A \langle B341\rangle - \chi_B \langle A341\rangle$, so any appearance of $\chi_{\hat{B}}$ in the three 5-brackets in (7.64) vanishes on the support of the $\delta^{(4)}$ in (7.67) under the $\int d^4 \chi_A d^4 \chi_B$-integral. In particular, the only contribution from χ_A, χ_B in $[\hat{B}, 1, 2, 3, \hat{4}]$ is through $\chi_{\hat{B}}$, so we immediately conclude that

$$\int d^4\chi_A d^4\chi_B \,[A, B, 3, 4, 1] \times [\hat{B}, 1, 2, 3, \hat{4}] = 0 \,. \qquad (7.68)$$

In the next case, $[\hat{B}, 1, 2, \hat{4}, A]$, we have $\delta^{(4)}(\chi_{\hat{B}}\langle 12\hat{4}A\rangle + \chi_A\langle \hat{B}12\hat{4}\rangle + \cdots)$. As before the $\chi_{\hat{B}}$-term can be dropped. Moreover, one can show that $\langle \hat{B}12\hat{4}\rangle$ vanishes (see Exercise 7.7 below), so we conclude

$$\int d^4\chi_A d^4\chi_B \,[A, B, 3, 4, 1] \times [\hat{B}, 1, 2, \hat{4}, A] = 0 \,. \qquad (7.69)$$

▶ **Exercise 7.7**

The 3-term Schouten identity (2.48) for angle and square spinors is the statement that three vectors in a plane are linearly dependent. As 4-component objects, the momentum twistors Z^I, $I = (\dot{a}, a)$, similarly satisfy a 5-term Schouten identity,

$$\langle i, j, k, l\rangle Z_m + \langle j, k, l, m\rangle Z_i + \langle k, l, m, i\rangle Z_j + \langle l, m, i, j\rangle Z_k + \langle m, i, j, k\rangle Z_l = 0 \,. \qquad (7.70)$$

Use (7.70) to derive the two identities

$$\langle \hat{B}12\hat{4}\rangle = 0 \,, \qquad \langle 234\hat{B}\rangle = -\frac{\langle 1234\rangle\langle 34AB\rangle}{\langle A341\rangle} \,. \qquad (7.71)$$

Hint: $\langle AB\hat{4}1\rangle = 0$; this identity follows from (7.65).

With the help of the second identity in (7.71), one finds that the result of integrating the $\delta^{(4)}$s in $[A, B, 3, 4, 1] \times [\hat{B}, 2, 3, \hat{4}, A]$ gives $\langle 1234\rangle^4\langle 34AB\rangle^4$.

In conclusion, after Grassmann integration, only the third 5-bracket in (7.64) contributes. After some simplifications one finds

$$\int d^4\chi_A d^4\chi_B \,[A, B, 3, 4, 1] \times [\hat{B}, 2, 3, \hat{4}, A] = \frac{\langle 1234\rangle\langle AB34\rangle}{\langle A234\rangle\langle B341\rangle} \times I_4(A, B) \,, \qquad (7.72)$$

where

$$I_4(A, B) = \frac{\langle 1234\rangle^2}{\langle AB12\rangle\langle AB23\rangle\langle AB34\rangle\langle AB41\rangle} \,. \qquad (7.73)$$

▶ **Exercise 7.8**

Derive (7.72).

$I_4(A, B)$ is the answer we expect, cf. (7.59), so now we have to massage its prefactor in (7.72). The recursion relations (7.63) instruct us to perform the forward limit by evaluating the $GL(2)$ integral:

$$\mathcal{A}_4^{\text{1-loop}} = \mathcal{A}_{4,\text{MHV}}^{\text{tree}} \int \frac{d^4 Z_A d^4 Z_B}{\text{Vol}[GL(2)]} \int_{GL(2)} I_4(A, B) \times \frac{\langle 1234\rangle\langle AB34\rangle}{\langle A234\rangle\langle B341\rangle} \,. \qquad (7.74)$$

This can be done by first making a $GL(2)$ rotation (7.55) of Z_A, Z_B and then integrating over the $GL(2)$ parameters. Since the integral is $GL(1)$ invariant, we can fix the scale in the $GL(2)$ matrix and set $c_{A'}{}^A = c_{B'}{}^B = 1$. With this "gauge fixing," we have

$$\begin{pmatrix} Z_A \\ Z_B \end{pmatrix} \rightarrow \begin{pmatrix} 1 & c_{A'} \\ c_{B'} & 1 \end{pmatrix} \begin{pmatrix} Z_A \\ Z_B \end{pmatrix}. \tag{7.75}$$

The result of this transformation on the 4-brackets is

$$\langle ABij \rangle \rightarrow \langle ABij \rangle \langle c_{A'} c_{B'} \rangle, \qquad \begin{aligned} \langle Aijk \rangle &\rightarrow \langle Aijk \rangle + c_{A'} \langle Bijk \rangle, \\ \langle Bijk \rangle &\rightarrow c_{B'} \langle Aijk \rangle + \langle Bijk \rangle, \end{aligned} \tag{7.76}$$

and the gauge fixing means that $\langle c_{A'} c_{B'} \rangle = 1 - c_{A'} c_{B'}$. Also, $\langle c_{A'} dc_{A'} \rangle = dc_{A'}$ and $\langle c_{B'} dc_{B'} \rangle = dc_{B'}$. So including the appropriate $GL(2)$ measure (7.57), we have

$$\mathcal{A}_4^{1\text{-loop}} = \mathcal{A}_{4,\text{MHV}}^{\text{tree}} \int \frac{d^4 Z_A d^4 Z_B}{\text{Vol}[GL(2)]} \, I_4(A, B) \, \langle 1234 \rangle \langle AB34 \rangle$$

$$\times \int \frac{dc_{A'} \, dc_{B'}}{\left(1 - c_{A'} c_{B'}\right) \left(\langle A234 \rangle + c_{A'} \langle B234 \rangle\right) \left(c_{B'} \langle A341 \rangle + \langle B341 \rangle\right)}. \tag{7.77}$$

Now the plan all along was that the $GL(2)$ integration was supposed to localize us on the forward limit. So consider the denominator factor $(c_{B'} \langle A341 \rangle + \langle B341 \rangle)$. The vanishing of this expression is the statement that $\hat{Z}_B = Z_B + c_{B'} Z_A$ is sent to the intersection point of the line (A, B) and the plane $(3, 4, 1)$: but this is exactly part of the forward limit $\hat{Z}_B \rightarrow Z_{\hat{B}} = (A, B) \bigcap (3, 4, 1)$. So to realize this, we take the contour (7.77) in the $c_{B'}$-plane to surround the pole $c_{B'*} = -\langle B341 \rangle / \langle A341 \rangle$. Now we also want to send A to the intersection point $\hat{B}$, but the integral (7.77) appears not to have a pole that achieves this. However, when we evaluate the $c_{B'}$-integral to localize $B \rightarrow \hat{B}$, the factor $(1 - c_{A'} c_{B'})$ actually develops the desired pole, namely $c_{A'*} = -\langle A341 \rangle / \langle B341 \rangle$. Just do it:

$$\int_{\mathcal{C}(c_{A'*})} dc_{A'} \int_{\mathcal{C}(c_{B'*})} dc_{B'} \frac{1}{\left(1 - c_{A'} c_{B'}\right) \left(\langle A234 \rangle + c_{A'} \langle B234 \rangle\right) \left(c_{B'} \langle A341 \rangle + \langle B341 \rangle\right)}$$

$$= \int_{\mathcal{C}(c_{A'*})} \frac{dc_{A'}}{\left(\langle A341 \rangle + c_{A'} \langle B341 \rangle\right) \left(\langle A234 \rangle + c_{A'} \langle B234 \rangle\right)}$$

$$= \frac{1}{\langle A234 \rangle \langle B341 \rangle + \langle A341 \rangle \langle B234 \rangle}$$

$$= -\frac{1}{\langle AB34 \rangle \langle 1234 \rangle}. \tag{7.78}$$

In the last line we used the 5-term Schouten identity (7.70). Plugging this result into (7.77), the factors $\langle AB34 \rangle \langle 1234 \rangle$ cancel and we are left with

$$\mathcal{A}_4^{1\text{-loop}} = -\mathcal{A}_{4,\text{MHV}}^{\text{tree}} \int \frac{d^4 Z_A d^4 Z_B}{\text{Vol}[GL(2)]} \, I_4(A, B). \tag{7.79}$$

This is the correct result, as we discussed in Section 7.4.

Note that in this derivation, the $GL(2)$ integral ended up localizing the integrand on the forward limit, where the poles that were used in this localization were given by the extra $[A, B, 3, 4, 1]$. This precisely realizes the idea we described around (7.60). Now you might be a little concerned that we could have chosen to localize on "non-forward" poles in the $GL(2)$ integral instead, but the answer would have been the same, as guaranteed by the large-$c_{A', B'}$ falloff of the integrand.

Finally, you may find it discouraging that it takes much more work and sophistication to calculate even the simplest of all $\mathcal{N} = 4$ SYM amplitudes with BCFW than it did with the generalized unitarity method, as we showed in Section 6.2. However, while it is not directly practical, it is morally encouraging – and perhaps even fascinating – that all information about the 4-point MHV 1-loop amplitude is encoded already in the 6-point NMHV tree-level amplitude. This is a realization of an interesting connection between amplitudes with different numbers of particles n, different N^KMHV levels, and different loop-orders L.

7.6 Higher loops

The planar loop-integrand recursion relations studied above can also be directly applied to higher-loop order in the planar limit of $\mathcal{N} = 4$ SYM. This was already shown in (7.61). However, in contrast to the unitarity approach, the integrands obtained from recursion generally contain spurious poles. Local poles (non-spurious) refer to propagator-like poles, these take the form of $1/\langle i, i-1, j, j-1 \rangle$ or $1/\langle A, B, i-1, i \rangle$ in momentum twistor space. Spurious poles, on the other hand, could take the form $1/\langle A, B, 4, 2 \rangle$; this is non-local in that it does not arise from a propagator in the loop-diagrams. At 4-point, since there is only one term in the recursion, such spurious poles must vanish by themselves, and indeed the final result is free of spurious poles. However, at higher-points the spurious poles cancel between various terms in the recursion relations, but their presence in the individual BCFW terms makes it difficult to carry out the integration of the loop-integrand to obtain the actual amplitude. Spurious poles are a hallmark of BCFW recursion relations – we discussed this for tree-level BCFW at the end of Section 3.2. BCFW builds in unitarity and gauge-invariance at the expense of manifest locality. While it provides us with a method to compute loop-integrands, it leads to complications as one eventually has to integrate these non-local functions in momentum space.

Given the large amount of symmetry enjoyed by $\mathcal{N} = 4$ SYM – superconformal symmetry and dual conformal symmetry as well as their enhancement to the Yangian – it is tempting to be ambitious and ask if it is possible to manifest *both* locality and dual conformal invariance at the same time. Certainly the unitary method discussed previously would suffice for this purpose, since the dual conformal invariant scalar integrals are local. However, when extended beyond 4-point, the number of dual conformal invariant integrals becomes large and not all of them may contribute to a given amplitude.

At 1-loop level, the 1-loop box integral in (7.59) is the only available dual conformal invariant local 4-point integral. In other words, dual conformal symmetry forces the triangle and bubble-contributions to be absent in $\mathcal{N} = 4$ SYM. Could it be that all amplitudes $\mathcal{N} = 4$ SYM are fixed by similar considerations? To study this involves *maximal cuts* and *Leading Singularities* – and some of the principles involved also extend beyond the planar limit and to SYM with less supersymmetry. We introduce the basic setup in the following chapter.

Leading Singularities and on-shell diagrams   **8**

Chapter 6 introduced unitarity cuts. In D-dimensions, we can at most cut D propagators per loop since each loop-momentum only has D components. When the maximum number of propagators, $D \times L$, is cut, the unitarity cut is called a ***maximal cut*** [117, 118]. The maximal cuts are very useful for determining the integrand, in particular in 4d planar $\mathcal{N} = 4$ SYM.

$\triangleright$ *Example.* As an example of a maximal cut, consider the quadruple cut of the 1-loop n-point amplitude in 4d:

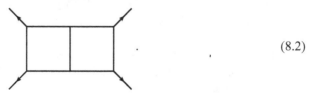

Four propagators are put on-shell:

$$\ell^2 = \left(\ell - K_1\right)^2 = \left(\ell - K_2\right)^2 = \left(\ell - K_3\right)^2 = 0 . \tag{8.1}$$

Here $K_1 = p_1 + \cdots + p_i$, $K_2 = p_{i+1} + \cdots + p_j$, $K_3 = p_{j+1} + \cdots + p_k$, and $K_4 = p_{k+1} + \cdots + p_n$. No further propagators can be put on-shell in 4d since the loop-momentum only has four components. $\triangleleft$

The result of a maximal cut is a product of on-shell tree amplitudes, $\mathcal{A}_{n_1}^{\text{tree}} \cdots \mathcal{A}_{n_j}^{\text{tree}}$, appropriately summed over all possible intermediate states, with the loop-momenta evaluated on the solutions to the cut constraints. For example, for the 1-loop box in 4d $\mathcal{N} = 4$ SYM, the value of the maximal cut (8.1) is $\int d^4 \eta_I \, \mathcal{A}_{n_1}^{\text{tree}} \mathcal{A}_{n_2}^{\text{tree}} \mathcal{A}_{n_3}^{\text{tree}} \mathcal{A}_{n_4}^{\text{tree}}$ evaluated on the two solutions to the quadratic loop-momentum constraints. At $L > 1$, the cut constraints generically have 2^L distinct solutions, however, there are situations where there are not enough propagators to cut; a simple example in 4d is the following 2-loop double-box integral

$$\tag{8.2}$$

It has only 7 propagators, but we need to take $2 \times 4 = 8$ propagators on-shell for a maximal cut. For such cases, the solution space for the loop-momenta is not a set of isolated points but rather a continuous multi-dimensional manifold. However, if we impose the cut constraints on one loop-momentum at a time, new poles appear and they can be used to fix the remaining degrees of freedom, again leaving us with a set of isolated solutions for the loop-momenta. We demonstrate this explicitly for the double-box (8.2) in Section 8.2.

The method of generalized unitarity is to find an integrand that reproduces all the unitarity cuts, including of course all the maximal cuts. But how exactly do we treat the distinct solutions to the maximal cut constraints? There are two ways to proceed:[1]

1. *Appropriate sampling over all solutions.* In one approach, we require that the correct integrand matches the maximal cut evaluated on a sampling of all 2^L solutions, with each solution given an appropriate weight. At 1-loop there are just two solutions and the proper weight is $1/2$ for both, thus in effect averaging over the two solutions, as in (6.31) and (6.32). For higher-loops, one starts with a set of integrals that integrate to zero. The appropriate weight for each solution is determined by the requirement that their contributions to the vanishing integrals need to sum to zero. Explicit examples and further discussions at 2-loop order can be found in [96].

2. *Match each solution.* In the other approach, we require the integrand to reproduce *each* of the cut solutions individually. The individual cut solutions are treated as independent entities and the resulting value for the maximal cut evaluated on each solution is called a ***Leading Singularity (LS)***. The name reflects that these objects are the residues of the most singular configuration of the loop-integrand (for generic external kinematics). Note that, despite the name, these contributions are finite.

We focus here on the second approach. A major motivation is that all planar loop-integrands of $\mathcal{N} = 4$ SYM can be written as a linear combination of dual conformal invariant "unit Leading Singularity integrands" (to be introduced below) [119]. The characterization of Leading Singularities turns out to be a quite interesting mathematical problem [120]. The Leading Singularities offer insight into the structure of planar $\mathcal{N} = 4$ SYM loop-integrands at all loop orders, but note that to obtain the actual amplitudes, one still needs to perform the challenging loop-integrations. While the notion of dual conformal invariance is only applicable in the planar limit, it is a well-defined (and open) question whether the full non-planar loop-integrands of $\mathcal{N} = 4$ SYM can also be determined by the Leading Singularities.

We begin our study of the Leading Singularities at 1-loop order. All amplitudes in this chapter are in 4d planar $\mathcal{N} = 4$ SYM.

8.1 1-loop Leading Singularities

To build intuition for the Leading Singularities at 1-loop order, we start with the simplest case, namely the 4-point integrand, then consider the new features at 5-point, and finally generalize to n-point.

---

[1] The maximal cut was formulated in [118] for 1-loop amplitudes of $\mathcal{N} = 4$ SYM and generalized to multi-loop amplitudes in [117]. A more detailed review is offered in [68].

4-point

For $n = 4$, the maximal cut conditions (8.1) are simply

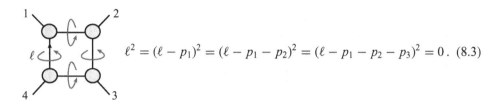

$$\ell^2 = (\ell - p_1)^2 = (\ell - p_1 - p_2)^2 = (\ell - p_1 - p_2 - p_3)^2 = 0 . \quad (8.3)$$

In dual variables, this is simply $y_{01}^2 = y_{02}^2 = y_{03}^2 = y_{04}^2 = 0$ (see Section 7.1). And translating that to momentum twistor space (as in the early part of Section 7.4), we have

$$\langle AB12 \rangle = \langle AB23 \rangle = \langle AB34 \rangle = \langle AB41 \rangle = 0 . \quad (8.4)$$

The geometric interpretation of the cut constraints (8.4) is that (A, B) is a line in $\mathbb{CP}^3$ that intersects each of the four lines $(1, 2)$, $(2, 3)$, $(3, 4)$, and $(4, 1)$. As anticipated from the quadratic nature of the constraint (8.3), there are two independent solutions. This is rather obvious geometrically:

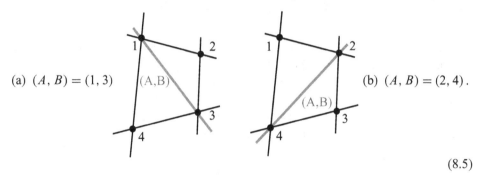

(a) $(A, B) = (1, 3)$

(b) $(A, B) = (2, 4)$.

$$(8.5)$$

Now, on the quadruple cut, the 1-loop integrand factorizes into a product of four tree-level 3-point amplitudes, which can be either MHV or anti-MHV. Recall that special kinematics, (2.58) and (2.59), apply to the 3-point amplitudes – we summarize it here:

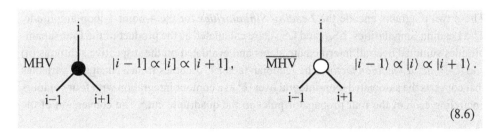

MHV $\quad |i-1] \propto |i] \propto |i+1]$, $\qquad \overline{\text{MHV}} \quad |i-1\rangle \propto |i\rangle \propto |i+1\rangle .$

$$(8.6)$$

We use a black blob to indicate an MHV 3-point subamplitude (or vertex), and a white blob for 3-point $\overline{\text{MHV}} = $ anti-MHV.

A configuration where two MHV subamplitudes with external legs are adjacent vanishes:

$$= 0 \, . \tag{8.7}$$

This is because special kinematics (8.6) implies $|1\rangle \propto |2\rangle$, so that $s_{12} = -(p_1 + p_2)^2 = \langle 12 \rangle [12] = 0$. Of course this does not hold true for generic momenta p_1 and p_2. Hence, for generic external momenta, we are not allowed to have helicity configurations, such as (8.7), where two MHV or two anti-MHV subamplitudes with external legs are adjacent.

The only helicity options for the 4-point quadruple cut are therefore

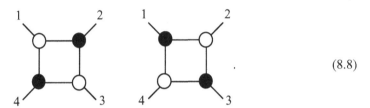

$$\tag{8.8}$$

Now these two diagrams have to be evaluated on the kinematic solutions (8.5). For the solution (a) where the line (A, B) is $(1, 3)$, we can simply pick the momentum twistor of the loop line ℓ to be $Z_1 = (|1\rangle, [\mu_1|)$. This means that $|\ell\rangle \propto |1\rangle$ and that selects the kinematics where the vertex that line 1 attaches to is anti-MHV, i.e. this picks the first helicity configuration in (8.8). Likewise, solution (b) with $(A, B) = (2, 4)$ selects the opposite helicity configuration. So we conclude that the maximal cuts have the two solutions:

(a) $(A, B) = (1, 3)$ (b) $(A, B) = (2, 4) \, .$ (8.9)

These two diagrams encode the **_Leading Singularities_** for the 4-point 1-loop amplitude. The Leading Singularities, $LS_{(a)}$ and $LS_{(b)}$, are calculated as the product of the four subamplitudes summed over all intermediate states and evaluated on the respective solutions (a) and (b), *times a Jacobian factor*. The Jacobian factor $1/J$ comes from a change of variables that converts the associated loop-integral over $\mathbb{R}^4$ to a contour integration with four contours encircling each of the four propagator poles in the quadruple cut.[2] The conversion of the

---

[2] Since the solutions to the cut conditions may be complex-valued, we should really consider the loop-integral as an integral over the real section in $\mathbb{C}^4$; that makes it more natural to convert to a contour integral encircling the poles corresponding to the on-shell propagators.

integral can be done via a change of variables $u_i = y_{0i}^2$, for $i = 1, 2, 3, 4$, giving

$$\int \frac{d^4 y_0}{y_{01}^2 y_{02}^2 y_{03}^2 y_{04}^2} = \int \frac{du_1}{u_1} \frac{du_2}{u_2} \frac{du_3}{u_3} \frac{du_4}{u_4} J , \qquad (8.10)$$

where $J = \det(\partial y_0^\mu / \partial u_i)$ is the Jacobian. As we show explicitly in the example below, the Jacobian is

$$J = \frac{1}{y_{13}^2 y_{24}^2} = -\frac{\langle 12 \rangle \langle 23 \rangle \langle 34 \rangle \langle 41 \rangle}{\langle 1234 \rangle^2} . \qquad (8.11)$$

The Leading Singularity for the 4-point 1-loop amplitude is then

$$\mathrm{LS}_{(a)} = J \int \left[\prod_{i=1}^4 d^4 \eta_{\ell_i} \right] \left(\mathcal{A}_3^{\overline{\mathrm{MHV}}}(-\ell_4, p_1, \ell_1) \, \mathcal{A}_3^{\mathrm{MHV}}(-\ell_1, p_2, \ell_2) \right.$$

$$\left. \times \, \mathcal{A}_3^{\overline{\mathrm{MHV}}}(-\ell_2, p_3, \ell_3) \, \mathcal{A}_3^{\mathrm{MHV}}(-\ell_3, p_4, \ell_4) \right) \Bigg|_{\ell = \ell^{(a)}} . \qquad (8.12)$$

A similar expression is found for $\mathrm{LS}_{(b)}$. Evaluating the $\mathrm{LS}_{(a)}$ and $\mathrm{LS}_{(b)}$, one finds

$$\mathrm{LS}_{(a)} = \mathrm{LS}_{(b)} = \mathcal{A}_4^{\mathrm{tree}} . \qquad (8.13)$$

▶ Exercise 8.1

Evaluate the RHS of (8.12) to show that $\mathrm{LS}_{(a)} = \mathcal{A}_4^{\mathrm{tree}}$.

Now before exploring the Leading Singularities further, let us illustrate how the Jacobian is obtained. It can of course be calculated brute-force, but in the example below we carry out the calculation via a tour to momentum twistor:

▷ *Example.* We calculate J in (8.10) via the momentum twistor formulation. From (7.40)–(7.43), we read off

$$\int \frac{d^4 y_0}{y_{01}^2 y_{02}^2 y_{03}^2 y_{04}^2} = \int \frac{d^4 Z_A d^4 Z_B}{\mathrm{vol}(GL(2))} \frac{\langle 12 \rangle \langle 23 \rangle \langle 34 \rangle \langle 41 \rangle}{\langle AB12 \rangle \langle AB23 \rangle \langle AB34 \rangle \langle AB41 \rangle} . \qquad (8.14)$$

The loop momentum twistors Z_A and Z_B can be expanded on the basis of four external line momentum twistors as

$$Z_A = a_1 Z_1 + a_2 Z_2 + a_3 Z_3 + a_4 Z_4 , \qquad Z_B = b_1 Z_1 + b_2 Z_2 + b_3 Z_3 + b_4 Z_4 . \qquad (8.15)$$

This linear transformation gives $d^4 Z_A \, d^4 Z_B = \langle 1234 \rangle^2 \, d^4 a_i d^4 b_i$. The 4-brackets evaluate to $\langle A, B, i-1, i \rangle = \langle 1234 \rangle M_{i+1}$, where M_j is the jth minor of the 2×4 matrix

$$\begin{pmatrix} a_1 & a_2 & a_3 & a_4 \\ b_1 & b_2 & b_3 & b_4 \end{pmatrix} . \qquad (8.16)$$

For example, $\langle AB34 \rangle = \langle 1234 \rangle M_1 = \langle 1234 \rangle (a_1 b_2 - a_2 b_1)$. Now, consider a $GL(2)$-transformation (7.55) of Z_A, Z_B. We can use it to set $a_4 = b_2 = 0$ and $a_2 = b_4 = 1$.

▶ **Exercise 8.2**

Convince yourself that a $GL(2)$ rotation of Z_A, Z_B allows you to make the above choice of parameters, but that setting $a_4 = b_4 = 0$ would be illegal.

In this gauge, we have $\langle AB12 \rangle = \langle 1234 \rangle\, a_3$, etc., and the integrand then has no dependence (obviously) on a_4, b_2, a_2, b_4. This means that the $GL(2)$-volume factor cancels and we are then left with

$$\int \frac{d^4 y_0}{y_{01}^2 y_{02}^2 y_{03}^2 y_{04}^2} = \int \frac{da_1}{a_1} \frac{db_1}{b_1} \frac{da_3}{a_3} \frac{db_3}{b_3} \frac{\langle 12 \rangle \langle 23 \rangle \langle 34 \rangle \langle 41 \rangle}{-\langle 1234 \rangle^2}. \tag{8.17}$$

This way we have brought the loop-integral to the form on the LHS of (8.10) and we see that the Jacobian is indeed (8.11).

Now our integration variables a_i and b_i in (8.17) are not exactly the $u_i = y_{0i}^2$ that we introduced above (8.10): for example $u_1 = y_{01}^2 = \frac{\langle 41AB \rangle}{\langle 41 \rangle \langle AB \rangle} = -\frac{\langle 1234 \rangle}{\langle 41 \rangle \langle AB \rangle} b_3$. So the u_is are proportional to the $a_{1,3}$ and $b_{1,3}$, but the factors of proportionality drop out of the du_i/u_i measure. ◁

Recall that we are studying the Leading Singularities in order to find an integrand that faithfully reproduces both $LS_{(a)}$ and $LS_{(b)}$. The integrand that we already know for the 4-point 1-loop amplitude does the job – let us see how. We have previously found (see (6.39)) that

$$\mathcal{A}_4^{\text{1-loop}} = \mathcal{A}_4^{\text{tree}}\, y_{13}^2 y_{24}^2 \int \frac{d^4 y_0}{y_{01}^2 y_{02}^2 y_{03}^2 y_{04}^2}. \tag{8.18}$$

When we convert this to the contour integral, the prefactor $y_{13}^2 y_{24}^2$ exactly cancels the Jacobian (8.11), so we are left with

$$\mathcal{A}_4^{\text{1-loop}} = \mathcal{A}_4^{\text{tree}} \int \frac{da_1}{a_1} \frac{db_1}{b_1} \frac{da_3}{a_3} \frac{db_3}{b_3}. \tag{8.19}$$

From this, we can directly read off the residue at the propagator poles $a_1 = b_1 = a_3 = b_3 = 0$. The result is independent of which of the two solutions (a) or (b) we use to localize the loop-integral, so the quadruple cuts of the integral match exactly with the Leading Singularities, $LS_{(a)} = LS_{(b)} = \mathcal{A}_4^{\text{tree}}$. This may not shock you, but once we venture beyond 4-point amplitudes, things are not so simple.

The result $LS_{(a)} = LS_{(b)}$ is special for the 4-point case. It can be represented diagrammatically as

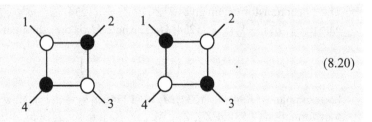

$$\tag{8.20}$$

This identity is called the **square move** and it will show up later in our discussions of higher-loop Leading Singularities and on-shell diagrams.

5-point

At this stage, we have constructed the 4-point 1-loop amplitude of $\mathcal{N} = 4$ SYM in three different ways: generalized unitarity, loop-level BCFW, and Leading Singularities. It is time to move ahead.

We consider a specific maximal cut of the 5-point 1-loop amplitude:

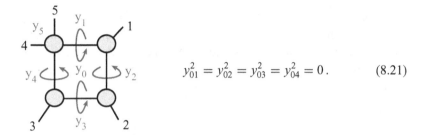

$$y_{01}^2 = y_{02}^2 = y_{03}^2 = y_{04}^2 = 0. \qquad (8.21)$$

The cut constraints for this maximal cut can be written in momentum twistor space as

$$\langle AB12 \rangle = \langle AB23 \rangle = \langle AB34 \rangle = \langle AB51 \rangle = 0. \qquad (8.22)$$

This says that (A, B) is a line that intersects the four lines $(1, 2)$, $(2, 3)$, $(3, 4)$, and $(5, 1)$. There are two solutions:

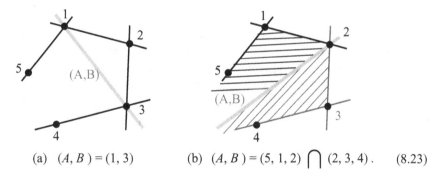

$$\text{(a)} \quad (A, B) = (1, 3) \qquad\qquad \text{(b)} \quad (A, B) = (5, 1, 2) \bigcap (2, 3, 4). \qquad (8.23)$$

It is straightforward to see that (a) is a solution to (8.22). As for (b), note that generically two planes in $\mathbb{CP}^3$ intersect in a line. Any points A and B on the intersection of the two planes in (b) will be linearly dependent with any two points in either plane. This establishes that (b) is a solution to (8.22).

Since intersections of planes in $\mathbb{CP}^3$ may not feel as natural to you as brushing your teeth (hopefully), let us make the solutions (8.23) explicit in momentum space. With $\ell = y_{10}$, the constraints are $\ell^2 = (\ell - p_1)^2 = (\ell - p_1 - p_2)^2 = (\ell - p_1 - p_2 - p_3)^2 = 0$ and it is not hard to verify that the two solutions for the loop-momentum ℓ can be written

$$\ell^{(1)} = -|1\rangle \left([1| + \frac{\langle 23 \rangle}{\langle 13 \rangle} [2| \right), \qquad \ell^{(2)} = -\left(|1\rangle + \frac{[23]}{[13]} |2\rangle \right) [1|. \qquad (8.24)$$

Note that even though $\ell^{(1)}$ is formally the complex conjugate of $\ell^{(2)}$, their geometric interpretations in momentum twistor space are quite different. This is because momentum twistors are chiral objects (only $|i\rangle$ appears, not $|i]$).

▶ **Exercise 8.3**

Show that $\ell^{(1)}$ and $\ell^{(2)}$ in (8.24) solve the cut constraints. Check little group scaling. Then show that two solutions, $\ell^{(1)}$ and $\ell^{(2)}$, correspond to the two geometric solutions (*a*) and (*b*) of (8.23), respectively, in momentum twistor space.

The solution $\ell^{(1)}$ has $|\ell\rangle \propto |1\rangle$ and by momentum conservation these are also proportional to the angle spinor of $(\ell - p_1)$. This means that the special 3-point kinematics forces the vertex where line 1 attaches to be anti-MHV. Likewise, the solution $\ell^{(2)}$ forces the same vertex to be anti-MHV. By (8.7), the rest of the helicity structure is then fixed, and we see that the two solutions (a) and (b) correspond to the two options

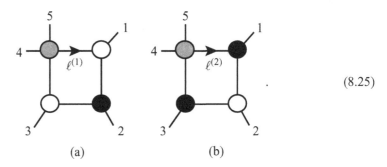

$$(8.25)$$

(a)   (b)

The gray blob for the 4-point vertex does not have specific helicity designation because 4-point on-shell amplitude is simultaneously MHV and anti-MHV.

Now, let us count the number of Grassmann ηs of these blob-diagrams. MHV has 8 ηs and 3-point anti-MHV has 4 ηs. For each of the four internal lines we have to do a $d^4\eta$-integral. For diagram (a), this then gives $8 + 4 + 4 + 8 - 4 \times 4 = 8$ corresponding to the 5-point MHV sector. For diagram (b): $8 + 8 + 8 + 4 - 4 \times 4 = 12$ which identifies it as belonging to the NMHV sector. Including the appropriate Jacobians, the results for the two diagrams (8.25) are the respective MHV and NMHV 5-point tree-level amplitudes, respectively.

The above discussion tells us that for a given MHV or NMHV 1-loop 5-point amplitude, only one of these solutions to the maximal cut conditions (8.21) is relevant. If we focus on the MHV sector, only diagram (a) in (8.25) matters, and it equals $\mathcal{A}_5^{\text{tree}}$ for solution $\ell^{(1)}$ and is zero when evaluated on solution $\ell^{(2)}$. This leads us to the crux of the problem we mentioned in the beginning of this chapter: whether the integral basis Ansatz we write for the integrand faithfully reproduces all Leading Singularities. Let us illustrate this explicitly. Consider the scalar box-integral whose propagators are those considered in the maximal cut (8.21):

$$I_{5,\text{box}}(1, 2, 3, 4) = \frac{\langle 5123\rangle\langle 1234\rangle}{\langle AB51\rangle\langle AB12\rangle\langle AB23\rangle\langle AB34\rangle}. \qquad (8.26)$$

We use the labels on the n-point box-integral $I_{n,\text{box}}(i, j, k, l)$ to specify the first external leg on each of the vertices. For example, for the arrangement in (8.1), the corresponding scalar box-integral would be labeled $I_n(1, i + 1, j + 1, k + 1)$.

When evaluating the quadruple cut for the integral $I_{5,\text{box}}(1, 2, 3, 4)$, the Jacobian cancels the numerator factor $\langle 5123 \rangle \langle 1234 \rangle$, and since there is no other dependence on the loop-momenta than the four propagators we are cutting, this integral produces the same answer, namely 1, no matter if we evaluate it on solution $\ell^{(1)}$ or $\ell^{(2)}$: i.e.

$$I_{5,\text{box}}\big|_{(1)} = I_{5,\text{box}}\big|_{(2)} = 1 \,. \tag{8.27}$$

On the other hand, we now know that for the MHV sector the corresponding Leading Singularities of diagram (a) in (8.25) are

$$\text{LS}_{(1)} = \mathcal{A}_{5,\text{MHV}}^{\text{tree}} \,, \qquad \text{LS}_{(2)} = 0 \,. \tag{8.28}$$

This means that the Ansatz

$$\mathcal{A}_{5,\text{MHV}}^{\text{1-loop}} = \mathcal{A}_{5,\text{MHV}}^{\text{tree}} \times \Big(I_{5,\text{box}}(1, 2, 3, 4) + \text{other box-integrals} \Big) \tag{8.29}$$

does *not* produce the Leading Singularities faithfully. However, it does produce the average of the two maximal cuts correctly because, when multiplied by a factor of $\mathcal{A}_{5,\text{MHV}}^{\text{tree}}$,

$$\frac{1}{2}\Big(I_{5,\text{box}}\big|_{(1)} + I_{5,\text{box}}\big|_{(2)} \Big) = \frac{1}{2}(1 + 1) = 1 \tag{8.30}$$

equals the sum of the Leading Singularities $\text{LS}_{(1)} + \text{LS}_{(2)}$. The message is that the integrand Ansatz (8.29) can produce the correct maximal cut when one averages over the two constraints (as is usually done in applications of the generalized unitarity method), but it does not produce each Leading Singularity honestly. If you just want an answer for the amplitude, you don't have to care. But let us try to be caring people and see where it takes us.

We have learned now that we need something else in the Ansatz (8.29) in order to match the Leading Singularities. That something else turns out to be the ***pentagon integral***

$$I_{5,\text{pentagon}} = \frac{\langle A, B|(1, 2, 3) \bigcap (3, 4, 5)\rangle \langle 2451 \rangle}{\langle AB12 \rangle \langle AB23 \rangle \langle AB34 \rangle \langle AB45 \rangle \langle AB51 \rangle} \,. \tag{8.31}$$

The numerator includes the bi-twistor $(1, 2, 3) \bigcap (3, 4, 5)$ that characterizes the line of intersection between the planes $(1, 2, 3)$ and $(3, 4, 5)$; the intersection formula was given in (5.67). When we evaluate the maximal cut (8.21) of the pentagon, the residue depends on the loop-momentum and hence on which solution (8.24) we choose to evaluate it. Including the Jacobian $J = \langle 5123 \rangle \langle 1234 \rangle$, one finds

$$I_{5,\text{pentagon}}\big|_{(a)} = 0, \qquad I_{5,\text{pentagon}}\big|_{(b)} = -1 \,. \tag{8.32}$$

This is good news, because now the improved Ansatz

$$\mathcal{A}_{5,\text{MHV}}^{\text{1-loop}} = \mathcal{A}_{5,\text{MHV}}^{\text{tree}} \times \Big(I_{5,\text{box}}(1, 2, 3, 4) + I_{5,\text{pentagon}} + \text{other} \Big) \tag{8.33}$$

(where "other" is assumed not to contribute to our quadruple cut) has the following maximal cut (8.21): on two solutions, $\ell^{(1)}$ and $\ell^{(2)}$, it gives

$$
\begin{aligned}
\mathcal{A}_{5,\text{MHV}}^{\text{tree}}\left(I_{5,\text{box}}\big|_{(1)} + I_{5,\text{pentagon}}\big|_{(1)}\right) &= \mathcal{A}_{5,\text{MHV}}^{\text{tree}} \times (1+0) = \mathcal{A}_{5,\text{MHV}}^{\text{tree}} = \text{LS}_{(1)}\,, \\
\mathcal{A}_{5,\text{MHV}}^{\text{tree}}\left(I_{5,\text{box}}\big|_{(2)} + I_{5,\text{pentagon}}\big|_{(2)}\right) &= \mathcal{A}_{5,\text{MHV}}^{\text{tree}} \times (1-1) = 0 = \text{LS}_{(2)}\,.
\end{aligned}
\tag{8.34}
$$

So it produces the correct Leading Singularities for the cut (8.21)!

Now, unfortunately we are not done yet, because we have to worry about all the other cuts: there are a total of $2 \times 5 = 10$ Leading Singularities for the 5-point 1-loop amplitude. With just the box diagram and the pentagon diagram in (8.33), there is no chance that this can be the full answer: the reason is simply that the sum of those two integrals is not cyclically invariant. It takes just one more integral to achieve cyclic invariance, namely the box integral $I_{5,\text{box}}(3, 4, 5, 1)$. Diagrammatically, we can express the final answer as

$$
\mathcal{A}_{5,\text{MHV}}^{\text{1-loop}} = \mathcal{A}_{5,\text{MHV}}^{\text{tree}} \times \left(I_{5,\text{box}}(1, 2, 3, 4) + I_{5,\text{pentagon}} + I_{5,\text{box}}(3, 4, 5, 1) \right)
$$

$$
= \mathcal{A}_{5,\text{MHV}}^{\text{tree}} \left(\cdots \right). \tag{8.35}
$$

The diagrammatic notation for the pentagon integral has a wavy line indicating that the bi-twistor $(1, 2, 3) \cap (3, 4, 5)$ goes in the numerator in (8.31). Our previous results plus cyclic invariance then guarantee that (8.35) produces all ten Leading Singularities correctly.

▶ **Exercise 8.4**

Show that (8.32) is true.

▶ **Exercise 8.5**

Show that (8.35) is invariant under cyclic permutations of the external lines.

We have introduced here box and pentagon 1-loop integrands, (8.26) and (8.31), whose quadruple cuts evaluate to either $+1$, -1 or 0. Such integrands are called ***unit Leading Singularity integrands***.

It has been shown [119] that all planar loop amplitudes of $\mathcal{N} = 4$ SYM can be obtained as a linear combination of unit Leading Singularity integrands (times a tree amplitude). The coefficients in front of the unit Leading Singularity integrands are determined by the Leading Singularity, thus knowing them is sufficient to determine the entire amplitude. For planar amplitudes, we need unit Leading Singularity integrands that are also dual conformal invariant and local, and this is a rather restrictive class of integrands. We have seen the Leading Singularity method at work for 4- and 5-point 1-loop amplitudes. The structure generalizes to higher points, as we now outline.

6-point and beyond

The 1-loop n-point MHV amplitude is given by a simple generalization of the 5-point result:

$$
A_{n,\text{MHV}}^{\text{1-loop}} = A_{n,\text{MHV}}^{\text{tree}} \left(\sum_{1<i<j<n} \vcenter{\hbox{}} \right)
$$

$$
= \int_{A,B} \sum_{1<i<j<n} \frac{\langle A,B|(i-1,i,i+1)\cap(j-1,j,j+1)\rangle \langle i,j,n,1\rangle}{\langle A,Bi,i-1\rangle\langle AB,i,i+1\rangle\langle A,B,j-1,j\rangle\langle A,Bj,j+1\rangle\langle A,B,n,1\rangle}.
$$

(8.36)

In the sum, there are two boundary cases: $i = 2$, $j = 3$ and $i = n - 2$, $j = n - 1$. These correspond to box integrals whose numerators are simply the Jacobian coming from cutting all four propagators. More precisely, we have

$$
\vcenter{\hbox{}} = \int_{A,B} \frac{\langle n123\rangle\langle 1234\rangle}{\langle AB12\rangle\langle AB23\rangle\langle AB34\rangle\langle ABn1\rangle}
$$

(8.37)

and

$$
\vcenter{\hbox{}} = \int_{A,B} \frac{\langle n-3, n-2, n-1, n\rangle\langle n-2, n-1, n, 1\rangle}{\langle A,B,n-3,n-2\rangle\langle A,B,n-2,n-1\rangle\langle A,B,n-1,n\rangle\langle A,B,n,1\rangle}.
$$

(8.38)

In conclusion, the two Leading Singularities of arbitrary 1-loop MHV amplitudes can be reproduced by including the simple combination of tensorial (due to the loop-momentum dependence in the numerator) pentagon integrals. These are all local unit Leading Singularity integrands. Note that these integrands can be used as part of the basis for 1-loop amplitudes in any massless quantum field theory. The special situation for $\mathcal{N} = 4$ SYM is that these integrals provide the entire answer, whereas for a generic QFT, one needs in addition the various lower-gon integrals that are not captured by the maximal cuts.

You may (and should) be puzzled that in the beginning of Section 6.2, we stated that the 1-loop amplitudes in a unitary 4d quantum field theory can be expanded on a basis of scalar box-, triangle-, and bubble-integrals with possible additional input from rational terms. This was summarized in equation (6.26), and we noted that in $\mathcal{N} = 4$ SYM, the only non-vanishing contributions were the box-integrals. There were no pentagons in that story! So what is the deal? The point of the pentagon integrals in the present section is that they allow us to write the $\mathcal{N} = 4$ SYM 1-loop integrand in a form that reproduces each Leading Singularity faithfully. On the other hand, (6.26) determines the 1-loop $\mathcal{N} = 4$ SYM amplitudes as a sum of box-integrals whose coefficients are evaluated by quadruple cuts, evaluated as the *average* of the two loop-constraint solutions. More box-diagrams contribute in (6.26) than in (8.36). So what *is* the deal? Well, the two representations of the integrand must yield the same answer for the amplitude. The integrals have to be regulated, and if one uses dimensional regularization $4 - 2\epsilon$, the difference between the two representations is only in the $O(\epsilon)$-terms. Specifically, the pentagon integrals contain the information about the "missing" boxes plus $O(\epsilon)$ [121]. Thus the two procedures yield the same integrated answer.

8.2 2-loop Leading Singularities

At the start of this chapter we noted that not all loop-diagrams appear to have enough propagators available for a maximal cut of $4L$-lines. A representative example is the double-box diagram of the 2-loop 4-point amplitude:

$$(8.39)$$

With seven propagators, we can only localize seven of the eight components of the two loop-momenta, leaving behind a 1-dimensional loop-integral. However, when the seven propagators are on-shell, the 4-point 1-loop analysis tells us that the lefthand box in (8.39) is a Leading Singularity that equals the 4-point tree amplitude, $A_4^{\text{tree}}[\ell_2, p_1, p_2, \ell_2 - p_1 - p_2]$, where ℓ_2 parameterizes the loop-momentum in the righthand box. But this tree amplitude has a propagator $1/(\ell_2 - p_1)^2$ that can now be used to localize the final component of the loop-momenta, thus providing a maximal cut. Moreover, on this pole, the 4-point tree amplitude factorizes into two 3-point amplitudes, and therefore the result is simply an on-shell 4-point 1-loop box diagram. The procedure is illustrated here:

$$(8.40)$$

The last step uses the fact that the 1-loop 4-point Leading Singularity is equal to the tree amplitude $A_4^{\text{tree}}[1234]$. In conclusion:

(i) the Leading Singularities are well-defined for the 2-loop 4-point amplitude, even though the double-box only has seven loop propagators, and

(ii) the 4-point "double-box Leading Singularity" equals the 4-point tree amplitude.

A Leading Singularity that involves a "pole-under-a-pole" is called a **composite Leading Singularity**. A prototype of such a composite object is the 3-variable contour integral

$$\oint dx\,dy\,dz\,\frac{1}{x(x+yz)}. \tag{8.41}$$

This integrand appears to have only two poles, insufficient to localize the 3d integral. However, if the x-contour encircles the pole at $x = 0$, then an additional pole emerges in the form $1/(yz)$ and this can then localize the remaining two integrals, giving the residue 1 (ignoring $2\pi i$s).

The idea of composite Leading Singularities resolves the subtlety about defining maximal cuts and Leading Singularities for higher-loop amplitudes. Henceforth we work with the understanding that the Leading Singularities of multi-loop amplitudes are always well-defined.

8.3 On-shell diagrams

We have found in Section 8.1 that the 4-point 1-loop Leading Singularity is equal to the 4-point tree amplitude:

$$\tag{8.42}$$

This looks rather peculiar since the LHS is a 1-loop diagram while the RHS is a tree amplitude. It actually turns out that the LHS can be interpreted as a super-BCFW diagram! We now show how.

Consider the top two vertices in the Leading Singularity diagram

$$|a\rangle \propto |c\rangle \propto |1\rangle,$$
$$|b] \propto |c] \propto |2]. \tag{8.43}$$

The MHV and anti-MHV designations imply the indicated special 3-particle kinematics. Up to an overall factor z, this determines p_c to be $p_c = -z|1\rangle[2|$. Momentum conservation then fixes p_a and p_b to be

$$p_a = -|1\rangle([1| + z[2|) \quad \text{and} \quad p_b = -(|2\rangle - z|1\rangle)[2| \,. \tag{8.44}$$

We recognize p_a and p_b as BCFW $[1, 2\rangle$-shifted momenta $\hat{p}_1$ and $\hat{p}_2$!

What about the Grassmann variables? Let us carry out the η_c-integral in the product of the Grassmann delta functions of the first two vertices,

$$\int d^4\eta_a \, d^4\eta_b \, d^4\eta_c \, \delta^{(4)}\big([1c]\eta_a + [ca]\eta_1 + [a1]\eta_c\big) \, \delta^{(8)}\big(|2\rangle\eta_2 - |b\rangle\eta_b - |c\rangle\eta_c\big)$$

$$\propto \int d^4\eta_a \, d^4\eta_b \, \delta^{(8)}\Big(|1\rangle\big(\eta_a - (\eta_1 - z\eta_b)\big) - |2\rangle\big(\eta_b - \eta_2\big)\Big) \,. \tag{8.45}$$

The last integral localizes η_a and η_b to be

$$\eta_a = \eta_1 - z\eta_2 \quad \text{and} \quad \eta_b = \eta_2 \,. \tag{8.46}$$

This is exactly the shift of the Grassmann variables associated with the supersymmetrization of the BCFW shift (8.44).

Finally, let us see how the internal line d in (8.43) fixes z. The on-shell condition is $0 = p_d^2 = (p_3 + p_b)^2 = (\langle 23\rangle - z\langle 13\rangle)[23]$, i.e. $z = \langle 23\rangle/\langle 13\rangle$. This corresponds exactly to the pole where the propagator $1/\hat{p}_{23}$ in the $[1, 2\rangle$-shifted 4-point tree amplitude goes on-shell. As we know from the super-BCFW calculation (4.56), this is exactly the factorization pole that allows us to determine the full 4-point tree amplitude in $\mathcal{N} = 4$ SYM from the $\text{MHV}_3 \times$ anti-MHV_3 super-BCFW diagram.

We have established the connection between the Leading Singularity diagram on the LHS of (8.42) and the 4-point super-BCFW diagram (4.56), and this allows us to understand why the 4-point 1-loop Leading Singularity is just the 4-point tree superamplitude. The connection is summarized diagrammatically as the ***BCFW bridge***

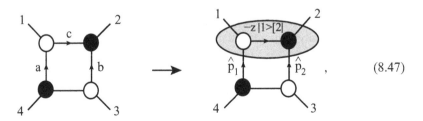

$$\tag{8.47}$$

where the upper two vertices, surrounded by the shaded region, are the "bridge." The bridge provides the BCFW super-shift.

Exchanging black and white dots in the BCFW bridge simply corresponds to the conjugate BCFW shift. This also gives another meaning to the *square move* (8.20):

$$\tag{8.48}$$

It says that the two BCFW super-shifts $[1, 2\rangle$ and $[2, 1\rangle$ give the same 4-point amplitude.

With the square move and the BCFW shift, it becomes fun to calculate Leading Singularities. Starting from the fundamental 3-point vertices, we can build *on-shell diagrams* that contain information about the higher-loop amplitudes. Each 3-point vertex represents the MHV or anti-MHV 3-point amplitude, along with the implication that the square or angle spinors of its legs are proportional. The vertices are glued together by "on-shell propagators" whose rules can be written

$$\frac{\quad I \quad}{} = \int \frac{d^2|I\rangle d^2|I]d^4\eta_I}{U(1)}. \tag{8.49}$$

The η_I integral is the usual state sum. The integration over the momentum variables will be localized by the momentum conservation delta function on both sides of the propagator.

In addition to the square move (8.48), there are two rules that help us simplify complicated on-shell diagrams. The first rule follows from the observation that each MHV 3-vertex imposes that the square spinors of the associated lines are proportional, so two consecutive MHV vertices imply that all four square spinors are proportional. This gives the rule

$$\tag{8.50}$$

There is of course an equivalent rule for anti-MHV. The black 4-vertex blob imposes, per definition, that the four lines have proportional square spinors. This blob does not represent a 4-point MHV tree amplitude; it is just a short-hand notation for the double-blob diagrams.

The second rule is

$$\tag{8.51}$$

The 3-particle kinematics forces the internal lines in the bubble to be collinear, and this collapses the bubble. This formally eliminates a loop-integral.

Combining the two rules (8.50) and (8.51) gives

$$
\tag{8.52}
$$

▶ Exercise 8.6

Show that the internal lines in the bubble (8.51) are collinear and that (8.52) follows from (8.50) and (8.51).

The point of these rules is to simplify the evaluation of on-shell diagrams. For the on-shell diagram of the 2-loop 4-point Leading Singularity, we first apply the square move (8.48) and then the collapse-moves (8.50) and (8.52) to get

$$
\tag{8.53}
$$

This shows that the 4-point 2-loop Leading Singularity equals the 4-point 1-loop Leading Singularity, which in turn is just the 4-point tree amplitude. We had already found this result in Section 8.2 by evaluating the composite Leading Singularity. The rules for the on-shell diagrams offer a simpler diagrammatical derivation.

▶ Exercise 8.7

Write down 3-loop on-shell diagrams for the 4-point MHV amplitude and show that they reduce to the 1-loop result.

Using the BCFW bridge, we can begin to build up more complicated on-shell diagrams. For example, we can use the BCFW bridge to interpret the on-shell diagram:

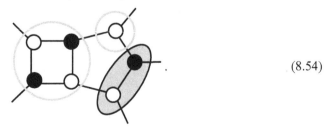

$$
\tag{8.54}
$$

The circles highlight the 4-point MHV tree amplitude and 3-point anti-MHV tree subamplitudes. The BCFW bridge, indicated with the shaded area, induces the BCFW shift on the two affected lines, so this on-shell diagram represents the BCFW diagram for the 5-point MHV tree-level superamplitude.

The super-BCFW recursion relations for the 6-point NMHV tree amplitude can be represented with on-shell diagrams as

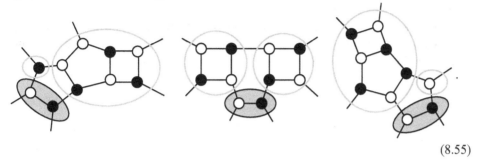

$$(8.55)$$

The left-hand diagram is the $MHV_3 \times$ anti-MHV_5 BCFW diagram, the middle diagram is the BCFW diagram with two 4-point MHV tree subamplitudes and the right-hand diagram is the $MHV_5 \times$ anti-MHV_3 BCFW diagram.

▶ **Exercise 8.8**

Do the η-counting to show that (8.54) represents an on-shell diagram for an MHV amplitude and (8.55) an NMHV amplitude.

▶ **Exercise 8.9**

Interpret the effect of the rule (8.50) on the on-shell diagram (8.54). Show that the 3-loop diagram

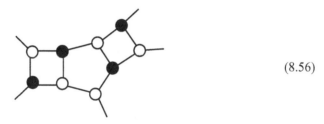

$$(8.56)$$

is equivalent to (8.54). What does that tell you about Leading Singularities?

The concept of the Leading Singularity is well-defined for lower-$\mathcal{N}$ SYM and for non-supersymmetric theories, and so are the on-shell diagrams. The only distinction is that the edges need arrows because the on-shell states split into two CPT conjugate multiplets, one with the positive helicity gluon and the other with the negative helicity gluon. ($\mathcal{N} = 4$ SYM is special in that its supermultiplet is CPT self-conjugate.) Opposite helicity multiplets must sit at different ends of the propagators, so they are dressed with arrows to indicate the assignment.

In the next section, we discuss a formula that reproduces all Leading Singularities in planar $\mathcal{N} = 4$ SYM. This formula will tell us that the number of distinct Leading Singularities for given n and given N^KMHV-level is fixed. For example, for $K = 0$ (MHV) there is only one Leading Singularity: that is why both the 1- and 2-loop Leading Singularities for the 4-point amplitude evaluate to the same value, namely the 4-point tree amplitude. That this pattern continues is rather remarkable, since it says that on-shell diagrams with four external lines when evaluated all give the 4-point tree amplitude, no matter how many hundreds of loops we add in.

The problem of determining and classifying all distinct on-shell diagrams, under the equivalence-moves, turns out to be an interesting mathematical problem that has become an exciting research topic [120].

PART III

TOPICS

The on-shell BCFW recursion formulas (7.39) and (7.61) have taught us that tree superamplitudes and loop-integrands of *planar* $\mathcal{N} = 4$ SYM can be written as

$$\mathcal{A}_n^{L\text{-loop}} = \mathcal{A}_{n,\text{MHV}}^{\text{tree}} \times Y_n^{L\text{-loop}}, \qquad (9.1)$$

where $Y_n^{L\text{-loop}}$ is a dual conformal invariant. In the N^KMHV sector, Y_n^{tree} is a sum of K products of 5-brackets (or R-invariants in Section 4.5), as found in Sections 5.3 and 5.4. At loop-level, $Y_n^{L\text{-loop}}$ is a linear combination of dual conformal invariant integrands.

The MHV tree amplitude prefactor in (9.1) serves an important purpose: since it is dual conformal *covariant* with homogeneous dual conformal inversion-weight of the external particles – as in (5.39) – it generates the necessary dual conformal "anomaly" that modifies the dual conformal generators in such a way that they become part of the level 1 generators of a Yangian symmetry. Hence, as described in Section 4.5, the MHV factor in (9.1) is essential for Yangian symmetry.

There are several interesting points to consider:

1. The color-ordered planar $\mathcal{N} = 4$ SYM superamplitudes have **cyclic symmetry** in the labels of the external states. However, in the dual conformal BCFW representations (7.39) and (7.61), the cyclic symmetry is completely obscured. This is not surprising, because these recursion formulas are based on shifts of two adjacent external lines: making two lines special breaks the cyclic symmetry. In the pursuit of happiness and manifest symmetries, we may ask if there is a formalism for the planar $\mathcal{N} = 4$ SYM superamplitudes in which both the (dual) conformal symmetry *and* the cyclic symmetry are manifest? This suggestion will guide us in Section 9.1.

2. In Section 8.3, we gave examples of how individual Leading Singularity diagrams can be understood as BCFW terms of tree amplitudes. Since each BCFW term is Yangian invariant, as indicated in the recursion formula (7.39), this implies that the Leading Singularities are also Yangian invariant. Thus, understanding the ***most general Yangian invariants*** is a step towards gaining control of the planar superamplitudes in $\mathcal{N} = 4$ SYM at *any* loop order. Of course, one still needs to understand how to put the Yangian invariants together to obtain a given superamplitude; read on.

3. Super-BCFW recursion for $\mathcal{N} = 4$ SYM amplitudes can be based on any choice of two shifted external momenta. Different choices can give very different representations of the same amplitude, in particular with distinct spurious poles. For the amplitude to be *local*, i.e. free of spurious poles, the residues of the spurious poles must cancel in the sum of BCFW diagrams. Thus, the ***equivalence of two different BCFW representations*** is intimately related to ***locality***. Each BCFW diagram is Yangian invariant, so by

understanding how to enforce locality in a Yangian invariant way, it turns out that the equivalence between the different BCFW representations can be trivialized.

Interestingly, the above three points can be addressed jointly. The strategy is to find a way to generate the most general Yangian invariant rational function from a formula in which cyclic symmetry is manifest. That's our job now, so let's get to work.

9.1 Yangian invariance and cyclic symmetry

The level 0 Yangian generators are the superconformal generators studied in Section 5.1. In Section 5.2, we introduced the supertwistors $\mathcal{W}_i^A = ([i|^a, |\tilde{\mu}_i\rangle^{\dot{a}}, \eta_{iA})$ in order to linearize the action of the superconformal generators. The Grassmann components of the supertwistor are simply the on-shell superspace coordinates η_{iA}, and $|\tilde{\mu}_i\rangle^{\dot{a}}$ is the Fourier conjugate coordinate of $|i\rangle^{\dot{a}}$. A function f of on-shell momentum space spinor-helicity variables is Fourier transformed to the (super)twistor space as

$$\int \left[\prod_{i=1}^n d^2|i\rangle \right] f([i|, |i\rangle, \eta_i) \, e^{i \sum_{j=1}^n \langle j\tilde{\mu}_j\rangle} \equiv \tilde{f}(\mathcal{W}_i^A). \qquad (9.2)$$

In supertwistor space, the level-0 superconformal generators are

$$G^A_{\ B} = \sum_{i=1}^n G_i^A_{\ B} = \sum_{i=1}^n \mathcal{W}_i^A \frac{\partial}{\partial \mathcal{W}_i^B}, \qquad (9.3)$$

and the level-1 generators can be written in bi-local form as (Section 5.3)

$$\sum_{i<j}^n (-1)^{|C|} \left[G_i^A_{\ C} \, G_j^C_{\ B} - (i \leftrightarrow j) \right]. \qquad (9.4)$$

Our aim is a cyclic invariant formula that generates Yangian invariant rational functions; these are the building blocks for N^KMHV superamplitudes of planar $\mathcal{N} = 4$ SYM. Let us try to motivate the construction, step by step. To start with, note that the level 0 generators (9.3) act on the supertwistor variables as $SL(2, 2|4)$ linear transformations. Any delta function $\delta^{4|4}$ whose argument is a linear combination of the supertwistors is invariant under the linear $SL(2, 2|4)$ transformation: for example

$$\delta^{4|4}\left(\sum_{i=1}^n C_i \, \mathcal{W}_i^A \right) \equiv \delta^2\left(\sum_{i=1}^n C_i [i|^a \right) \delta^2\left(\sum_{i=1}^n |i\rangle^{\dot{a}} C_i \right) \delta^{(4)}\left(\sum_{i=1}^n C_i \, \eta_{iA} \right), \qquad (9.5)$$

with some arbitrary auxiliary coefficients $C_i \in \mathbb{C}$. For a level 0 generator $G^A_{\ B}$ with $A \neq B$, this is because it transforms the argument of one of the delta functions to that of another delta function; schematically

$$x \frac{\partial}{\partial y} \delta(x)\delta(y) = x \, \delta(x) \, \delta'(y) = 0. \qquad (9.6)$$

For "diagonal" generators, such as $G^1{}_1$, invariance follows only after using the tracelessness condition, i.e. writing the generators in the form analogous to (5.19).

For an $N^K MHV$ superamplitude, we need Yangian invariants that are Grassmann polynomials of degree $4(K+2)$; so it is natural to take

$$k \equiv K + 2 \qquad (9.7)$$

products of (9.5). Note that k counts the number of negative helicity gluons in the pure gluon amplitude. To avoid having k identical delta functions, we introduce k sets of the auxiliary variables C_{ai} labeled by a new index $a = 1, 2, \ldots, k$. So now we have a Grassmann degree $4(K+2)$ object

$$\prod_{a=1}^{k} \delta^{4|4}\left(\sum_{i=1}^{n} C_{ai} \mathcal{W}_i^A\right) \qquad (9.8)$$

that is $SL(2, 2|4)$ invariant. The parameters C_{ai} are sometimes called **link variables** [23].

The $n \times k$ parameters C_{ai} are arbitrary, so to remove the dependence on them let us integrate (9.8) over all C_{ai}. This has the further benefit of making the integrated result cyclically invariant: a permutation of the $\mathcal{W}_i$s is compensated by a permutation of the C_{ai}s. This is a change of integration variable with unit Jacobian. However, it is not clear what measure we should use when integrating over the C_{ai}s. So let us allow for a general cyclically invariant function $f(C)$ and write our candidate "generating function" as

$$\int d^{k \times n} C \; f(C) \prod_{a=1}^{k} \delta^{4|4}\left(\sum_{i=1}^{n} C_{ai} \mathcal{W}_i^A\right) . \qquad (9.9)$$

The integral of the $k \times n$ complex parameters is intended to be carried out as a contour integral. The choice of contour is a very important and physically relevant aspect that will be discussed in Section 9.3.

When the level 1 generators are considered, it turns out that there is a unique choice of $f(C)$ such that (9.9) is Yangian invariant. We will not repeat the argument here, but refer you to [122, 123]. The unique function that gives (9.9) full Yangian symmetry is

$$f(C) = \frac{1}{M_1 M_2 \cdots M_n} , \qquad (9.10)$$

where M_i is the ith ordered minor of the $k \times n$ matrix C_{ai}: this is the determinant of the $k \times k$ submatrix whose first column is the ith column of C_{ai}, specifically

$$M_i \equiv \epsilon^{a_1 a_2 \cdots a_k} C_{a_1 i} C_{a_2, i+1} \cdots C_{a_k, i+k-1} , \qquad (9.11)$$

with $i = 1, 2, \ldots, n$. One goes around cyclically when reaching the end of the C-matrix. For example, the $n = 5$ and $k = 2$ matrix

$$C = \begin{pmatrix} C_{11} & C_{12} & C_{13} & C_{14} & C_{15} \\ C_{21} & C_{22} & C_{23} & C_{24} & C_{25} \end{pmatrix} \qquad (9.12)$$

gives $M_1 = C_{11} C_{22} - C_{12} C_{21}$ and $M_5 = C_{15} C_{21} - C_{11} C_{25}$.

At this stage, we have learned that the integral

$$\int \frac{d^{k \times n} C}{M_1 M_2 \cdots M_n} \prod_{a=1}^{k} \delta^{4|4} \left(\sum_{l=1}^{n} C_{al} \mathcal{W}_l^A \right) \tag{9.13}$$

is Yangian invariant and has cyclic symmetry.

Before declaring victory, there are some loose ends to tie. *First*, a minor issue (yes, that's a pun) is that if M_i contains columns that are not strictly increasing due to cyclicity (for example $M_{n-1} = \cdots C_{an} C_{a1} \cdots$) then the proof of Yangian invariance goes through only on the support of the bosonic delta functions in (9.13). *Second*, a major issue is that the integral we so proudly wrote down in (9.13) is not at all well-defined – it is divergent. To see this, note that the product of delta functions is invariant under a $GL(k)$ rotation of the k a-indices. The minors only respect $SL(k)$ transformations: $GL(1)$ takes $C_{ai} \to t C_{ai}$, hence $M_i \to t^k M_i$, but this excess weight is canceled by the Jacobian of $d^{k \times n} C$. Thus the integral has $GL(k)$ symmetry. To define a proper integral we need to "gauge fix" the $GL(k)$ redundancy. We indicate the need to gauge fix $GL(k)$ by writing

$$\tilde{\mathcal{L}}_{n,k}(\mathcal{W}_i) = \int \frac{d^{n \times k} C_{ai}}{GL(k) \prod_{j=1}^{n} M_j} \prod_{a=1}^{k} \delta^{4|4} \left(\sum_{l=1}^{n} C_{al} \mathcal{W}_l^A \right). \tag{9.14}$$

It turns out [122, 123] that for given n and k, $\mathcal{L}_{n,k}$ *is the unique cyclically invariant integral-expression that generates all Yangian invariants!* We are going to give examples in the following sections. The formula (9.14) was first introduced by Arkani-Hamed, Cachazo, Cheung, and Kaplan [23], who at the time conjectured that it produces all Leading Singularities of planar $\mathcal{N} = 4$ SYM. A similar integral formula was presented by Mason and Skinner [60] based on momentum supertwistors $\mathcal{Z}$, as opposed to the "regular" supertwistors $\mathcal{W}$, thus interchanging the role of the ordinary superconformal and dual superconformal symmetries. The twistor and momentum twistor versions of the Grassmannian integral are directly related (Arkani-Hamed *et al.* [60]). For a streamlined proof and more recent review of the Grassmannian formulation, see Elvang *et al.* [60].

We used supertwistors $\mathcal{W}_i^A = ([i|^a, |\tilde{\mu}_i\rangle^{\dot{a}}|\, \eta_{iA})$ to emphasize superconformal and Yangian symmetry in the construction above. However, since we are more familiar with scattering amplitudes in momentum space ($[i|^a, |i\rangle^{\dot{a}}|\, \eta_{iA}$), we are going to inverse-Fourier transform all $|\tilde{\mu}_i\rangle$ in (9.14) back to $|i\rangle$. This is conveniently done in a gauge-fixing of the $GL(k)$ symmetry where the first $k \times k$ block of C_{ai} is the unit matrix. For example for $n = 7$ and $k = 3$, we have

$$C = \begin{pmatrix} 1 & 0 & 0 & c_{14} & c_{15} & c_{16} & c_{17} \\ 0 & 1 & 0 & c_{24} & c_{25} & c_{26} & c_{27} \\ 0 & 0 & 1 & c_{34} & c_{35} & c_{36} & c_{37} \end{pmatrix}. \tag{9.15}$$

In this gauge, (9.14) becomes

$$\int \frac{d^{(n-k) \times k} c}{\prod_{j=1}^{n} M_j} \prod_{a=1}^{k} \delta^2 \left([a| + \sum_{l=k+1}^{n} c_{al} [l| \right) \delta^2 \left(|\tilde{\mu}_a\rangle + \sum_{l=k+1}^{n} |\tilde{\mu}_l\rangle c_{al} \right) \delta^{(4)} \left(\eta_a + \sum_{l=k+1}^{n} c_{al} \eta_l \right). \tag{9.16}$$

Performing the inverse-Fourier transform $\int d^2|\tilde{\mu}_j\rangle\, e^{-i\langle j\,\tilde{\mu}_j\rangle}$ for each $j = 1, \ldots, n$ gives

$$
\mathcal{L}_{n,k}([i|, |i\rangle), \eta_i) = \int \frac{d^{(n-k)\times k}c}{\prod_{j=1}^n M_j}\left[\prod_{a=1}^k \delta^2\left([a| + \sum_{l=k+1}^n c_{al}[l|\right)\delta^{(4)}\left(\eta_a + \sum_{l=k+1}^n c_{al}\,\eta_l\right)\right]
$$
$$
\times \left[\prod_{i=k+1}^n \delta^2\left(|i\rangle - \sum_{a=1}^k |a\rangle c_{ai}\right)\right]. \qquad (9.17)
$$

▶ **Exercise 9.1**

Fill out the details of the inverse Fourier transformation to derive (9.17).

The representation (9.17) is central in the next section where we study the geometric interpretation of $\mathcal{L}_{n,k}$. In Section 9.3, we show that familiar amplitude expressions can be derived from $\mathcal{L}_{n,k}$.

9.2 The Grassmannian

It is very convenient to view the $n \times k$ matrices C_{al} in (9.14) as k n-component vectors that define a k-plane in $\mathbb{C}^n$. The space of all k-planes in an n-dimensional space is called the **Grassmannian** $\mathrm{Gr}(k, n)$. The formula (9.14) for $\tilde{\mathcal{L}}_{n,k}$ is therefore naturally viewed as a cyclic invariant integral over all k-planes in the Grassmannian. Since any non-degenerate linear transformation of the k n-vectors gives the same plane, there is a natural $GL(k)$ invariance. It is precisely the same $GL(k)$ redundancy we encountered previously in the discussion of the integral (9.14): the Grassmannian integral (9.14) is well-defined only when "gauge fixing" the $GL(k)$ redundancy. Because of the $GL(k)$ redundancy, the dimensions of the Grassmannian $\mathrm{Gr}(k, n)$ are $k \times n - k^2 = k(n - k)$.

With this geometric picture in mind, let us now examine the bosonic delta functions in the gauge-fixed expression (9.17) for $\mathcal{L}_{n,k}$. They enforce the constraints

$$
\sum_{i=1}^n C_{ai}\,[i|^a = 0, \qquad \sum_{i=1}^n \tilde{C}_{a'i}\langle i| = 0, \qquad (9.18)
$$

where C and $\tilde{C}$ are $k \times n$ and $(n - k) \times n$ matrices respectively, so $a' = k + 1, \ldots, n$. They are explicitly given as

$$
C = \begin{pmatrix} 1 & 0 & \cdots & 0 & c_{1,k+1} & \cdots & c_{1n} \\ 0 & 1 & \cdots & 0 & c_{2,k+1} & \cdots & c_{2n} \\ \vdots & \vdots & \vdots & \vdots & \vdots & \vdots & \vdots \\ 0 & \cdots & 0 & 1 & c_{kk+1} & \cdots & c_{kn} \end{pmatrix}, \quad
\tilde{C} = \begin{pmatrix} -c_{1,k+1} & \cdots & -c_{k,k+1} & 1 & 0 & \cdots & 0 \\ -c_{1,k+2} & \cdots & -c_{k,k+2} & 0 & 1 & \cdots & 0 \\ \vdots & & \vdots & \vdots & \vdots & \vdots & \vdots \\ -c_{1n} & \cdots & -c_{kn} & 0 & 0 & \cdots & 1 \end{pmatrix}.
$$
$$
(9.19)
$$

An important feature is that

$$C \tilde{C}^T = \sum_{i=1}^{n} C_{ai} \, \tilde{C}_{a'i} = 0 \, . \tag{9.20}$$

▶ **Exercise 9.2**

Construct $\tilde{C}$ associated with (9.15) and check that (9.20) holds.

We can view $\tilde{C}$ as $(n-k)$ n-vectors spanning an $(n-k)$-plane in n dimensions. The condition (9.20) states that the $(n-k)$-plane defined by $\tilde{C}$ is the orthogonal complement of the k-plane defined by C.

In this notation, we can reinstate the $GL(k)$ redundancy and write our momentum space Grassmannian integral (9.17) as

$$\mathcal{L}_{n,k} = \int \frac{d^{n \times k} C}{GL(k) \prod_{j=1}^{n} M_j} \left[\prod_{a=1}^{k} \delta^2 \Big(\sum_i C_{ai} \, [i|\Big) \delta^{(4)} \Big(\sum_i C_{ai} \, \eta_{iA} \Big) \right]$$
$$\times \left[\prod_{a'=k+1}^{n} \delta^2 \Big(\sum_i \tilde{C}_{a'i} \langle i| \Big) \right] , \tag{9.21}$$

with the understanding that $\tilde{C}$ is defined as the complement to C in the sense of (9.20).

Now a geometric picture is emerging of the meaning of the constraints (9.18). The collection of the n $|i\rangle$s defines a 2-plane in an n-dimensional space,

$$\begin{pmatrix} |1\rangle^{\dot{1}} & |2\rangle^{\dot{1}} & \cdots & |n\rangle^{\dot{1}} \\ |1\rangle^{\dot{2}} & |2\rangle^{\dot{2}} & \cdots & |n\rangle^{\dot{2}} \end{pmatrix} . \tag{9.22}$$

Similarly the $[i|$s define a 2-plane in an n-dimensional space.

The constraints (9.18) say that the 2-plane spanned by the $[i|$s is orthogonal to the k-plane C and that the 2-plane defined by the $|i\rangle$s is orthogonal to the $(n-k)$-plane $\tilde{C}$. This is illustrated in Figure 9.1. Since $\tilde{C}$ and C are orthogonal complements, we immediately conclude that $\tilde{C}$ must contain the 2-plane $[i|$ while C must contain $|i\rangle$. This in turn tells us that the 2-plane of $[i|$ must be orthogonal to the 2-plane of $|i\rangle$, i.e.

$$\sum_{i=1}^{n} |i\rangle [i| = 0 \, . \tag{9.23}$$

This is just the statement that the external momenta satisfy momentum conservation. Thus, seemingly out of nowhere, the cyclic- and Yangian-invariant generating function $\mathcal{L}_{n,k}$ "knows" about momentum conservation!

As we have just shown, the bosonic delta functions in (9.21) give non-vanishing results only on the support of momentum conservation, $\delta^4(P)$. With this in mind, let us count the number of "free" integration variables in (9.21), i.e. the number of c_{ai}s *not* localized by the bosonic delta functions. After gauge fixing $GL(k)$, we have a total of $k \times (n-k)$

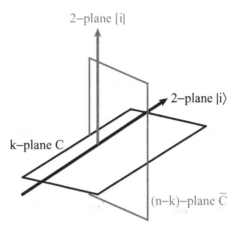

The geometry of planes in the Grassmannian. The k-plane C and $(n - k)$-plane $\tilde{C}$ are orthogonal complements, so the constraint $\sum_i C_{ai}[i| = 0$ in (9.18) means that the 2-plane spanned by the n $[i|$s is orthogonal to C and hence must be contained in $\tilde{C}$. Similarly, C contains the 2-plane spanned by the $|i\rangle$s. It follows from the geometry that the 2-planes spanned by $[i|$ and $|i\rangle$, respectively, are orthogonal, but that is exactly the statement of momentum conservation $\sum_{i=1}^{n} |i\rangle[i| = 0$.

Fig. 9.1

c_{ai}-variables. There are $[2k + 2(n - k)] = 2n$ bosonic delta functions, but this includes momentum conservation $\delta^4(P)$, so four of the delta functions do not localize any c_{ai}-variables. Therefore, a total of

$$\#(\text{integration variables}) = [k \times (n - k) - (2n - 4)] = (k - 2)(n - k - 2) \quad (9.24)$$

c_{ai}-variables are left to be integrated.

In the MHV sector, $k = 2$ so we learn from (9.24) that the integral (9.21) is fully localized by the bosonic delta functions. For generic k and n, $\mathcal{L}_{n,k}$ is a multi-dimensional integral that localizes on the poles in the minors; we work out an explicit example in the next section. For $k = 0$, the integral vanishes since it is proportional to $\delta^2(|i\rangle)$ which does not have support for generic momenta. This is simply the statement that the "all-plus" gluon amplitude vanishes in $\mathcal{N} = 4$ SYM. For $k = 1$, the last delta function in (9.21) forces all of the $|i\rangle$s to be proportional to each other, but this lacks support for generic momenta, with the exception of special kinematics for $n = 3$. Not surprisingly, this says that the all-plus-and-one-minus gluon amplitudes vanish in $\mathcal{N} = 4$ SYM for $n > 3$.

The counting of integration variables in (9.24) is invariant under $k \to (n - k - 2)$. This corresponds to a flip of what we identify as positive and negative helicity, i.e. which states are associated with highest/lowest Grassmann weight. Indeed, for $k = n - 2$, the (super)amplitude is anti-MHV, so it makes sense that the Grassmannian integral is localized completely by the bosonic delta functions, just as it is for the MHV sector $k = 2$.

Perhaps you have noticed that the dimension $(k - 2)(n - k - 2)$ of the $\mathcal{L}_{n,k}$-integral is also the dimension of the Grassmannian $\text{Gr}(k - 2, n - 4)$. This is not a coincidence. The bosonic delta functions enforce that C_{ai} contains the 2-plane $|i\rangle$, so using $GL(k)$ redundancy we can choose the first two rows of C_{ai} to be $|i\rangle$. The $2k - 4$ delta functions (where the -4 comes from momentum conservation) impose that the remaining $(k - 2)$-plane in

n-dimensions is orthogonal to $|j]$. This places $2(k-2)$ constraints on C_{ai}. Taking the $GL(k-2)$ redundancy into account, the number of free variables in C_{ai} is therefore $n(k-2) - (k-2)^2 - 2(k-2) = (k-2)(n-k-4)$. This is the dimension of the Grassmannian $\text{Gr}(k-2, n-4)$.

So far we have constructed a cyclic and Yangian invariant integral $\mathcal{L}_{n,k}$ in the Grassmannian, but we have not really done anything with it. In fact, other than showing that it captures momentum conservation, we have given you little reason to believe that there is any connection to scattering amplitudes in planar $\mathcal{N} = 4$ SYM. So now we had better show how it works.

9.3 Yangian invariants as residues in the Grassmannian

We carry out the Grassmannian integral (9.21) in the simplest cases to illustrate how the familiar MHV and NMHV superamplitudes appear.

9.3.1 MHV amplitudes

By the counting (9.24), the bosonic delta functions completely localize the integral (9.21) for the MHV sector ($k = 2$). Furthermore, the geometric description in the previous section tells us that for MHV, the bosonic delta functions exactly encode conservation of 4-momentum on the n external states: $\delta^4(P)$. What about the Grassmann delta function in (9.21)? Well, when $k = 2$, C defines a 2-plane, and since the 2-plane $|i\rangle$ must be contained in C, we can simply identify the two 2-planes; up to a $GL(2)$ transformation we therefore have

$$\begin{pmatrix} C_{11} & C_{12} & \cdots & C_{1n} \\ C_{21} & C_{22} & \cdots & C_{2n} \end{pmatrix} = \begin{pmatrix} |1\rangle^{\dot{1}} & |2\rangle^{\dot{1}} & \cdots & |n\rangle^{\dot{1}} \\ |1\rangle^{\dot{2}} & |2\rangle^{\dot{2}} & \cdots & |n\rangle^{\dot{2}} \end{pmatrix}. \tag{9.25}$$

Using this explicit representation of the C_{ai}s, the Grassmann delta function becomes the familiar statement of supermomentum conservation,

$$\prod_{a=1}^{2} \delta^{(4)}\Big(\sum_i C_{ai}\, \eta_i\Big) = \prod_{\dot{a}=1}^{2} \delta^{(4)}\Big(\sum_i |i\rangle^{\dot{a}}\, \eta_i\Big) = \delta^{(8)}(\tilde{Q}). \tag{9.26}$$

Finally, consider the minors in (9.21):

$$M_i = \epsilon^{ab}\, C_{ai}\, C_{b,i+1} = -\epsilon_{\dot{a}\dot{b}}\, |i\rangle^{\dot{a}}\, |i+1\rangle^{\dot{b}} = -\langle i, i+1\rangle. \tag{9.27}$$

Putting everything together, we find

$$\mathcal{L}_{n,2} = (-1)^n\, \frac{\delta^{(8)}(\tilde{Q})\, \delta^4(P)}{\prod_{i=1}^{n} \langle i, i+1\rangle} = (-1)^n\, \mathcal{A}_{n,\text{tree}}^{\text{MHV}}. \tag{9.28}$$

So for $k = 2$, the cyclic invariant integral $\mathcal{L}_{n,2}$ nicely produces the MHV tree amplitude (up to an overall convention-dependent sign).

▷ *Example.* Did the argument above go a little fast? Fair enough, let us evaluate $\mathcal{L}_{n,2}$ in full detail, starting with the gauged-fixed expression (9.17), which for $k = 2$ gives

$$\mathcal{L}_{n,2} = \int \frac{d^{(n-2)\times 2} c}{M_1 \cdots M_n} \left[\prod_{a=1}^{2} \delta^2 \left([a| + \sum_{l=3}^{n} c_{al}[l| \right) \delta^{(4)} \left(\eta_a + \sum_{l=3}^{n} c_{al}\,\eta_l \right) \right] \\ \times \left[\prod_{i=3}^{n} \delta^2 \left(|i\rangle - |1\rangle c_{1i} - |2\rangle c_{2i} \right) \right]. \tag{9.29}$$

The last set of delta functions localize the $2(n-2)$ components c_{1i} and c_{2i}:

$$\delta^2 \left(|i\rangle - |1\rangle c_{1i} - |2\rangle c_{2i} \right) = \frac{1}{\langle 12 \rangle} \delta \left(c_{1i} - \frac{\langle i2 \rangle}{\langle 12 \rangle} \right) \delta \left(c_{2i} - \frac{\langle i1 \rangle}{\langle 21 \rangle} \right). \tag{9.30}$$

Thus, on the support of these delta functions, the first four bosonic delta functions in (9.29) give

$$\delta^2 \left([1| + \sum_{l=3}^{n} c_{1l}[l| \right) \delta^2 \left([2| + \sum_{l=3}^{n} c_{2l}[l| \right) = \langle 12 \rangle^2 \, \delta^4(P). \tag{9.31}$$

This is how the momentum conservation delta function appears.

Likewise for the Grassmann delta function: on the support of (9.30) it gives

$$\prod_{a=1}^{2} \delta^{(4)} \left(\eta_a + \sum_{l=3}^{n} c_{al}\,\eta_l \right) = \frac{1}{\langle 12 \rangle^4} \, \delta^{(8)}(\tilde{Q}). \tag{9.32}$$

Finally, we evaluate the minors M_i. With the help of the Schouten identity we find

$$M_1 = 1, \quad M_2 = \frac{\langle 23 \rangle}{\langle 12 \rangle}, \quad M_3 = -\frac{\langle 34 \rangle}{\langle 12 \rangle}, \quad M_4 = -\frac{\langle 45 \rangle}{\langle 12 \rangle}, \quad \ldots, \quad M_n = -\frac{\langle n1 \rangle}{\langle 12 \rangle}, \tag{9.33}$$

and hence

$$\prod_{i=1}^{n} M_i = \frac{(-1)^n}{\langle 12 \rangle^n} \left(\prod_{i=1}^{n} \langle i, i+1 \rangle \right). \tag{9.34}$$

Inserting everything into (9.29) we indeed obtain $(-1)^n$ times the MHV tree superamplitude, just as in (9.28). ◁

▶ **Exercise 9.3**

Derive (9.30)–(9.34). Then plug the results into (9.29) to verify that all powers of $\langle 12 \rangle$ cancel.

The Grassmannian integral $\mathcal{L}_{n,k}$ has given a unique result, $\mathcal{A}_{n,\text{tree}}^{\text{MHV}}$, for $k = 2$. Given that the Grassmannian integral produces all possible Yangian invariants [122, 123], it must be that all MHV superamplitudes have the same Leading Singularities in planar $\mathcal{N} = 4$ SYM, up to a sign, *to all loop-orders*. We have already seen a non-trivial manifestation of this in Section 8.2 (and again in Section 8.3), where the Leading Singularities of the 1-loop and 2-loop 4-point superamplitudes were found to be the MHV tree superamplitude. We now know that even if we were to take the 234-loop MHV superamplitude and solve the on-shell constraints that localize the $234 \times 4 = 936$ loop-momenta, the result of the Leading Singularities will again be MHV tree superamplitudes! No other Yangian invariants are available at MHV order.

9.3.2 6-point NMHV amplitudes

Let us now move on to a slightly more complicated – hence more exciting – example, the 6-point NMHV amplitude.[1] With $k = 3$ and $n = 6$, the counting formula (9.24) reveals that the Grassmannian integral $\mathcal{L}_{6,3}$ involves just one non-trivial integration. To evaluate it, we choose the gauge

$$
\begin{pmatrix}
c_{21} & 1 & c_{23} & 0 & c_{25} & 0 \\
c_{41} & 0 & c_{43} & 1 & c_{45} & 0 \\
c_{61} & 0 & c_{63} & 0 & c_{65} & 1
\end{pmatrix} .
\tag{9.35}
$$

The c-variables are labeled such that the bosonic delta functions in (9.21) can be written as:

$$
\delta^2 \left([\bar{i}| + \sum_j c_{\bar{i}j} [j| \right), \quad
\delta^2 \left(|j\rangle - \sum_{\bar{i}} |\bar{i}\rangle c_{\bar{i}j} \right),
\tag{9.36}
$$

where $\bar{i} = 2, 4, 6$ and $j = 1, 3, 5$. Since the integral is 1-dimensional, there must be a 1-parameter family of solutions that solve the delta function constraints (9.36). Indeed, if $c_{\bar{i}j}^*$ is a solution, then

$$
\hat{c}_{\bar{i}j}(\tau) = c_{\bar{i}j}^* + \tfrac{1}{4} \tau \, \epsilon_{\bar{i}\bar{j}\bar{k}} \, \langle \bar{j}\bar{k} \rangle \, \epsilon_{jkl} [kl] ,
\tag{9.37}
$$

is also a solution for any τ. Here $\epsilon_{\bar{i}\bar{j}\bar{k}}$ is a Levi-Civita symbol for the indices $\bar{i} = 2, 4, 6$, and similarly for ϵ_{jkl}. There are implicit sums over repeated labels in (9.37). That $\hat{c}_{\bar{i}j}(\tau)$ is a solution can be seen from the result that the τ dependence drops out from the constraints in (9.36) due to the Schouten identity:

$$
[\bar{i}|^a + \sum_j \hat{c}_{\bar{i}j}(\tau) [j|^a = \tfrac{1}{4} \tau \, \epsilon_{\bar{i}\bar{j}\bar{k}} \langle \bar{j}\bar{k} \rangle \sum_j \epsilon_{jkl} [j|^a [kl]
$$

$$
= \tfrac{1}{2} \tau \, \epsilon_{\bar{i}\bar{j}\bar{k}} \langle \bar{j}\bar{k} \rangle \big([1|^a[35] + [3|^a[51] + [5|^a[13] \big) = 0 .
\tag{9.38}
$$

We can now remove the bosonic delta functions by localizing the integral on the solution to the constraints (9.36) such that the remaining integral is over the 1-dimensional parameter

---

[1] For a much simpler evaluation of the NMHV residues in momentum twistor space, see Elvang *et al.* [60].

τ. That gives

$$\mathcal{L}_{6,3} = \int \frac{d^9 c_{\bar{i}j}}{M_1 \cdots M_n} \left[\prod_j \delta^2 \left(|j\rangle - \sum_{\bar{i}} |\bar{i}\rangle c_{\bar{i}j} \right) \right]$$

$$\times \left[\prod_{\bar{i}} \delta^2 \left([\bar{i}| + \sum_j c_{\bar{i}j} [j| \right) \delta^{(4)} \left(\eta_{\bar{i}} + \sum_j c_{\bar{i}j} \eta_j \right) \right]$$

$$= \delta^4(P) \int \frac{d^9 c_{\bar{i}j} d\tau}{M_1 \cdots M_n} \delta^9 \left(c_{\bar{i}j} - \hat{c}_{\bar{i}j}(\tau) \right) \prod_{\bar{i}} \delta^{(4)} \left(\eta_{\bar{i}} + \sum_j c_{\bar{i}j} \eta_j \right)$$

$$= \delta^4(P) \int \frac{d\tau}{\hat{M}_1 \cdots \hat{M}_n} \prod_{\bar{i}} \delta^{(4)} \left(\eta_{\bar{i}} + \sum_j \hat{c}_{\bar{i}j} \eta_j \right). \tag{9.39}$$

The "hat" indicates dependence on τ via (9.37). In the gauge (9.35), the minors are

$$\hat{M}_1 = \hat{c}_{43} \hat{c}_{61} - \hat{c}_{41} \hat{c}_{63}, \quad \hat{M}_3 = \hat{c}_{23} \hat{c}_{65} - \hat{c}_{25} \hat{c}_{63}, \quad \hat{M}_5 = \hat{c}_{21} \hat{c}_{45} - \hat{c}_{25} \hat{c}_{41},$$
$$\hat{M}_2 = -\hat{c}_{63}, \qquad\qquad \hat{M}_4 = -\hat{c}_{25}, \qquad\qquad \hat{M}_6 = -\hat{c}_{41}. \tag{9.40}$$

At this stage, there appears to be no a priori prescription of which contour to pick in the τ-plane. Each minor $\hat{M}_i$ has a simple pole in τ, so there are six different residues that we denote $\{M_i\}$. Let us focus on the pole in $\hat{M}_4$. This means that τ is evaluated at τ_* such that $\hat{c}_{25}(\tau_*) = 0$. We can make the calculation simpler by choosing the origin for τ such that $\hat{M}_4 = 0$ for $\tau = 0$; in other words, we choose $\hat{c}_{25}^* = 0$. Let us use the constraints (9.36) to solve for the eight other $\hat{c}_{\bar{i}j}^*$s. From

$$|5\rangle - |4\rangle c_{45}^* - |6\rangle c_{65}^* = 0, \qquad [2| + c_{21}^*[1| + c_{23}^*[3| = 0, \tag{9.41}$$

we deduce

$$c_{45}^* = \frac{\langle 56 \rangle}{\langle 46 \rangle}, \quad c_{65}^* = \frac{\langle 45 \rangle}{\langle 46 \rangle}, \quad c_{21}^* = -\frac{[23]}{[13]}, \quad c_{23}^* = -\frac{[12]}{[13]}. \tag{9.42}$$

And this in turn allows us to solve

$$[4| + c_{41}^*[1| + c_{43}^*[3| + c_{45}^*[5| = 0, \qquad [6| + c_{61}^*[1| + c_{63}^*[3| + c_{65}^*[5| = 0, \tag{9.43}$$

to find

$$c_{41}^* = -\frac{\langle 6|4+5|3]}{\langle 46 \rangle [13]}, \quad c_{43}^* = \frac{\langle 6|4+5|1]}{\langle 46 \rangle [13]}, \quad c_{61}^* = \frac{\langle 4|5+6|3]}{\langle 46 \rangle [13]}, \quad c_{63}^* = -\frac{\langle 4|5+6|1]}{\langle 46 \rangle [13]}. \tag{9.44}$$

▶ **Exercise 9.4**

Use the above results for $c_{\bar{i}j}^*$ to show that the unused constraints in (9.36) give $\delta^4(P)$.

We can now substitute the solutions $c_{\bar{i}j}^*$ into the minors (9.40) and the Grassmann delta functions to obtain the residue of the integral (9.39) of the pole $1/M_4$, denoted by $\{M_4\}$. For simplicity, consider a particular component amplitude, namely the gluon amplitude with helicity assignments $(+, -, +, -, +, -)$. For this amplitude, the coefficient from the

Grassmann delta functions is just 1. Taking into account the extra factor of $\langle 46 \rangle[13]$ coming from $\hat{c}_{25} = -\tau\langle 46 \rangle[13]$, we find

$$\{M_4\} = \frac{\langle 46 \rangle^4[13]^4}{\langle 4|5+6|1]\langle 6|4+5|3][21][23]\langle 54 \rangle\langle 56 \rangle P_{456}^2}. \qquad (9.45)$$

▶ **Exercise 9.5**

Show that $\hat{M}_1\big|_{\tau=0} = \frac{P_{456}^2}{\langle 46 \rangle[13]}$. Evaluate the other minors at $\tau = 0$ and use them to derive the result (9.45) for the residue at $\tau = 0$.

We could calculate the residues associated with each of the other minors similarly. $\{M_6\}$ and $\{M_2\}$ are just cyclic permutations of $\{M_4\}$ by two sites, so we have

$$\{M_6\} = \frac{\langle 62 \rangle^4[35]^4}{\langle 6|1+2|3]\langle 2|6+1|5][43][45]\langle 16 \rangle\langle 12 \rangle P_{612}^2},$$

$$\{M_2\} = \frac{\langle 24 \rangle^4[51]^4}{\langle 2|3+4|5]\langle 4|2+3|1][65][61]\langle 32 \rangle\langle 34 \rangle P_{234}^2}. \qquad (9.46)$$

For the residues $\{M_1\}$, $\{M_3\}$, $\{M_5\}$, it is convenient to choose the gauge

$$\begin{pmatrix} 1 & c_{12} & 0 & c_{14} & 0 & c_{16} \\ 0 & c_{32} & 1 & c_{34} & 0 & c_{36} \\ 0 & c_{52} & 0 & c_{54} & 1 & c_{56} \end{pmatrix}. \qquad (9.47)$$

Then following the same steps as before we find that the $\{M_1\}$ residue for the $(+, -, +, -, +, -)$ amplitude is

$$\{M_1\} = \frac{-\langle 6|2+4|3]^4}{\langle 1|5+6|4]\langle 5|6+1|2][23][34]\langle 56 \rangle\langle 61 \rangle P_{561}^2}. \qquad (9.48)$$

The other residues, $\{M_3\}$ and $\{M_5\}$, are obtained by relabeling the external states in (9.48).

We have now extracted six residues $\{M_i\}$ from (9.39) for a projection that corresponds to the helicity configuration $(+, -, +, -, +, -)$ of a gluon amplitude. But it is not yet clear what the residues have to do with the amplitude. Each of the $\{M_i\}$s contains spurious poles, such as $\langle 4|5+6|1]$ in $\{M_2\}$ and $\{M_4\}$. However, in the sum $\{M_2\} + \{M_4\}$, this spurious pole cancels. In fact, in the sum $\{M_2\} + \{M_4\} + \{M_6\}$ all three spurious poles $- \langle 4|5+6|1]$, $\langle 6|4+5|3]$, and $\langle 2|6+1|5]$ – cancel, so this is a local object. Your brain may even be tingling with the sensation that you have seen this combination before. Go back to look at Exercise 3.9: there we calculated the 6-point tree amplitude $A_6[1^+2^-3^+4^-5^+6^-]$ from a $[2, 3\rangle$-BCFW shift and found that it was exactly

$$A_6[1^+2^-3^+4^-5^+6^-] = \{M_2\} + \{M_4\} + \{M_6\}. \qquad (9.49)$$

In the BCFW construction, each of the three terms in (9.49) corresponds exactly to a BCFW diagram. Now we have also seen that each term can be understood as the residue of a pole

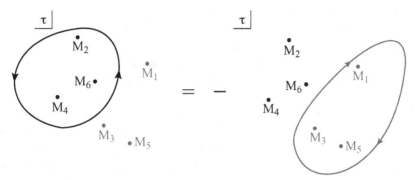

The "tree contour" in the Grassmannian. It circles the residues of the poles $\{M_2\}$, $\{M_4\}$, $\{M_6\}$. Through contour deformation, the result is equivalent to minus the sum of $\{M_1\}$, $\{M_3\}$, $\{M_5\}$. **Fig. 9.2**

associated with the minor M_i in the cyclically invariant Grassmannian integral. So for $\mathcal{N} = 4$ SYM, individual BCFW terms are in one-to-one correspondence with the residues of the Grassmannian integral. Since the Grassmannian integral was constructed to produce Yangian invariants, we now understand that each super-BCFW term is a Yangian invariant.

Consider the contour that encircles the minors $\{M_2\}$, $\{M_4\}$, $\{M_6\}$. It is this contour that gives a Yangian invariant rational function that is local and free of spurious singularities. The statement of locality has become a choice of contour.

Through contour deformation, illustrated schematically in Figure 9.2, we have

$$\{M_2\} + \{M_4\} + \{M_6\} = -\{M_1\} - \{M_3\} - \{M_5\}. \tag{9.50}$$

This means that the tree amplitude $A_6[1^+2^-3^+4^-5^+6^-]$ can also be represented by (minus) the sum of $\{M_1\}$, $\{M_3\}$, and $\{M_5\}$. Indeed, this is the representation that one obtains from the BCFW shift $[3, 2\rangle$, the "parity conjugate" of the shift $[2, 3\rangle$ that produced the $\{M_2\}$, $\{M_4\}$, $\{M_6\}$ representation. Actually, this is a little too quick, because the shift $[3, 2\rangle$ would be an illegal $[+, -\rangle$-shift for the component-amplitude $A_6[1^+2^-3^+4^-5^+6^-]$. The correct statement is that the $\{M_1\}$, $\{M_3\}$, $\{M_5\}$ representation is the result of a $[3, 2\rangle$ BCFW *super*-shift recursion relation with a subsequent projection to the $(+, -, +, -, +, -)$ gluon helicity states.

The insight gained here is that the mysterious six-term identity (9.50) that arises from the equivalence of the two conjugate BCFW super-shifts $[2, 3\rangle$ and $[3, 2\rangle$ is simply a consequence of the residue theorem of the Grassmannian integral $\mathcal{L}_{n,k}$! Actually, the identity (9.50) is the 5-bracket six-term identity (5.63) projected to the $(+, -, +, -, +, -)$ gluon helicity states. In Chapter 10 we expose an underlying geometric interpretation of such identities.

The Grassmannian, the tree contour, and the twistor string

In Witten's *twistor string* [54], mentioned briefly in Section 5.2, the N^KMHV superamplitudes in $\mathcal{N} = 4$ SYM are calculated as open string current algebra correlators integrated over the moduli space of degree $(K+1)$ curves in supertwistor space. It turns out that the so-called *RSV connected prescription* [124] for the twistor string has a direct relation to the BCFW recursion relations. Moreover, different BCFW representations are related via (higher-dimensional versions of) Cauchy's theorem [125]. Sounds similar to the properties of the Grassmannian integral, does it not? In fact, it can be shown that the tree contour in the Grassmannian precisely gives the RSV connected prescription for the twistor string.

Let us be more explicit about how this works. Recall the procedure for carrying out the Grassmann integral $\mathcal{L}_{n,k}$ in (9.21). Gauge fixing $GL(k)$ leaves $nk - k^2$ integration variables of which the bosonic delta functions fix $2k + 2(n - k) - 4 = 2n - 4$. (The -4 is momentum conservation.) So this leaves $nk - k^2 - (2n - 4) = (k - 2)(n - k - 2)$ variables and the strategy is to fix those by localizing all of them on zeroes of the minors in the denominator of $\mathcal{L}_{n,k}$. Reversing the procedure, we ignore the delta functions to begin with and first localize on the minors: this leaves an integral with $nk - k^2 - (k - 2)(n - k - 2) = 2n - 4$ variables. These will of course be fixed by the bosonic delta functions, but note that $2n - 4$ is the dimension of the Grassmannian Gr(2, n). Thus, after localizing on the minors, the integral $\mathcal{L}_{n,k}$ naturally lives in Gr(2, n). Now the punchline is that precisely with the tree contour, reducing the Gr(k, n)-integral to a Gr(2, n)-integral results in the twistor string formula after a suitable Fourier transform [126]. Thus the underlying property that makes the tree contour special is that it localizes the Gr(k, n) Grassmannian integral to Gr(2, n) in a particular fashion that is intimately tied to locality and the twistor string. We will see this story repeat itself when we consider the Grassmannian formula for the 3-dimensional ABJM theory in Section 11.3. There are several papers in the literature on the relationship between amplitudes and the twistor string, see for example [54, 124–129].

The Grassmannian picture is interesting and has given us insight about locality, but there is perhaps a small stone in our shoe: we have not yet been able to see how the cancellation of spurious poles takes place within the tree superamplitude in a manifest fashion. There is a geometric story about how that happens and we draw it in Chapter 10.

9.4 From on-shell diagrams to the Grassmannian

The terms appearing in the BCFW expansion of the 6-point NMHV amplitude have now appeared in two distinct guises. We have seen in the previous section that they are given as residues of an integral over a Gr(3,6) Grassmannian manifold. In Section 8.3, they were the result of gluing on-shell cubic vertices together into on-shell diagrams. So can we make a connection between the Grassmannian and the on-shell diagrams? Yes, we can!

In Section 8.3, we did not explicitly compute any of the on-shell diagrams beyond the simplest box-diagram. The reason is simple: explicitly solving all momentum conservation constraints at each vertex is a complicated task because it is quadratic in spinor variables. In this section, we have seen that momentum conservation can be converted into a linear constraint with the aid of the Grassmannian. This means that if we convert all 3-point vertices in an on-shell diagram into Grassmannian integrals, then the momentum conservation constraints are just a set of linear equations. Let us see how this is done in practice. For the MHV 3-point amplitude, the Grassmannian integral is simply Gr(2,3):

$$
\mathcal{A}_3^{\text{MHV}} = \int \frac{d^{2\times 3}C}{M_1 M_2 M_3} \, \delta^{2\times 2}\big(C_i[i|\big) \, \delta^{(4\times 2)}\big(C_i \eta_i\big) \, \delta^{2\times 1}\big(\tilde{C}_i \langle i|\big),
\tag{9.51}
$$

where we have used the momentum space representation (9.21): C is a 2×3 matrix, and $\tilde{C}$ is its 1×3-dimensional orthogonal complement. For the anti-MHV 3-point amplitude, the analogue Grassmannian integral in Gr(1,3) is

$$
\mathcal{A}_3^{\text{anti-MHV}} = \int \frac{d^{1\times 3}C}{M_1 M_2 M_3} \, \delta^{2\times 1}\big(C_i[i|\big) \, \delta^{(4\times 1)}\big(C_i \eta_i\big) \, \delta^{2\times 2}\big(\tilde{C}_i \langle i|\big),
\tag{9.52}
$$

where now C is a 1×3-dimensional matrix and $\tilde{C}$ is its 2×3-dimensional orthogonal complement. To see that this indeed gives the correct 3-point amplitude, note that the first bosonic delta function requires that C is orthogonal to the 2-plane $[i|$, so this localizes the integral (up to an irrelevant overall rescaling) to

$$
C = \Big([23] \,, \; [31] \,, \; [12] \Big).
\tag{9.53}
$$

Substituting this into (9.52), one indeed recovers the anti-MHV 3-point amplitude.

▶ Exercise 9.6

Given (9.53), determine a representation of $\tilde{C}_{ai}$. Substitute the result into (9.52) to recover the 3-point amplitude. Note that all potential Jacobian factors can be fixed by dimension-counting and symmetry analysis.

In summary, the MHV and anti-MHV 3-point amplitudes can be viewed as providing 2×2 and 1×2 linear constraints respectively for the $[i|$s. For later convenience, we parameterize the Gr(2, 3) and Gr(1, 3) Grassmannians as follows:

$$
\begin{array}{ccc} b & c & a \end{array} \\
\begin{pmatrix} 1 & 0 & \alpha_b \\ 0 & 1 & \alpha_c \end{pmatrix} \qquad
\tag{9.54}
$$

$$
\begin{array}{ccc} a & b & c \end{array} \\
\begin{pmatrix} 1 & \beta_b & \beta_c \end{pmatrix} \qquad
\tag{9.55}
$$

The arrows represent the particular gauge that we have chosen for each Grassmannian integral. Incoming lines on the 3-point vertex indicate that the corresponding columns in the Grassmannian are $GL(k)$-gauge-fixed to be the identity matrix. Outgoing lines correspond to unfixed columns. In this gauge, the $[i|$ parts of the bosonic delta functions are:

$$\text{MHV:}\quad \delta^2\big([b| + \alpha_b[a|\big)\,\delta^2\big([c| + \alpha_c[a|\big),\qquad \text{anti-MHV:}\quad \delta^2\big([a| + \beta_b[b| + \beta_c[c|\big).$$
$$(9.56)$$

For each vertex, the spinors of the incoming lines are expressed as a linear combination of those of the outgoing lines. One can perform a similar analysis for the fermionic delta functions and the bosonic delta functions of $\langle i|$. The analyses are exactly parallel, so we leave them implicit.

▶ **Exercise 9.7**

What does the bosonic delta function for $\langle i|$ look like? What constraints does it impose?

We are now ready to start gluing 3-point on-shell vertices together. Recall from (8.49) that each internal line in the on-shell diagram corresponds to an integral over the set of internal variables,

$$\int \frac{d^2|I\rangle\,d^2|I]\,d^4\eta_I}{U(1)}.$$
$$(9.57)$$

Since the spinors $\big(|I\rangle,\;|I]\big)$ also appear in the vertices on each end of the line, the bosonic delta functions of these vertices can be used to localize the integral (9.57). This can be made manifest in a graphical way. For each on-shell diagram, we decorate the lines with arrows following the rule that for each black vertex, there should be two incoming lines and one outgoing line, while for each white vertex there should be one incoming line and two outgoing lines, just as in (9.54), (9.55). One might wonder if it is always possible to find such decoration consistent throughout the on-shell diagram. For diagrams of *physical relevance* the answer is yes, since one can interpret the outgoing lines as $+$ helicity, incoming lines as $-$ helicity, and a consistent decoration is equivalent to consistent helicity assignments. For example, consider gluing six vertices together to form a double box diagram. We can have a consistent decoration if there is at least one different color vertex, but not if they are all the same:[2]

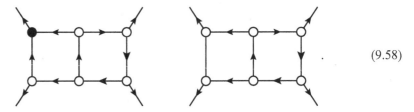

$$(9.58)$$

The lines of the second diagram cannot be consistently oriented. It follows from the previous discussion that the spinors of the internal line are completely determined by the outgoing

---

[2] You may find it a little peculiar that we show examples corresponding to on-shell diagrams for all-plus amplitudes; the point is just that they give simple illustrations of the consistency conditions.

lines of one of the vertices. For example, decorations of the diagram

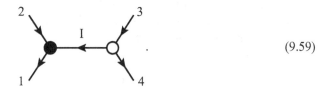

$$(9.59)$$

dictate the bosonic delta functions for the square spinors to be

$$\delta^2\big([2| + \alpha_2[1|\big), \quad \delta^2\big([I| + \alpha_I[1|\big), \quad \delta^2\big([3| + \beta_4[4| + \beta_I[I|\big).$$

$$(9.60)$$

The second bosonic delta function localizes the $\int d^2|I]$ integral, while the remaining two delta functions become

$$\delta^2\big([2| + \alpha_2[1|\big), \quad \delta^2\big([3| + \beta_4[4| - \beta_I\alpha_I[1|\big).$$

$$(9.61)$$

The delta functions in (9.61) can be combined to the form $\prod_{a=1}^{2}\delta^2(C_{ai}[i|)$ with the Gr(2,4) Grassmannian given as

$$C_{ai} = \begin{pmatrix} \overset{1}{\alpha_2} & \overset{2}{1} & \overset{3}{0} & \overset{4}{0} \\ -\beta_I\alpha_I & 0 & 1 & \beta_4 \end{pmatrix}.$$

$$(9.62)$$

So gluing the two 3-point vertices together now gives a new Grassmannian integral

$$\int \frac{d\alpha_2}{\alpha_2} \frac{d\alpha_I}{\alpha_I} \frac{d\beta_4}{\beta_4} \frac{d\beta_I}{\beta_I} \frac{1}{U(1)} \, \delta^{2\times2}\big(C_i[i|\big) \, \delta^{(4\times2)}\big(C_i\eta_i\big) \, \delta^{2\times2}\big(\tilde{C}_i\langle i|\big),$$

$$(9.63)$$

where the C_{ai} is identified in (9.62).

Note the leftover $1/U(1)$ in (9.63). We have been treating $(|I\rangle, |I])$ as independent variables, each being fixed by the bosonic delta functions. However, there remains a gauge-fixing functional that is present to remove the little-group redundancy. This functional is represented by this $1/U(1)$ factor. We do not need its explicit form, just remember that it can be used to fix an additional degree of freedom.

The above simple example generalizes to an arbitrary decorated on-shell diagram. A diagram with n_b black vertices and n_w white vertices contains $2 \times (2n_b + n_w)$ constraints on the $|i]$s. If it has n_I internal lines, then the $2 \times n_I$ integrations over the internal $|I]$s can be localized by these bosonic delta functions. Finally, there will be $2 \times (2n_b + n_w - n_I)$ constraints left and they can be conveniently grouped into a degree $2 \times k$ delta function $\prod_{a=1}^{k}\delta^2(C_{ai}[i|)$ where $k = (2n_b + n_w - n_I)$ and $n = 3(n_w + n_b) - 2n_I$. Note that by counting the Grassmann degrees of the black (2×4) and white (1×4) blobs minus the n_I internal Grassmann integrations ($n_I \times 4$), this k is exactly the same k as in the N^{k+2}MHV classification. Thus each on-shell diagram corresponds to the following Grassmannian integral:

$$\int \left(\prod_{i=1}^{n_b} \frac{d\alpha_{i1}}{\alpha_{i1}} \frac{d\alpha_{i2}}{\alpha_{i2}}\right) \left(\prod_{i=1}^{n_w} \frac{d\beta_{i1}}{\beta_{i1}} \frac{d\beta_{i2}}{\beta_{i2}}\right) \left(\prod_{i=1}^{n_I} \frac{1}{U(1)_i}\right) \delta^{2\times k}\big(C_i[i|\big) \, \delta^{(4\times k)}\big(C_i\eta_i\big) \, \delta^{2\times(n-k)}\big(\tilde{C}_i\langle i|\big).$$

$$(9.64)$$

As we have seen, this $Gr(k, n)$ Grassmannian integral is parameterized by an on-shell diagram decorated with arrows consistently throughout the diagram. The Grassmannian integral (9.64) is in a $GL(k)$-gauge-fixed form. Each vertex contains two degrees of freedom, but the n_I internal lines each leave a $1/U(1)$ gauge-fixing function, so the dimension of this Grassmannian integral is

$$\dim(C) = 2 \times n_v - n_I \,, \tag{9.65}$$

where the number of vertices is $n_v = n_b + n_w$. Using Euler's formula for a planar diagram $(n_f - n) - n_I + n_v = 1$, where n_f is the number of faces in a diagram, we find that the dimension of the Grassmannian integral corresponding to a particular on-shell diagram is

$$\dim(C) = n_f - 1 \,. \tag{9.66}$$

The total number of bosonic delta functions is $2 \times (k + n - k) - 4 = 2n - 4$. If $\dim(C) = 2n - 4$, then all the degrees of freedom in the integral are completely localized by the bosonic delta functions. This is the case for the on-shell diagrams that correspond to the BCFW terms in Section 8.3. We saw this in Section 9.3.2 for $Gr(3,6)$: prior to solving the bosonic delta functions, each BCFW term is obtained by localizing on the zeroes of one of the minors. This is precisely the $9 - 1 = 8$-dimensional Grassmannian manifold indicated by the on-shell diagrams.

If $\dim(C) < 2n - 4$, then the bosonic delta functions over-constrain the external data and can only be satisfied in special kinematics. This is precisely the scenario for the example that led to (9.62), where $\dim(C) = 3 < 4$. From (9.62) one can readily read off what the special kinematics is: $[1| \sim [2|$.

What happens if $\dim(C) > 2n - 4$? It corresponds to a term in the BCFW representation of a loop amplitude, where the remaining integrations can be translated into the loop-momentum integration! For example, consider attaching a BCFW bridge on the forward limit of (8.55). In our calculation of the loop-recursion in Section 7.5, only the BCFW term that corresponds to the middle on-shell diagram of (8.55) has a non-vanishing forward limit contribution. If we attach a BCFW bridge to this middle diagram and perform a series of equivalence moves, we find:

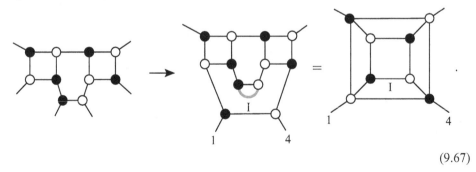

$$\tag{9.67}$$

The final diagram is precisely the 1-loop 4-point amplitude. One can count that with $n = 4$, $\dim(C) = 9 - 1 = 4 + (2n - 4)$, i.e. four integrals remain after solving all bosonic delta functions. These four extra integrals correspond to the integration over the four components of the loop-momentum. We can readily identify these extra components in the on-shell diagram: since the original tree-diagram contains no extra integration variable, the new degrees of freedom must arise from the procedure of taking the forward limit. This introduces a factor of $\int d^2|I\rangle\, d^2|I]\, d^4\eta_I\, /\, U(1)$ in (9.67); that is three integrations because the $U(1)$ is mod'ed out. The presence of the BCFW bridge introduces the fourth integration $\int \frac{dz}{z}$. The loop-momentum ℓ can then be written

$$\ell = |I\rangle[I| + z|1\rangle[4| \,. \tag{9.68}$$

It is quite remarkable that starting out with a fully on-shell construction of "on-shell diagrams" leads to a loop-integrand construction in which the loop-momentum is off-shell.

▶ **Exercise 9.8**

Use equivalence moves to prove the last two diagrams of (9.67) are equivalent to each other.

There is much more information in the connection between the Grassmannian and the on-shell diagrams, but we also have other fish to fry and birds to scare. If we have awoken your appetite for blob diagrams and Grassmannians and you are interested in learning more about their relations to permutations, stratifications, amalgamation, dimers, bipartite graphs, and quivers, you should take a look at reference [120] for more details.

In Chapter 9, we learned that the individual terms in a BCFW expansion of an N^{k-2}MHV n-point superamplitude are residues of a cyclically invariant integral-formula in the Grassmannian $Gr(k, n)$. In this language, we found that the different BCFW representations of the 6-point NMHV amplitude $A_6[1^+2^-3^+4^-5^+6^-]$ are related by a simple contour deformation. This was encoded in the 6-term identity (9.50). In *momentum supertwistor space* [60], the 6-term identity is promoted to the 5-bracket relation

$$[2, 3, 4, 6, 1] + [2, 3, 4, 5, 6] + [2, 4, 5, 6, 1] = [3, 1, 6, 5, 4] + [3, 2, 1, 6, 5]$$
$$+ [3, 2, 1, 5, 4]. \qquad (10.1)$$

We have already encountered this version of the 6-term identity in (5.63) when we discussed the equivalence of the BCFW recursion relations based on $[2, 3\rangle$ and $[3, 2\rangle$ super-shifts. The 5-brackets were defined in Section 5.4 as

$$[i, j, k, l, m] \equiv \frac{\delta^4(\chi_{iA}\langle jklm\rangle + \text{cyclic})}{\langle ijkl\rangle\langle jklm\rangle\langle klmi\rangle\langle lmij\rangle\langle mijk\rangle}, \qquad (10.2)$$

with 4-brackets $\langle ijkl\rangle \equiv \epsilon_{IJKL}Z_i^I Z_j^J Z_k^K Z_l^L$ involving the bosonic components $Z_i^I = (|i\rangle, [\mu_i|)$ of the $SU(2, 2|4)$ momentum supertwistors $\mathcal{Z}_i^A \equiv (|i\rangle^{\dot{a}}, [\mu_i|^a \mid \chi_{iA})$, $A = (\dot{a}, a, A)$. The $[\mu_i|$ are defined by the incidence relations (5.44).

Recall that momentum conservation is automatic for momentum twistors, so the 6-term identity (10.1) must hold for any six momentum twistors; it is not specific to the NMHV 6-point amplitude but is an intrinsic property of the 5-brackets (10.2). Hence it seems worthwhile to try to understand the structure of (10.1) better. Using cyclic and reflection symmetry of the 5-brackets, we can rewrite (10.1) as

$$[1, 2, 3, 4, 5] - [2, 3, 4, 5, 6] + [3, 4, 5, 6, 1] - [4, 5, 6, 1, 2]$$
$$+ [5, 6, 1, 2, 3] - [6, 1, 2, 3, 4] = 0. \qquad (10.3)$$

Pretend for a moment that we do not know what the 5-brackets are. Consider a fully antisymmetric 5-bracket $\langle i, j, k, l, m\rangle$ defined as the contraction of five 5-component vectors $Z_i^{\mathcal{I}}$ with a 5-index Levi-Civita tensor. Such an object would satisfy the 6-term Schouten identity

$$\langle 1, 2, 3, 4, 5\rangle Z_6^{\mathcal{I}} - \langle 2, 3, 4, 5, 6\rangle Z_1^{\mathcal{I}} + \langle 3, 4, 5, 6, 1\rangle Z_2^{\mathcal{I}}$$
$$- \langle 4, 5, 6, 1, 2\rangle Z_3^{\mathcal{I}} + \langle 5, 6, 1, 2, 3\rangle Z_4^{\mathcal{I}} - \langle 6, 1, 2, 3, 4\rangle Z_5^{\mathcal{I}} = 0. \qquad (10.4)$$

Since this looks quite similar to (10.3), including relative signs, we might be tempted to think that (10.3) somehow arises as a Schouten identity. This is of course too speculative: the 5-brackets $[i, j, k, l, m]$ really represent rational functions of the 4-component momentum twistors Z_i, not some 5-index objects contracted with a 5-index Levi-Civita tensor. However, we can entertain the idea a little further. Could the 5-bracket (10.2) be written in terms of some new 5-vectors? The fermionic variables χ_{iA} appear in (10.2) on a different footing than their bosonic counterparts. In the name of democracy, let us define the following purely bosonic 5-component vector

$$Z_i^{\mathcal{I}} = \begin{pmatrix} Z_i^I \\ \chi_i \cdot \psi \end{pmatrix}, \qquad \mathcal{I} = 1, \ldots, 5, \tag{10.5}$$

where $\chi_i \cdot \psi = \chi_i^A \psi_A$ is $SU(4)$-invariant and ψ_A is an auxiliary Grassmann variable common for all external particles $i = 1, 2, \ldots, n$. If we define $\langle i, j, k, l, m \rangle$ as the contraction of five of these 5-vectors with a 5-indexed Levi-Civita tensor, then they will satisfy the Schouten identity (10.4) – but that is not what we are after, so read on.

To write the 5-bracket $[i, j, k, l, m]$ in terms of the 5-vectors (10.5), we must remove the auxiliary variable ψ_A. Since it is fermionic, this can be done via a Grassmann-integration: one finds that the 5-bracket (10.2) can be written as

$$[i, j, k, l, m]$$
$$= \frac{1}{4!} \int d^4\psi \, \frac{\langle i, j, k, l, m \rangle^4}{\langle 0, i, j, k, l \rangle \langle 0, j, k, l, m \rangle \langle 0, k, l, m, i \rangle \langle 0, l, m, i, j \rangle \langle 0, m, i, j, k \rangle}, \tag{10.6}$$

where $\langle ijklm \rangle \equiv \epsilon_{IJKLM} Z_i^I Z_j^J Z_k^K Z_l^L Z_m^M$ and we have introduced the auxiliary reference 5-vector

$$Z_0^{\mathcal{I}} = \begin{pmatrix} 0 \\ 0 \\ 0 \\ 0 \\ 1 \end{pmatrix}. \tag{10.7}$$

The representation (10.6) is certainly not just contractions of five 5-vectors with a Levi-Civita tensor, so the origin of the identity in (10.3) is not a Schouten identity. But let us not give up just yet, for it will be worthwhile to examine (10.6) further.

Since the integral $\int d^4\psi$ is universal for all 5-brackets, we ignore it for the time being and focus on the integrand of (10.6). Each $Z_i^{\mathcal{I}}$ appears an equal number of times in the numerator and the denominator, so the integrand is invariant under $Z_i^{\mathcal{I}} \to t_i Z_i^{\mathcal{I}}$ for each $i = 1, 2, \ldots, n$. In other words, the 5-vectors $Z_i^{\mathcal{I}}$ appear projectively in (10.6), and therefore we can think of the $Z_i^{\mathcal{I}}$ as homogeneous coordinates of points in projective space $\mathbb{CP}^4$. The presence of the reference vector $Z_0^{\mathcal{I}}$ in the denominator breaks projective invariance, but only at this particular point.

There is an analogous case where we have encountered something similar. The momentum twistors Z_i^I in Section 5.4 are defined projectively and are elements in $\mathbb{CP}^3$. The map in Figure 5.1 shows how to relate momentum twistors Z_i^I with points y_i in dual space. Specifically, the distance between two points y_i and y_j in dual space is

$$y_{ij}^2 = \frac{\langle i-1, i, j-1, j \rangle}{\langle i-1, i \rangle \langle j-1, j \rangle} = \frac{\langle i-1, i, j-1, j \rangle}{\langle I_0, i-1, i \rangle \langle I_0, j-1, j \rangle}. \tag{10.8}$$

The first equality is simply (5.52): it has a momentum twistor 4-bracket in the numerator and regular angle brackets in the denominator. In the second equality we have rewritten the denominator in a more suggestive form involving only 4-brackets, at the cost of introducing a reference bi-twistor I_0^{IJ} defined as

$$I_0^{IJ} = \begin{pmatrix} 0 & 0 \\ 0 & \epsilon_{\dot{a}\dot{b}} \end{pmatrix}. \tag{10.9}$$

In the literature, I_0 is often referred to as the **infinity twistor**,[1] and its role is to break $SL(4)$ conformal invariance and provide a preferred metric for the definition of distance.

The expression (10.8) is similar to the integrand of (10.6): both are projectively defined, except for the reference bi-twistor/vector. The reference bi-twistor appears twice in the denominator of (10.8), reflecting the fact that this expression gives the distance between the two points i and j. An analogous expression involving three points and a reference vector appearing thrice defines the area of a triangle. And so on. The appearance of the reference vector $Z_0^{\mathcal{I}}$ five times in the denominator of (10.6) gives us a hint that *the rational integrand in (10.6) is the volume of a geometric figure defined by five points in* $\mathbb{CP}^4$! In the following, we pursue the interpretation of 5-brackets as volumes of simplices and their sum – the superamplitudes – as volumes of **polytopes**. Definitions and explanations follow next.

10.1 Volume of an n-simplex in $\mathbb{CP}^n$

Let us begin with the concepts of polytopes and simplices before reconnecting with the motivation above.

Polytopes and simplices: definitions and examples

We are all familiar with polygons: triangles, squares (or more generally quadrilaterals), pentagons, hexagons, chiliagons, star-shapes, etc. These are figures in the plane bounded by a finite number of straight line-segments. Their 3-dimensional analogues – tetrahedrons, cubes, prisms, dodecahedrons, etc – are solids whose faces are polygons. The n-dimensional versions of polygons and polyhedrons are called **polytopes** or **n-polytopes**. A 2-polytope is a polygon and a 3-polytope is a polyhedron.

A simplex is in a sense the simplest example of an polytope. To define it, recall first that a **convex set** C (in, for example, $\mathbb{R}^n$ or $\mathbb{CP}^n$) has the property that the line segment between any two points in C lies entirely in C. In the plane, triangles are convex, but star-shaped polygons are not. Given a set of points S, the **convex hull** of S is the intersection of all convex sets containing S. Examples from the plane: 1) the convex hull of a circle is the closed disk bounded by the circle; 2) the convex hull of three points is a triangle. Adding

---

[1] The infinity twistor in (10.9) corresponds to a flat space metric. For AdS$_4$ it is given as $I_0^{IJ} = \begin{pmatrix} \epsilon_{ab}\Lambda & 0 \\ 0 & \epsilon_{\dot{a}\dot{b}} \end{pmatrix}$, where Λ is the cosmological constant.

a fourth point that lies inside the triangle, the convex hull of the four points is the same triangle. A fourth point outside the triangle (but in the same plane) gives a convex hull that is a convex quadrilateral, or in pictures:

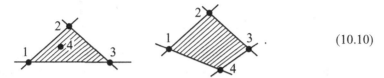

$$(10.10)$$

An n-**simplex** is the convex hull of a set of $n+1$ points. Examples:

$$
\begin{aligned}
&\text{0-simplex} = \text{a point,}\\
&\text{1-simplex} = \text{line segment,}\\
&\text{2-simplex} = \text{triangle,}\\
&\text{3-simplex} = \text{tetrahedron.}
\end{aligned}
\qquad (10.11)
$$

An n-simplex is bounded by $n+1$ $(n-1)$-simplices which intersect each other in $\binom{n+1}{2}$ $(n-2)$-simplices. For $n+1$ generic points in $\mathbb{R}^n$, an n-simplex has an n-dimensional volume. (For $\mathbb{CP}^n$ it will be n-complex dimensional.) The volume of a polytope can be calculated by "tessellating" it into simplices, whose volumes are easier to calculate.

Now that we know what simplices and polytopes are, let us progress towards understanding how the integrand in (10.6) represents the volume of a 4-simplex in $\mathbb{CP}^4$, as claimed. As a warm-up, we begin in 2-dimensions with a 2-simplex (a triangle).

Area of a 2-simplex in $\mathbb{CP}^2$

The area of a triangle in a 2-dimensional plane can be computed as

$$
\text{Area}\left[\;\parbox{2cm}{\rule{0pt}{1cm}}\;\right] = \frac{1}{2}\begin{vmatrix} x_1 & x_2 & x_3 \\ y_1 & y_2 & y_3 \\ 1 & 1 & 1 \end{vmatrix}, \qquad (10.12)
$$

where the (x_i, y_i) are the coordinates of the three vertices.

▶ **Exercise 10.1**

If the area formula (10.12) is not familiar, you should derive it by showing that it is equivalent to the "$\frac{1}{2} \times$ base $\times$ height" formula that was imprinted on your brain in elementary school.

The 1s in the last row of (10.12) are redundant as we can write the same formula as a sum of the 2×2 minors. In physics, when faced with a redundancy we can choose to eliminate it or promote it to a feature. Choosing the latter, we define three 3-vectors along with a reference vector:

$$
W_{iI} = \begin{pmatrix} x_i \\ y_i \\ 1 \end{pmatrix}, \qquad Z_0^I = \begin{pmatrix} 0 \\ 0 \\ 1 \end{pmatrix}, \qquad I = 1, 2, 3. \qquad (10.13)
$$

The area can now be written

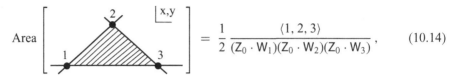

$$\text{Area}\left[\ \ \right] = \frac{1}{2}\frac{\langle 1,2,3\rangle}{(Z_0\cdot W_1)(Z_0\cdot W_2)(Z_0\cdot W_3)}\,, \qquad (10.14)$$

where the 3-bracket is the contraction of a 3-index Levi-Civita tensor with the three W_i vectors: $\langle 1,2,3\rangle = \epsilon^{IJK}W_{1I}W_{2J}W_{3K}$. Using (10.13), the $\langle 1,2,3\rangle$-numerator exactly equals the 3×3-determinant in (10.12), so you might consider the trivial dot-products $Z_0\cdot W_i = Z_0^I W_{iI} = 1$ in the denominator a provocation of your sense of humor. However, written in the form (10.14), the redundancy has been promoted to projective symmetry: the new area-formula (10.14) is invariant under scalings $W_i \to t_i\,W_i$. When the 3-vectors W_i are "gauge fixed" to the canonical form in (10.13), we immediately recover the original area formula. Since the triangle vertices are specified in terms of projectively defined 3-vectors, we can think of the triangle as an object in $\mathbb{CP}^2$ and the W_is as the homogeneous coordinates of the vertices.

The area formula in (10.14) involves an antisymmetric 3-bracket as well as the inner product of 3-vectors. To make contact with (10.6) and (10.8), we would like to have a representation that is given solely in terms of 3-brackets. To achieve this, it is useful to characterize the triangle by its edges instead of its vertices. We define a "dual space" whose points Z_a^I are associated with lines in W-space: a given line is defined as the set of points W_I satisfying the incidence relations

$$Z^I W_I = 0\,. \qquad (10.15)$$

Since Z^I is a vector in the 2-dimensional space $\mathbb{CP}^2$, the constraint indeed defines a 1-dimensional subspace, i.e. a line.

Now, to define the triangle in terms of three lines in dual space, note that each W_{iI} is characterized by lying simultaneously on two lines. Labeling the three edges of the triangle as a, b, and c, the vertex W_{1I} is the intersection of lines a and c, so that $Z_a^I W_{1I} = Z_c^I W_{1I} = 0$. These two constraints are easily solved and we have

$$\rightarrow\quad \begin{aligned} W_1 &= \langle *, Z_c, Z_a\rangle \\ W_2 &= \langle *, Z_a, Z_b\rangle \\ W_3 &= \langle *, Z_b, Z_c\rangle\,, \end{aligned} \qquad (10.16)$$

where the $*$ indicates the free index, e.g. $W_{1I} = \langle *, Z_c, Z_a\rangle = \epsilon_{IJK}Z_c^J Z_a^K$.

Plugging the map (10.16) into (10.14), we find that the area is now given as

$$\text{Area}\left[\ \ \right] = \frac{1}{2}\frac{\langle a,b,c\rangle^2}{\langle 0,b,c\rangle\langle 0,a,b\rangle\langle 0,c,a\rangle} \equiv \big[a,b,c\big], \quad (10.17)$$

where $\langle a,b,c\rangle = \epsilon_{IJK}Z_a^I Z_b^J Z_c^K$ and "0" indicates the reference vector Z_0^I introduced in (10.13).

▶ **Exercise 10.2**

Show that (10.17) follows from (10.14).

As advertised earlier, we now see that the "volume" (i.e. area) of a 2-simplex is given by a rational function whose denominator is the product of all 3-brackets involving two of the edge variables Z_i^I, $i = a, b, c$, and a reference vector Z_0^I. Requiring projective invariance for each Z_i^I with $i = a, b, c$ uniquely fixes the numerator. For later convenience, we have introduced the notation $[a, b, c]$ to denote the volume (10.17). Note that the area-formula (10.17) comes with an "orientation" in the sense that $[a, b, c]$ is fully antisymmetric in a, b, c.

Volume of an n-simplex in $\mathbb{CP}^n$

The expression in (10.17) can be generalized to the volume of an n-simplex in $\mathbb{CP}^n$: we denote it by an antisymmetric $(n+1)$-bracket

$$[Z_{i_1}, \dots Z_{i_{n+1}}] = \frac{1}{n!} \frac{\langle i_1, i_2, \dots, i_{n+1} \rangle^n}{\langle 0, i_1, \dots, i_n \rangle \langle 0, i_2, \dots, i_{n+1} \rangle \cdots \langle 0, i_{n+1}, i_1, \dots, i_{n-1} \rangle}, \quad (10.18)$$

where the angle-brackets are the contractions of $n+1$ $\mathbb{CP}^n$-vectors Z_i with an $(n+1)$-index Levi-Civita. The $n+1$ variables $Z_{i_1}, \dots, Z_{i_{n+1}} \in \mathbb{CP}^n$ carry the information about the $n+1$ boundaries of the n-simplex as follows. For a given vector Z_i^I, the set of W_Is satisfying the incidence relation $Z_i^I W_I = 0$ span an $(n-1)$-dimensional subspace of $\mathbb{CP}^n$. These subspaces contain the $n+1$ faces of the n-simplex; each face is an $(n-1)$-simplex.

In the example of the 2-simplex in $\mathbb{CP}^2$, the 3 Z_is label the 1-dimensional lines a, b, c bounding the triangle. Each pair of lines intersect in a point that is a vertex of the triangle: we can label the vertices (a, b), (b, c), and (c, a). They are defined in terms of the Z_is in (10.16) and the denominator of the volume formula (10.17) is the dot-product of all vertex point vectors with the reference vector.

As a second example, consider a 3-simplex (tetrahedron) in $\mathbb{CP}^3$. The four 4-component homogeneous coordinates of (dual) $\mathbb{CP}^3$ – Z_a^I, Z_b^I, Z_c^I, Z_d^I – have 2-dimensional orthogonal complements spanned by the W_Is satisfying $Z_i^I W_I = 0$. Pairwise, these generic 2-planes intersect in a line: this gives six lines, (a, b), (b, c), etc., that define the 1-simplex edges of the tetrahedron. Three generic 2-planes in $\mathbb{CP}^3$ intersect in a point: this defines the four vertices of the tetrahedron and we label them (a, b, c), (b, c, d), (c, d, a), and (d, a, b), as illustrated here:

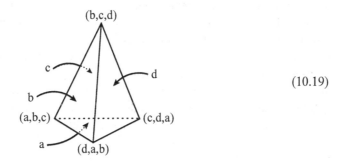

$$(10.19)$$

Just as in the case of the triangle, the denominator of the volume formula (10.18) is the product of each vertex coordinate dotted into the reference vector Z_0. The numerator compensates the scaling of each "face"-variable $Z_a^I, Z_b^I, Z_c^I, Z_d^I$ to make the volume formula projective.

Now, here comes the highlight of your day: for $n = 4$ the 5-bracket volume expression (10.18) for a 4-simplex is identical to the integrand in the amplitude 5-bracket expression (10.6)! This verifies our statement at the beginning of the chapter that the rational function in (10.6) is indeed the volume of a 4-simplex in $\mathbb{CP}^4$. The denominator factors in (10.18) involve the five vertices of the 4-simplex. Since an NMHV tree superamplitude is a sum of 5-brackets, we are led to view the amplitude as a volume of a polytope in $\mathbb{CP}^4$. We realize this expectation in the next section and discuss its consequences.

10.2 NMHV tree superamplitude as the volume of a polytope

The simplest NMHV case is the 5-point (anti-MHV)superamplitude

$$A_5^{\mathrm{NMHV}}[1, 2, 3, 4, 5] = A_5^{\mathrm{MHV}} \times [1, 2, 3, 4, 5]. \tag{10.20}$$

Thus, up to the MHV factor, A_5^{NMHV} is the volume of a 4-simplex in $\mathbb{CP}^4$.

Next, for the NMHV 6-point superamplitude, consider the $[2, 3\rangle$ super-BCFW representation on the LHS of (10.1):

$$A_6^{\mathrm{NMHV}}[1, 2, 3, 4, 5, 6] \propto \underbrace{[2, 3, 4, 6, 1]}_{\langle 4,6,1,2\rangle,\ \langle 2,3,4,6\rangle} + \underbrace{[2, 3, 4, 5, 6]}_{\langle 6,2,3,4\rangle,\ \langle 4,5,6,2\rangle} + \underbrace{[2, 4, 5, 6, 1]}_{\langle 2,4,5,6\rangle,\ \langle 6,1,2,4\rangle}. \tag{10.21}$$

Apart from the overall MHV factor, the 6-point NMHV superamplitude is the sum of the volumes of three 4-simplices in $\mathbb{CP}^4$; we expect this to be the volume of a polytope obtained by somehow gluing the three simplices together. But how exactly does this work? To address this question, it is useful to examine the poles in the 5-brackets.

Recall from (5.50) and (5.52) that momentum twistor 4-brackets $\langle i - 1, i, j - 1, j \rangle$ in the denominator give local poles, whereas other 4-brackets, such as $\langle 1, 2, 4, 6 \rangle$, give spurious "non-local" poles. Examining the denominator terms of the 5-brackets in (10.21), we find that each of them has two spurious poles; they are listed under each 5-bracket. The spurious poles come in pairs – for example $\langle 4, 6, 1, 2 \rangle$ and $\langle 6, 1, 2, 4 \rangle$ in the first and third 5-brackets – and cancel in the sum (10.21), as required by locality of the physical amplitude. In the geometric description of a 5-bracket as a 4-simplex in $\mathbb{CP}^4$, each of the five factors in the denominator of the volume-expression (i.e. the 5-bracket) is determined by a vertex of the associated 4-simplex. In particular, spurious poles must be associated with vertices in $\mathbb{CP}^4$ that somehow "disappear" from the polytope whose volume equals the sum of the simplex-volumes in (10.21). We now discuss how the "spurious" vertices disappear in the sum of simplices. Let us start in $\mathbb{CP}^2$ where the polytopes are easier to draw.

Polytopes in $\mathbb{CP}^2$

In 2-dimensions, consider the 4-edge polytope

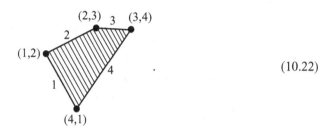

$$(10.22)$$

All vertices for this $\mathbb{CP}^2$ "amplitude" are defined by adjacent edges and are in this sense local. We would like to compute the area of the 2-polytope (10.22) using 2-simplex volumes $[a, b, c]$. There are several different ways to do this, corresponding to different triangulations of the polytope. As an example, introduce a "non-local" point $(1, 3)$ as the intersection of lines 1 and 3. The resulting triangulation is

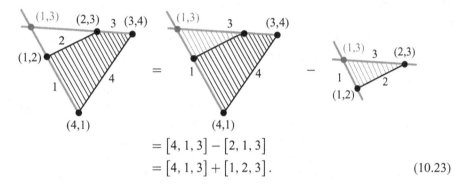

$$= [4, 1, 3] - [2, 1, 3]$$
$$= [4, 1, 3] + [1, 2, 3]. \qquad (10.23)$$

The area of the 4-edge polytope is given by the difference of two triangular areas. The non-local vertex $(1, 3)$ appears in both triangles. Comparing the last two lines, the sign of the 3-bracket indicates the orientation of the triangle with respect to a particular predetermined ordering of all edges (or, in higher dimensions, boundaries).

It is useful to also consider another triangulation, so introduce the point $(2, 4)$:

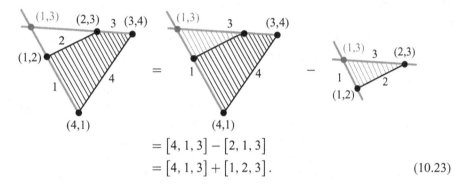

$$= [1, 2, 4] - [2, 4, 3]$$
$$= [1, 2, 4] + [2, 3, 4]. \qquad (10.24)$$

The two triangulations (10.23) and (10.24) compute the same area ("amplitude"), so we have a $\mathbb{CP}^2$ version of the identity (10.1), namely $[4, 1, 3] + [1, 2, 3] = [1, 2, 4] + [2, 3, 4]$ which can also be written

$$[2, 3, 4] - [1, 3, 4] + [1, 2, 4] - [1, 2, 3] = 0. \qquad (10.25)$$

▶ **Exercise 10.3**

Suppose the 4-vertex polytope in the example above was not convex as drawn in (10.24): show that the volume of a non-convex 4-vertex polytope can also be written $[1, 2, 4] + [2, 3, 4]$.

Polytopes in $\mathbb{CP}^4$

Extending the simple $\mathbb{CP}^2$ example to $\mathbb{CP}^4$, one finds that the BCFW representation of a 6-point NMHV tree superamplitude corresponds to a triangulation of the associated polytope by introduction of three new auxiliary vertices. This allows one to use the given external data, the boundaries of the polytope, to efficiently construct the corresponding triangulation. Efficiency here means using a minimum number of 4-simplices; we come back to this point in Section 10.4. Different BCFW constructions simply correspond to different choices of auxiliary vertices. As an example, the auxiliary vertices for the BCFW representation in (10.21) are $(2, 4, 5, 6)$, $(6, 1, 2, 4)$ and $(2, 3, 4, 6)$. Note that these exactly label the spurious poles in (10.21).

Let us now see how the removal of an auxiliary vertex works in $\mathbb{CP}^4$. Since it can be slightly challenging to draw a 4-dimensional object on paper, we go to the 3d boundary of the 4-dimensional polytope. Specifically, at the 3d boundary defined by $Z_1 \cdot W = 0$, only the two simplices $[2, 3, 4, 6, 1]$ and $[2, 4, 5, 6, 1]$ in (10.21) contribute, and their projections to the boundary are the tetrahedrons defined by the faces Z_2, Z_3, Z_4, and Z_6 and, respectively, by Z_2, Z_4, Z_5, and Z_6. The two boundary tetrahedrons share the non-local vertex $(1, 2, 4, 6)$, which we simply label $(2, 4, 6)$ on the boundary. Since this vertex does not appear in other terms of (10.21), we should be able to visualize its cancellation geometrically on the boundary defined by Z_1. On the 3d subspace, the superamplitude contains the combination

$$[2, 4, 5, 6, 1] + [2, 3, 4, 6, 1] \xrightarrow{Z_1 \text{ bdr}} [6, 2, 4, 5] - [6, 2, 4, 3] = \text{vol(bdr polytope)}. \qquad (10.26)$$

On the RHS we have arranged the common faces, Z_6, Z_2, and Z_4, to appear in the same order to facilitate the geometrical interpretation. The pictorial representation of (10.26) is

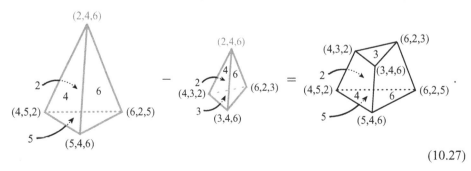

$$(10.27)$$

The "non-local" auxiliary vertex $(2, 4, 6)$ indeed "cancels" in the sum and leaves behind the volume of the 3-dimensional polytope with five faces and six *local* vertices! To see that the remaining vertices are local, remember that we are in the subspace $Z_1 \cdot W = 0$, so each vertex is really represented as $(1, *, *, *)$ in $\mathbb{CP}^4$. Thus each of the six vertices in (10.27), namely

$$(1, 2, 3, 4), \quad (1, 2, 3, 6), \quad (1, 3, 4, 6), \quad (1, 4, 5, 2), \quad (1, 4, 5, 6), \quad (1, 2, 5, 6), \quad (10.28)$$

involves two pairs of adjacent labels, and therefore – by (5.52) – they correspond to local poles. Thus we conclude that on the subspace $Z_1 \cdot W = 0$, which involves only two of the simplices in (10.21), the amplitude is free of non-local vertices. One can similarly understand the cancellation of the two other spurious poles in (10.21).

We have found that the 6-point NMHV tree superamplitude is given by the volume of a polytope in $\mathbb{CP}^4$. It is defined as the sum of the three 4-simplices in (10.21) and its six boundaries are in 1-to-1 correspondence with the momentum supertwistors Z_i^I, $i = 1, \ldots, 6$. Different BCFW representations correspond to the different tessellations of the polytope into 4-simplices; each representation requires introduction of "spurious" vertices and the associated spurious poles cancel because they are absent in the original polytope. The vertices of the polytopes are all local and can be characterized as the nine quadruple intersections $(i, i + 1, j, j + 1)$ of the six boundaries determined by Z_i^I.

The polytope interpretation of the amplitudes was first presented by Hodges [24] with the goal of geometrizing the cancellation of spurious poles in the BCFW expansion. Building on Hodges' work, the authors of [130] constructed the representation of the NMHV superamplitude where both dual superconformal symmetry and locality are manifest.

10.3 The boundary of simplices and polytopes

We have studied the volumes of simplices and polytopes, but their boundaries also have interesting properties. Let us again start with a simple triangle in $\mathbb{CP}^2$,

$$(10.29)$$

The subspace defined by $Z_1 \cdot W = 0$ contains part of the boundary of the triangle, namely the line segment bounded by the intersections of lines 2 and 3 with line 1. The length of the line segment is just the projection to the subspace defined by Z_1, namely $[2, 3]$. Note that since the "volumes" (i.e. lengths) of the line segments are defined with a choice of sign, we have to pick an orientation for each: here and in the following, we pick the orientation of the faces to point into the volume of the polytope that they are bounding. With this choice of orientation, the circumference is $[12] + [23] + [31]$.

To see a little more structure, consider the tetrahedron in $\mathbb{CP}^3$,

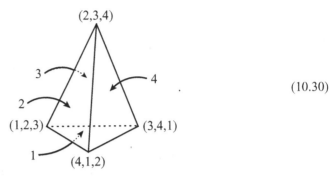

$$(10.30)$$

The volume is $[1, 2, 3, 4]$. The 2-plane defined by Z_1 contains the face bounded by the intersections of the plane 1 with the planes 2, 3, and 4. Therefore the area of this face is $[2, 3, 4]$. Keeping careful track of the orientations of the faces, we find that the total area of boundary of the tetrahedron is $[2, 3, 4] + [1, 4, 3] + [2, 4, 1] + [2, 1, 3]$.

We can summarize the results for the boundary "volumes" so far as

$$\text{bdr of } \mathbb{CP}^2 \text{ triangle}: \quad \partial[1, 2, 3] = [2, 3] - [1, 3] + [1, 2],$$

$$\text{bdr of } \mathbb{CP}^3 \text{ tetrahedron}: \quad \partial[1, 2, 3, 4] = [2, 3, 4] - [1, 3, 4] + [1, 2, 4] - [1, 2, 3].$$

$$(10.31)$$

This motivates us to define the **boundary operation** for any $(n-1)$-simplex:

$$\partial[123\ldots n] = \sum_{i=1}^{n}(-1)^{i+1}\left[1, 2, \ldots, i-1, i+1, \ldots, n\right]. \qquad (10.32)$$

▶ **Exercise 10.4**

The boundary of a boundary vanishes, so our definition (10.32) had better yield a nilpotent operator, $\partial^2 = 0$. Show that the action of ∂^2 on any n-simplex is zero.

At this stage you might have noticed the similarity between the RHS of the tetrahedron boundary identity (10.31) and the $\mathbb{CP}^2$ vanishing identity (10.25). This is easy to understand: in $\mathbb{CP}^2$, we cannot construct a 3-simplex $[1, 2, 3, 4]$ with a 3d volume, so in particular the boundary of such a formal object must vanish:

$$\mathbb{CP}^2: \quad 0 = \partial[1, 2, 3, 4] = [2, 3, 4] - [1, 3, 4] + [1, 2, 4] - [1, 2, 3]. \qquad (10.33)$$

This gives another geometric interpretation of the $\mathbb{CP}^2$ BCFW identity (10.25).

Similarly, a 5-simplex $[1, 2, 3, 4, 5, 6]$ in $\mathbb{CP}^4$ has vanishing boundary:

$$\mathbb{CP}^4: \quad 0 = \partial[1, 2, 3, 4, 5, 6] \equiv [2, 3, 4, 5, 6] - [3, 4, 5, 6, 1] + [4, 5, 6, 1, 2]$$
$$- [5, 6, 1, 2, 3] + [6, 1, 2, 3, 4] - [1, 2, 3, 4, 5]. \qquad (10.34)$$

The RHS is exactly the 6-term identity (10.3) which originated from the equivalence of different super-BCFW shifts (10.1). This was the identity that motivated our study at the beginning of the chapter: we understand now that it is not a Schouten identity, but that it has an interpretation as the vanishing boundary of a formal 5-simplex in $\mathbb{CP}^4$.

Let us now take a look at the action of the boundary operation on a superamplitude. As per usual, we start with $\mathbb{CP}^2$ to get intuition for the problem. Consider the triangulation used in (10.23) to calculate the volume of a 4-sided polygon in $\mathbb{CP}^2$,

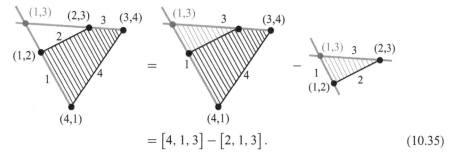

$$= [4, 1, 3] - [2, 1, 3]. \tag{10.35}$$

We apply the boundary operator to each 2-simplex and find:

$$\begin{aligned} \partial[4, 1, 3] &= [1, 3] - [4, 3] + [4, 1], \\ \partial[2, 1, 3] &= [1, 3] - [2, 3] + [2, 1]. \end{aligned} \tag{10.36}$$

The RHS of each equation is the circumference of the respective triangles. Let us try to interpret $\partial([4, 1, 3] - [2, 1, 3])$. It contains $([4, 1] - [2, 1])$: this is a difference of lengths of two line segments in the subspace defined by Z_3 and thus it is the length of the side labeled 3 in the 4-sided polygon on the LHS of (10.35). Similarly, $([4, 3] - [2, 3])$ is the length of side 1 of the polygon. Now we are left with two terms that are both labeled $[1, 3]$ in (10.36). It is tempting to cancel these two terms, but this is not quite correct: $[1, 3]$ in $\partial[4, 1, 3]$ lives in the subspace defined by Z_4, while $\partial[2, 1, 3]$ is in the Z_2-subspace. So the two $[1, 3]$s are the lengths of the sides 2 and 4 in the polygon (10.35). Why does their difference show up in $\partial([4, 1, 3] - [2, 1, 3])$ instead of their sum? Easy: that is because they have the opposite orientations: in our conventions, side 4 in the big triangle in (10.35) is oriented to point into the polygon, but side 2 in the small triangle points out of the 4-sided polygon. Flipping the orientation and labeling the 2-brackets by the subspace Z_i they live on, we see that the difference of the two terms in (10.36) exactly calculates the circumference of the 4-sided polygon on the LHS of (10.35):

$$\partial([4, 1, 3] - [2, 1, 3]) = ([4, 3] - [2, 3])_{Z_1} + [3, 1]_{Z_2} + ([4, 1] - [2, 1])_{Z_3} + [1, 3]_{Z_4}. \tag{10.37}$$

The boundary operator ∂ was introduced [130] as a formal operation useful for studying the cancellation of spurious poles in the BCFW expansion, without emphasis on the interpretation as the "boundary volume" we have presented here. So let us now comment on the application of ∂ in [130]. Note that for a simplex, there is a unique point "opposite" each face: in particular in the triangles in (10.35) the point labeled $(1, 3)$ is the non-local "spurious" point that sits across from the line segments $[1, 3]_{Z_2}$ and $[3, 1]_{Z_4}$, respectively. So since the two $[1, 3]$s define the same point $(1, 3)$, one can in a *vertex-interpretation* of the boundary operation cancel them in $\partial([4, 1, 3] - [2, 1, 3])$: one can think of this as the cancellation of the spurious point in the $(1, 3)$ in this particular triangulation. To distinguish the vertex-interpretation from the boundary volume, we include a V (for vertex) with each term; then we write

$$\partial\big([4, 1, 3] - [2, 1, 3]\big) = V[3, 4] + V[4, 1] + V[2, 3] + V[1, 2]. \qquad (10.38)$$

Note how each term on the RHS is of the form $V[i, i + 1]$ indicating that the polytope has only local vertices; the non-local vertex $V[1, 3]$ canceled. This is the interpretation of the boundary operation given in [130].

▶ Exercise 10.5

As an example in $\mathbb{CP}^3$, consider the dissection (10.27) of the 5-faced polytope into two tetrahedrons. Keep careful track of the orientations of the boundaries to show that $\partial([6, 2, 4, 5] - [6, 2, 4, 3])$ calculates the surface area of the 5-faced 3-polytope on the RHS of (10.27). Next use the vertex-interpretation discussed above to show that the spurious poles cancel. Lift the example back to $\mathbb{CP}^4$ (remember that (10.27) was the projection on the subspace defined by Z_1) to see that each boundary vertex term is of the form $V[i, i + 1, j, j + 1]$ as in (10.28).

Enough of toy-examples! Let us compute the boundary of the NMHV 6-point tree super-amplitude in the BCFW representation

$$A_6^{\mathrm{NMHV}}[1, 2, 3, 4, 5, 6] = A_6^{\mathrm{MHV}} \times \big([1, 3, 4, 5, 6] + [3, 5, 6, 1, 2] + [5, 1, 2, 3, 4]\big). \qquad (10.39)$$

Using the vertex-interpretation, ∂ acts on the first two 4-simplices to give

$$\partial\big([1, 3, 4, 5, 6] + [3, 5, 6, 1, 2]\big) = V[3, 4, 5, 6] + V[4, 5, 6, 1] + V[6, 1, 3, 4]$$
$$+ V[5, 6, 1, 2] + V[2, 3, 6, 1] + V[2, 3, 5, 6]$$
$$+ V[1, 3, 4, 5] + V[1, 2, 3, 5]. \qquad (10.40)$$

All vertices on the RHS are local except the last two. Including the boundary of the third 5-bracket in (10.39), the non-local vertices cancel completely and we have

$$\partial\big([1, 3, 4, 5, 6] + [3, 5, 6, 1, 2] + [5, 1, 2, 3, 4]\big)$$
$$= \sum_{i=1}^{6} V[i, i + 1, i + 2, i + 3] + \sum_{i=1}^{3} V[i, i + 1, i + 3, i + 4], \qquad (10.41)$$

where, as usually, the arguments are understood cyclically. This shows that, indeed, the boundary of the polytope that corresponds to our tree superamplitude A_6^{NMHV} contains only local vertices. The power of the boundary operation is that it makes the cancellation of spurious points clear without a need to draw any polyhedrals.

▶ Exercise 10.6

How many local vertices are there in the polytope corresponding to the 7-point NMHV tree superamplitude?

10.4 Geometric aftermath

The BCFW triangulation is an efficient representation of the tree-level NMHV superamplitudes in the sense that it involves only relatively few terms. However, we can imagine other triangulations. For example, consider the 4-sided polygon in $\mathbb{CP}^2$. Introducing an auxiliary point W^* inside the polygon, we can triangulate it as

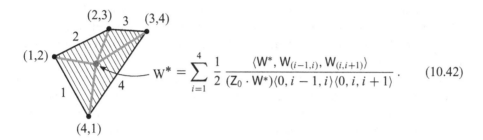

$$W^* = \sum_{i=1}^{4} \frac{1}{2} \frac{\langle W^*, W_{(i-1,i)}, W_{(i,i+1)} \rangle}{(Z_0 \cdot W^*)\langle 0, i-1, i \rangle \langle 0, i, i+1 \rangle}. \qquad (10.42)$$

This gives a 4-term expression for the volume of the polygon, as opposed to the 2-term BCFW triangulations in (10.23) or (10.24). In this sense, BCFW is more efficient. The representation (10.42) may remind you of another representation of scattering amplitudes, namely the CSW expansion (or MHV vertex expansion) of Section 3.4. It is actually not quite the same; CSW in momentum twistor space involves a reference supertwistor $\mathcal{Z}^* = (0, |X], 0)$ instead of W^*; see [131] for details.

As yet another way to calculate the amplitudes, we might ask if there is a triangulation that does not give spurious poles? A prescription for such a representation was given in [130] for the tree-level NMHV superamplitudes. To give a hint of how it works, consider the boundary $Z_1 \cdot W = 0$ that we also analyzed in (10.27). Instead of the 2-term triangulation applied in (10.27) at the cost of a non-local vertex, we can triangulate the 5-sided polytope as

$$(10.43)$$

The manifestly local tessellation of the polytope gives more terms than the BCFW representation. You can find the general expression in [130].

We have argued that each n-point tree-level NMHV superamplitude of $\mathcal{N} = 4$ SYM can be interpreted as the volume of a polytope in $\mathbb{CP}^4$. It should be rather obvious by now that the reverse is not true: not all polytopes in $\mathbb{CP}^4$ correspond to tree-level superamplitudes in $\mathcal{N} = 4$ SYM. An example is the polytope obtained by gluing together two of the three simplices in the BCFW representation (10.39): that is a perfectly fine polytope, but it has non-local poles so does not correspond to a physical amplitude. The color-ordering plays a key role in interpreting the superamplitudes polytopes. It is of course an interesting question if this geometric picture can be extended beyond the leading color level – or if other polytopes might have interpretations in terms of scattering processes.

The current discussion of tree superamplitudes utilizes the dual superconformal invariance of planar $\mathcal{N} = 4$ SYM, focusing on the 5-brackets invariants. Since the tree amplitudes of pure Yang–Mills theory can be projected out from $\mathcal{N} = 4$ SYM superamplitudes, a similar analysis can be applied to pure YM as well. In fact, it was in pure Yang–Mills theory that Hodges first realized the polytope picture [24].

The polytope interpretation described here is valid for NMHV n-point tree superamplitudes as well as 1-loop n-point MHV integrands in planar $\mathcal{N} = 4$ SYM [130]. The generalization is not obvious. Tree-level N^KMHV superamplitudes involve sums of products of K 5-brackets, so a geometric interpretation in terms of simplex-volumes is not straightforward. There is nonetheless a geometrization of all N^KMHV tree superamplitudes and loop-integrands in planar $\mathcal{N} = 4$ SYM: it goes under the name of the ***amplituhedron*** [132]. This is a polytope defined in a space whose coordinates are a union of momentum supertwistors and Grassmannian coordinates that extend the $\chi_i \cdot \psi$-construction in (10.5). For tree-level NMHV, the amplituhedron reduces to a dual of the polytope discussed here, with vertices and faces interchanged. The amplituhedron description makes locality manifest, while unitarity is an emergent property. BCFW arises as a particular triangulation. Loops appear from integrating out pairs of "hidden" points, in a somewhat similar way to the description of the loop-integrands in Chapter 7.

It is curious that in connection to amplitudes, polytopes can appear in different guises. An example, different from our discussion so far, is the observation [133] that a 1-loop box integral can be interpreted as the volume of a tetrahedron in AdS$_5$. The vertices of the tetrahedron are the four dual region variables y_i (in the embedding formalism) and the edges are geodesics in AdS$_5$. Since all $\mathcal{N} = 4$ SYM 1-loop amplitudes are given by an expansion in box integrals, the amplitudes can be interpreted as sums of volumes of such AdS$_5$ tetrahedrons, weighted by the appropriate box-coefficients. Non-planar 1-loop amplitudes can be given as linear combinations of planar ones, so the same conclusion extends to non-planar amplitude as well [134].

We have encountered many different representations of the tree amplitude and have now seen a unifying geometric interpretation. But there is yet another representation of amplitudes that we had a glimpse of in Section 2.5, namely the "BCJ representation" in which the color- and kinematic-structures enter on a dual basis. For such representations the amplitude can be manifestly local. We discuss BCJ further in Chapter 13, but note here that the color-ordering is crucial for the relation between polytopes and amplitudes: it allows us to relate the polytope to the momentum space representation of an amplitude and this is key for the statements about locality. Polytope or amplituhedron interpretations for the planar BCJ representation, non-planar, or non-color-ordered amplitudes are still open problems.

Amplitudes in dimensions $D \neq 4$

Just in case it slipped your mind, our discussion up to now has focused on scattering amplitudes in $D=4$ spacetime dimensions. There is a good reason for this: for one, this review was written in 3+1 dimensions (as far as we know) and this is where our particle physics experiments take place. And secondly, the power of the $D=4$ spinor helicity formalism and its extensions to twistors and momentum twistors allows us to explore the rich and exciting mathematical structure of 4d scattering amplitudes, especially those in planar $\mathcal{N} = 4$ SYM. However, there are also interesting quantum field theories in other dimensions; in this chapter we take a look at their scattering amplitudes. We discuss $D=6$ briefly, but otherwise our eyes are on $D=3$, particularly on the $\mathcal{N} = 8$ and $\mathcal{N} = 6$ superconformal theories known as BLG and ABJM.

11.1 Helicity formalism in $D \neq 4$

We have often emphasized that the modern on-shell approach relies heavily on having a "good" set of variables to parameterize the on-shell degrees of freedom: "good" typically means that the variables trivialize the kinematic constraints and transform linearly under the global symmetries of the theory. This is realized strikingly by the supertwistors and momentum supertwistors of planar $\mathcal{N} = 4$ SYM, but the trivialization of the massless on-shell condition $p_i^2 = 0$ in the spinor helicity formalism with $|i\rangle$ and $|i]$ was our starting point. So this is also where we begin for $D \neq 4$.

To parameterize massless kinematics in D-dimensions, consider bosonic spinors that carry a spinor index A of the Lorentz group $\mathrm{Spin}(1, D-1)$ and a fundamental index a of the little group $SO(D-2)$:

$$\lambda_{i\mathsf{a}}^A \begin{array}{l} \leftarrow \text{Lorentz} \\ \leftarrow \text{little grp} \end{array}. \tag{11.1}$$

As per usual, $i = 1, 2, \ldots, n$ is a particle label. The spinor-type (Weyl, Majorana, etc.) will be specified when we specialize to a given dimension D. If the spinors are complex, there will be a conjugate spinor $\tilde{\lambda}$ whose A and a indices are in the appropriate conjugate representations.

The 4d Lorentz group is $\mathrm{Spin}(1,3) = SL(2, \mathbb{C})$ and the little group is $SO(2) = U(1)$. The spinors

$$D = 4: \qquad \tilde{\lambda}_{i-}^{\dot{a}} = |i\rangle^{\dot{a}}, \qquad \lambda_{i+}^a = [i|^a, \tag{11.2}$$

are Weyl-spinors, so the index A is the familiar $SL(2, \mathbb{C})$ indices $a, \dot{a}$. The little group index a is $+$ or $-$ depending on how the spinors transform under the $SO(2) = U(1)$ little group transformations. The $D = 4$ light-like momentum is written as the bi-spinor as the familiar relation

$$D = 4: \qquad p_i^{\dot{a}a} = -\tilde{\lambda}_{i-}^{\dot{a}} \lambda_{i+}^{\dot{a}} = -|i\rangle^{\dot{a}} [i|^a . \qquad (11.3)$$

As discussed in Section 2.2, p_i^μ is real when the spinors (11.2) are complex conjugate.

The relation (11.3) implies that the 2×2 matrix $p_i^{\dot{a}a}$ has rank 1 and therefore solves the $D = 4$ massless constraint $p_i^2 = -\det(p_i) = 0$. To see if a similar construction could be available in D dimensions, we count degrees of freedom. A real light-like vector has $D-1$ degrees of freedom with the -1 from the condition $p_i^2 = 0$.[1] So the strategy is to find a Spin$(1,D-1)$ spinor representation that allows forming a little group invariant bi-spinor with $D-1$ degrees of freedom. Here is how the counting works in $D = 4$. The complex 2-component spinor $\lambda_{i+}^a = [i|^a$ has four real degrees of freedom, and when combined with its complex conjugate $\tilde{\lambda}_{i-}^{\dot{a}} = |i\rangle^{\dot{a}}$, the resulting bi-spinor (11.3) is invariant under the $U(1)$ little group rotation. Thus subtracting out the $U(1)$ redundancy, we indeed have $4 - 1 = 3$ degrees of freedom, matching that of a real light-like vector in 4d. Now let us look at how the counting works in other dimensions.

▶ **Exercise 11.1**

Do the above counting for the case where the spacetime signature is $(--++)$.

For $D = 3$, the Lorentz group is Spin$(1,2)=SL(2, \mathbb{R})$ and the minimal spinor representation is a 2-component Majorana spinor λ_i^a, where $a = 1, 2$ is an $SL(2, \mathbb{R})$ index. The null momentum is given by

$$D = 3: \qquad p_i^{ab} = \lambda_i^a \lambda_i^b . \qquad (11.4)$$

For p_i real, the spinors λ_i^a may be either real or purely imaginary. Either way, they encode two real degrees of freedom. So the RHS of (11.4) has two degrees of freedom, the correct count for a 3d light-like vector. Note that no little group index was included on the spinors λ_i^a because the little group $\mathbb{Z}_2$ is discrete. It acts as $\lambda_i^a \to -\lambda_i^a$, indeed leaving the momentum (11.4) invariant.

For $D = 6$, we have Spin$(1,5)=SU^*(4)$ and the little group is $SO(4) = SU(2) \times SU(2)$. The $*$ on the $SU^*(4)$ indicates that it is pseudo-real.[2] We pick a chiral spinor λ_{ia}^A in the fundamental of $SU^*(4)$, so $A = 1, 2, 3, 4$. The spinor is chiral, as opposed to anti-chiral, because it is in the fundamental, not anti-fundamental, representation of $SU^*(4)$. The two $SU(2)$-factors of the little group belong to the chiral and anti-chiral spinors, respectively, so λ_{ia}^A carries a little group index a $= 1, 2$ of the chiral $SU(2)$ factor. A candidate for the

---

[1] An on-shell massive momentum has $p_i^2 = -m_i^2$, but we view the constraint as imposed on $D + 1$ degrees of freedom, p_i^μ and m_i.

[2] Pseudo-real means that for each group element g, the complex conjugate g^* is related to g via a similarity transformation $g = \Omega g^* \Omega^{-1}$, where Ω is an antisymmetric matrix. (If Ω is symmetric, then the representation is a real.)

light-like momentum can now be formed as the little group invariant bi-spinor

$$D = 6: \qquad p_i^{AB} = \lambda_i^{A\mathsf{a}} \lambda_{i\mathsf{a}}^B . \qquad (11.5)$$

This works to give the right number of degrees of freedom, namely five, for a massless momentum in 6d: the spinor $\lambda_i^{A\mathsf{a}}$ has 4×2 degrees of freedom, but we have to mod out by the little group $SU(2)$-factor, giving $4 \times 2 - 3 = 5$.

The results for $D = 3, 4, 6$ can be summarized as follows:

	Spin$(1, D-1)$	little group	$p^2 = 0$
$D = 3$	$SL(2, \mathbb{R})$	$\mathbb{Z}_2$	$p_i^{ab} = \lambda_i^a \lambda_i^b$
$D = 4$	$SL(2, \mathbb{C})$	$SO(2) = U(1)$	$p_i^{a\dot{a}} = -\lambda_i^a \tilde{\lambda}_i^{\dot{a}}$
$D = 6$	$SU^*(4)$	$SO(4) = SU(2) \times SU(2)$	$p_i^{AB} = \lambda_i^{A\mathsf{a}} \lambda_{i\mathsf{a}}^B$

$$(11.6)$$

How about general D dimensions? The strategy is to introduce a bosonic spinor $\lambda_{i\mathsf{a}}^A$ (where a transforms under the little group, or a subgroup as in the 6d example) and use it (and possibly its conjugate spinor) to form a (real) light-like vector as a little group invariant bi-spinor, e.g. $\lambda_{i\mathsf{a}}^A \tilde{\lambda}_i^{\mathsf{a}B}$. However, for this to encode a null momentum, the number of real degrees of freedom of the bi-spinor, modulo the number of little group generators, has to match that of a light-like vector:

$$\mathrm{DOF}\big[\lambda_{i\mathsf{a}}^A \tilde{\lambda}_i^{\mathsf{a}B}\big] - \#(\text{little group generators}) = D - 1 . \qquad (11.7)$$

This is a non-trivial constraint because the bi-spinor typically has more than $D - 1$ degrees of freedom. One has to find a minimal spinor representation with maximal little group redundancy; this was particularly clear in the 6d example above. Indeed, we know solutions to these constraints only for $D = 3, 4, 6$.[3]

▶ **Exercise 11.2**

What is the smallest possible number of degrees of freedom for a little group invariant bi-spinor in $D = 10$? [Hint: In $D = 10$, the minimum spinor representation is a Majorana–Weyl spinor; it has 16 real components.]

It is possible to reduce the number of independent spinor degrees of freedom further by imposing the equations of motion, i.e. the zero-mass Dirac equation. Now you may be puzzled, because back in Chapter 2 we set up the $D = 4$ spinor helicity formalism by requiring at the starting point that the spinors $|i\rangle$ and $|i]$ satisfied the Dirac–Weyl equation. For $D = 3, 4, 6$, this approach is equivalent: the Lorentz contraction of the $(D - 1)$-component bi-spinor with one of its spinors gives zero, so the momentum space form of the massless Dirac equation is automatic.

For $D \neq 3, 4, 6$, setting up a spinor helicity formalism is possible but the resulting spinors are constrained in the sense that the Dirac equation is imposed as a non-trivial condition [135–137]. Constrained spinors are more difficult to work with, especially if one wants to construct symmetry generators in order to study symmetries of the D-dimensional

---

[3] $D = 3, 4, 6$ are precisely the dimensions in which twistor constructions that describe conformal symmetry are known; see Section II.C.5 of Siegel's "Fields" [57].

amplitudes. For this reason, we focus on $D = 3, 6$ in this chapter: we describe $D = 6$ briefly, then offer more details about the interesting structure of $D = 3$ amplitudes.

11.2 Scattering amplitudes in $D = 6$

Oh, who cares about 6d scattering amplitudes!!? Don't we live in 4d? Well, the 6d massless condition

$$-p_0^2 + p_1^2 + p_2^2 + p_3^2 + p_4^2 + p_5^3 = 0 \tag{11.8}$$

can be viewed from 4d spacetime as the on-shell condition for a massive 4d momentum vector: take $p_4^2 + p_5^2 = m^2$ (or $= m\tilde{m}$ if you are willing to accept complex masses). Then $p_{4d}^2 = -m^2$ follows from (11.8) with p_{4d} denoting the first four components of the 6d momentum. This makes the 6d formalism very useful for studies of 4d amplitudes with massive particles. An example of an explicit application to Higgs production processes can be found in [138]. See [87, 139, 140] for other examples. So let's care about 6d scattering amplitudes, at least for one section.

The 6d spinor helicity formalism was first developed by Cheung and O'Connell [141] and its supersymmetrization was carried out in [142]. It has been applied to tree- and loop-level scattering amplitudes in maximal super Yang–Mills theory in 6d [139, 143, 144] and also used in other 6d theories [145–148].

In 4d, we used $(\sigma^\mu)_{a\dot{b}}$ and $(\bar{\sigma}^\mu)^{\dot{a}b}$ to define the 2×2 matrices $p_{a\dot{b}} = p_\mu(\sigma^\mu)_{a\dot{b}}$ and $p^{\dot{a}b} = p_\mu(\bar{\sigma}^\mu)^{\dot{a}b}$. Similarly, the 6d Lorentz group $SO(1, 5) \sim SU^*(4)$ has antisymmetric 4×4 matrices $(\sigma^\mu)_{AB}$ and $(\tilde{\sigma}^\mu)^{AB}$, $A, B = 1, 2, 3, 4$, that allow us to define

$$p_{AB} = p_\mu(\sigma^\mu)_{AB}, \qquad p^{AB} = p_\mu(\tilde{\sigma}^\mu)^{AB}. \tag{11.9}$$

The explicit form of the $(\sigma, \tilde{\sigma})$ matrices as well as their relation to the 6d 8×8 γ-matrices can be found in Appendix A of [141].

The pseudo-real property of $SU^*(4)$, implies that p^{AB} and p_{AB} are related as

$$p^{AB} = \frac{1}{2}\epsilon^{ABCD}p_{CD}. \tag{11.10}$$

In this notation, the $SO(1, 5)$ invariant product $p^\mu p_\mu$ can be written as the manifestly $SU^*(4)$-invariant contraction

$$p^\mu p_\mu = -\frac{1}{4}p^{AB}p_{AB} = -\frac{1}{8}\epsilon_{ABCD}p^{AB}p^{CD}. \tag{11.11}$$

Now, in momentum space, the 6d Dirac equation for massless spinors is

$$p_{AB}\lambda_i^{Ba} = 0, \qquad p^{AB}\tilde{\lambda}_{iB\dot{a}} = 0, \tag{11.12}$$

where λ^{Ba} and $\tilde{\lambda}_{B\dot{a}}$ are chiral and anti-chiral spinors, and $a = 1, 2$ and $\dot{a} = 1, 2$ are fundamental indices of the two $SU(2)$s of the little group $SO(4) = SU(2) \times SU(2)$. The two pairs of Weyl spinors

$$\lambda_i^{Aa} = \langle i^a|^A = {}^A|i^a\rangle \quad \text{and} \quad \tilde{\lambda}_{iB\dot{a}} = [i_{\dot{a}}|_B = {}_B|i_{\dot{a}}] \tag{11.13}$$

are the building blocks of the 6d spinor helicity formalism. There is no distinction between bras and kets because there is no raising or lowering of the $SU^*(4)$ indices.

The little group indices can be raised/lowered using the $SU(2)$ Levi-Civita symbol as $\lambda_{\mathsf{a}} = \epsilon_{\mathsf{ab}} \lambda^{\mathsf{b}}$ and $\tilde{\lambda}^{\dot{\mathsf{a}}} = \epsilon^{\dot{\mathsf{a}}\dot{\mathsf{b}}} \tilde{\lambda}_{\dot{\mathsf{b}}}$. This allows us to form little group invariants, as in the bi-spinor construction (11.5). Indeed, the massless momentum is given as

$$p_i^{AB} = \lambda_i^{A\mathsf{a}} \lambda_{i\mathsf{a}}^{B}, \qquad p_{iAB} = \tilde{\lambda}_{iA\dot{\mathsf{a}}} \tilde{\lambda}_{iB}{}^{\dot{\mathsf{a}}}. \tag{11.14}$$

Due to the antisymmetric contraction of the $SU(2)$ indices, the bi-spinors in (11.14) are automatically antisymmetric in the $SU^*(4)$ indices A and B. By (11.14), the 4×4 matrix p_i^{AB} has rank 2, so $p_i^2 \sim \epsilon_{ABCD} \, p_i^{AB} \, p_i^{CD}$ is zero. Hence the massless on-shell condition $p_i^2 = 0$ is satisfied. Thus this realizes the construction (11.5).

Reverting the momentum in (11.14) from matrix form to vector form, we have

$$p_i^{\mu} = -\frac{1}{4} \langle i^a | \sigma^{\mu} | i_a \rangle = -\frac{1}{4} [i_{\dot{\mathsf{a}}} | \tilde{\sigma}^{\mu} | i^{\dot{\mathsf{a}}}]. \tag{11.15}$$

These expressions are the 6d versions of the 4d relation $k^{\mu} = \frac{1}{2} \langle k | \gamma^{\mu} | k]$ you derived in Exercise 2.4.

The Dirac equation (11.12) implies that $\lambda_i^{A\mathsf{a}} \tilde{\lambda}_{iA\dot{\mathsf{a}}} = 0$, so the chiral and anti-chiral spinors are related. Construction of symmetry generators using these variables must take such constraints into account. However, if only chiral spinors are needed, we can still work with unconstrained variables.

To get a better feeling for the 6d 4×2 spinors – and to facilitate reduction to 4d – consider the embedding of our good old 4d spinors in the new 6d spinors. Choosing $\mu = 0, 1, 2, 3$ to be the 4d subspace and setting $p_4 = p_5 = 0$, the 4d spinors appear in the 6d ones as

$$\lambda_{i\mathsf{a}}^A = \begin{pmatrix} 0 & \langle i|_{\dot{a}} \\ [i|^a & 0 \end{pmatrix}, \qquad \tilde{\lambda}_{iA\dot{\mathsf{a}}} = \begin{pmatrix} 0 & |i\rangle^{\dot{a}} \\ -|i]_a & 0 \end{pmatrix}. \tag{11.16}$$

Thus the constraint $\lambda_i^{A\mathsf{a}} \tilde{\lambda}_{iA\dot{\mathsf{a}}} = 0$ becomes nothing but the familiar $\langle i\, i \rangle = [i\, i] = 0$.

In 6d massless kinematics, the basic Lorentz invariant spinor products are:

- $\langle i^{\mathsf{a}} | j_{\dot{\mathsf{b}}}] \equiv \lambda_i^{A\mathsf{a}} \tilde{\lambda}_{jA\dot{\mathsf{b}}} = [j_{\dot{\mathsf{b}}} | i^{\mathsf{a}} \rangle$,
- $\langle i^{\mathsf{a}} j^{\mathsf{b}} k^{\mathsf{c}} l^{\mathsf{d}} \rangle \equiv \epsilon_{ABCD} \lambda_i^{A\mathsf{a}} \lambda_j^{B\mathsf{b}} \lambda_k^{C\mathsf{c}} \lambda_l^{D\mathsf{d}}$,
- $[i_{\dot{\mathsf{a}}} j_{\dot{\mathsf{b}}} k_{\dot{\mathsf{c}}} l_{\dot{\mathsf{d}}}] \equiv \epsilon^{ABCD} \tilde{\lambda}_{iA\dot{\mathsf{a}}} \tilde{\lambda}_{jB\dot{\mathsf{b}}} \tilde{\lambda}_{kC\dot{\mathsf{c}}} \tilde{\lambda}_{lD\dot{\mathsf{d}}}$.

In particular, the Mandelstam variable $s_{ij} = -(p_i + p_j)^2$ is

$$s_{ij} = -\frac{1}{2} \epsilon^{\mathsf{ab}} \epsilon^{\dot{\mathsf{a}}\dot{\mathsf{b}}} \langle i_{\mathsf{a}} | j_{\dot{\mathsf{a}}}] \langle i_{\mathsf{b}} | j_{\dot{\mathsf{b}}}] = -\det \langle i_{\mathsf{a}} | j_{\dot{\mathsf{a}}}]. \tag{11.17}$$

We have outlined the 6d spinor helicity formalism, so now it is time to apply it to amplitudes. Let us begin with 3-point amplitudes; this involves **special 3-particle kinematics** because all s_{ij} vanish. In 4d, we got around this by working with complex kinematics such that $\langle ji \rangle \neq ([ij])^*$ and that allowed us to choose either all the angle- or the square-brackets to be non-vanishing, but not both. In 6d 3-particle kinematics, the only Lorentz invariants available for s_{ij} are the brackets $\langle i_a | j_{\dot{a}}]$. But since $0 = s_{ij} = -\det \langle i_a | j_{\dot{a}}]$, the 2×2 matrix $\langle i_a | j_{\dot{a}}]$ must be rank 1. We have encountered 2×2 matrices of rank 1 before, namely the 4d massless $p_{a\dot{a}}$, and by now it should be a simple reflex to introduce two 2-component spinors, u_{ia} and $\tilde{u}_{j\dot{a}}$ such that $\langle i_a | j_{\dot{a}}] = u_{ia}\tilde{u}_{j\dot{a}}$ [141], so the 3-point amplitudes in 6d are written in terms of these "auxiliary" 2-component spinors.

Just as in 4d, the 6d **3-point amplitudes** are highly constrained by the little group and Lorentz invariance. For example, one finds that the 3-vector amplitudes only come in two types, one is generated by the $AA\partial A$ vertex of the Yang–Mills action while the other is generated by the operator $F_\mu{}^\nu F_\nu{}^\rho F_\rho{}^\mu$ [141]. A wide class of possible 3-point interactions was categorized in [146]. In particular, for 6d self-dual antisymmetric tensors – which are part of the $(2,0)$ supermultiplet that describes the degrees of freedom of M5-membranes in M-theory – one can demonstrate [146] that a 3-point amplitude cannot be both Lorentz invariant and carry the correct little group indices to describe scattering of three self-dual tensors; so it does not exist.

The 6d **4-point Yang–Mills amplitude** is given by:

$$A_4(1, 2, 3, 4) = \frac{\langle 1^a 2^b 3^c 4^d \rangle [1_{\dot{a}} 2_{\dot{b}} 3_{\dot{c}} 4_{\dot{d}}]}{s\, u}. \tag{11.18}$$

The 6d gluons are not labeled by the 4d concept of helicity: instead, a 6d massless spin-1 particle has four physical states labeled by the little group indices $^a{}_{\dot{a}}$.

▶ **Exercise 11.3**

Use the map in (11.16) to reduce the 6d amplitude (11.18) to 4d. You should find the usual suspect, the MHV gluon amplitude. But that is not all: identify the other possibilities and describe their origin.

In maximal SYM in 6d, the 4-point superamplitude takes the simple form

$$\mathcal{A}_4(1, 2, 3, 4) = \delta^6(P)\, \delta^{(4)}(Q)\, \delta^{(4)}(\tilde{Q})\, \frac{1}{y_{13}^2 y_{24}^2}. \tag{11.19}$$

We have used dual space to write $s = -y_{13}^2$ and $u = -y_{24}^2$. The supermomentum delta functions are defined in [142]. If we write the n-point superamplitude as $\mathcal{A}_n = \delta^6(P)\, \delta^{(4)}(Q)\, \delta^{(4)}(\tilde{Q})\, f_n$, we note from (11.19) that $I[f_4] = y_1^2 y_2^2 y_3^2 y_4^2\, f_4$ under dual conformal inversion (5.37), i.e. f_4 inverts in exactly the same way as the 4d 4-point superamplitude of $\mathcal{N} = 4$ SYM. Using a 6d version super-BCFW recursion, it was proven [144] for all n that

$$I[f_n] = \left[\prod_{i=1}^n y_i^2 \right] f_n. \tag{11.20}$$

In 4d, it was essential for dual superconformal symmetry of planar superamplitudes in $\mathcal{N} = 4$ SYM that the inversion weights of the bosonic and fermionic delta functions canceled, as shown in (5.38). This, however, does not happen in 6d maximal SYM: $\delta^6(P)$ inverts with weight 6, while the Grassmann delta functions have weight $-(4 + 4)/2$. Therefore, the planar superamplitudes of 6d maximal SYM do not have uniform inversion weight. Nonetheless, as is often the case with scattering amplitudes, even if a symmetry is not exact, it is still useful if it is broken in a predetermined fashion, as is the case here. Remarkably, using generalized unitarity methods it has been shown [144] that the planar L-loop integrands of 6d maximal SYM theory have the same dual conformal inversion weight as in 4d. (A similar result was found for 10d SYM [136].) While the origin of this form of dual superconformal symmetry is not clear (and the 6d and 10d SYM theories are *not* (super)conformal), it has non-trivial implications in 4d for the structure of (super)amplitudes on the Coulomb branch of $\mathcal{N} = 4$ SYM [140, 149].

11.3 Scattering amplitudes in $D = 3$

Scattering amplitudes in $D = 3$ turn out to have very interesting properties. After introducing the necessary kinematic tools and basic examples of amplitudes, we focus on scattering processes in the 3d $\mathcal{N} = 8$ and $\mathcal{N} = 6$ superconformal theories called BLG and ABJM.

11.3.1 $D = 3$ kinematics

We construct 3d kinematics by reduction from 4d using the fact that the 4d massless condition, $-p_0^2 + p_1^2 + p_2^2 + p_3^2 = 0$, is equivalent to a 3d massive constraint. It is convenient to identify the p_2-component with the 3d mass as $p_2^2 = m^2$ so that the 3d momentum p^μ with $\mu = 0, 1, 3$ satisfies $p^\mu p_\mu = -m^2$.

Recall that in 4d, the momentum can be given as

$$D = 4: \qquad p_{ab} = \begin{pmatrix} -p^0 + p^3 & p^1 - ip^2 \\ p^1 + ip^2 & -p^0 - p^3 \end{pmatrix}. \tag{11.21}$$

We restrict this to 3d by removing p_2 and writing

$$D = 3: \qquad p_{ab} = \begin{pmatrix} -p^0 + p^3 & p^1 \\ p^1 & -p^0 - p^3 \end{pmatrix}. \tag{11.22}$$

Then $\det p_{ab} = -(-p_0^2 + p_1^2 + p_3^2) = m^2$.

The 2×2 matrix p_{ab} is symmetric. If the 3d momentum p^μ, $\mu = 0, 1, 3$, is real, p_{ab} is also real.[4] A generic real symmetric 2×2 matrix can be written as [150][5]

$$m \neq 0: \qquad p_{ab} = \lambda_a \bar{\lambda}_b + \lambda_b \bar{\lambda}_a, \tag{11.23}$$

where $\bar{\lambda}_a = (\lambda_a)^*$ when p^μ is real.

---

[4] This contrasts with the 4d case, where p_{ab} is complex valued, and it reflects the different Lorentz groups, $SO(1, 2) = SL(2, \mathbb{R})$ in 3d and $SL(2, \mathbb{C})$ in 4d.

[5] We could also have written $p_{ab} = \lambda_a \lambda_b + \mu_a \mu_b$, where μ_a is some real 2-component spinor not parallel to λ_a, but this is equivalent to (11.23) by a linear redefinition.

► **Exercise 11.4**

If p^μ is complex, we take λ_a and $\bar\lambda_a$ to be independent. For each case, p^μ real or complex, count the number of degrees of freedom on each side of (11.23).

By direct calculation of the determinant of (11.23), we find that $m^2 = \det p_{ab} = -\langle\lambda\bar\lambda\rangle^2$, where $\langle\lambda\bar\lambda\rangle = \lambda^a\bar\lambda_a$ and spinor indices are raised and lowered with the 2-index Levi-Civita of the $SL(2,\mathbb{R})$ Lorentz group. For 3d massless kinematics, $m = 0$, we must therefore have $\langle\lambda\bar\lambda\rangle = 0$, implying that $\bar\lambda \propto \lambda$. Thus, we can write

$$m = 0: \qquad p_{ab} = \lambda_a\lambda_b = \langle p|_a\,\langle p|_b\,, \tag{11.24}$$

where $\lambda = \langle p|$ was rescaled such that the prefactor is just 1. Note that $\langle p|$ must be either purely real or purely imaginary for p_{ab} to be real.

It follows from (11.24) that in *3d massless kinematics, all Lorentz invariants are built out of one kind of angle brackets, namely* $\langle ij\rangle = \lambda_i^a\lambda_{j\,a}$. For example, since $2p_i.p_j = -\langle ij\rangle^2$, the Mandelstams s_{ij} are

$$D = 3: \qquad s_{ij} = -(p_i + p_j)^2 = \langle ij\rangle^2\,. \tag{11.25}$$

Momentum conservation $\sum_{i=1}^n p_i^\mu = 0$ can be written

$$D = 3: \qquad \sum_{i=1}^n |i\rangle\langle i| = 0\,. \tag{11.26}$$

Our 3d kinematics is ready, so let us see some amplitudes. As it is our style, we start with **3-particle amplitudes**. These are particularly easy in 3d, because 3-particle kinematics requires all $s_{ij} = \langle ij\rangle^2 = 0$ and hence there are no Lorentz invariants available for a massless 3-point amplitude. *Thus for massless kinematics, there are no 3-point on-shell amplitudes in 3d.*

The **little group** for massless kinematics in 3d is the discrete group $\mathbb{Z}_2$; it acts on the spinor variables as $|i\rangle \to -|i\rangle$. The homogeneous scaling of the scattering amplitudes distinguishes only two types of particles in 3d: *scalar particles* scale with $+1$ and *fermions* scale with -1. And spin-1 vector particles? A massless vector in D-dimensions has $D-2$ degrees of freedom, so in $D = 3$ this is just 1, the same as a scalar.

Tree-level scattering amplitudes of 3d super Yang–Mills theory can be obtained directly from 4d ones using dimensional reduction. For example, the dimensional reduction of the 4-point gluon amplitude $A_4[1^-2^+3^-4^+]$ of 4d Yang–Mills theory gives

$$A_4[1^-2^+3^-4^+] = \frac{\langle 13\rangle^4}{\langle 12\rangle\langle 23\rangle\langle 34\rangle\langle 41\rangle} \xrightarrow{4d\,\to\,3d} -\frac{\langle 13\rangle^4}{\langle 12\rangle^2\langle 23\rangle^2}\,. \tag{11.27}$$

We have used 3d momentum conservation $\langle 34 \rangle \langle 41 \rangle = -\langle 32 \rangle \langle 21 \rangle$ to simplify the result. The two helicity states of the 4d gluon become two degrees of freedom in 3d that we can organize as a 3d "gauge boson" and a scalar.

11.3.2 3d SYM and Chern–Simons theory

The 3-dimensional Yang–Mills action

$$\mathcal{L}_{\mathrm{YM}} = \frac{1}{g^2} \int d^3x \, \mathrm{Tr} \, F_{\mu\nu} F^{\mu\nu} \tag{11.28}$$

has a coupling g^2 of mass-dimension (mass)^1. We are particularly interested in theories with extra symmetry (after all, we keep getting marathonic mileage out of planar $\mathcal{N} = 4$ SYM), but a superconformal theory needs dimensionless couplings.

In 3d, the gauge field can be introduced with a dimensionless coupling via the **Chern–Simons Lagrangian**

$$\mathcal{L}_{\mathrm{CS}} = \frac{\kappa_{\mathrm{CS}}}{4\pi} \, \epsilon^{\mu\nu\rho} \, \mathrm{Tr} \left(A_\mu \partial_\nu A_\rho + \frac{2i}{3} A_\mu A_\nu A_\rho \right). \tag{11.29}$$

The coupling κ_{CS} is an integer and is called the **Chern–Simons level**.

The equation of motion derived from varying $\mathcal{L}_{\mathrm{CS}}$ with respect to the gauge field is

$$\partial_{[\mu} A_{\nu]} + i[A_\mu, A_\nu] = F_{\mu\nu} = 0 \,. \tag{11.30}$$

The solution to this equation is simply $A_\mu = g \partial_\mu g^{-1}$, where g is an arbitrary element in the gauge group. This means that the gauge field is pure gauge, or a flat connection. For us, the relevant implication is that the Chern–Simons gauge field does not carry any physical degrees of freedom, since one can always choose a gauge such that $A_\mu = 0$. This is an important difference between a gauge field whose dynamics is governed by $\mathcal{L}_{\mathrm{CS}}$ versus the usual Yang–Mills Lagrangian $\mathcal{L}_{\mathrm{YM}}$: the gauge boson scattering amplitudes of 3d Yang–Mills theory are non-trivial, but for a theory with just a Chern–Simons term the scattering amplitudes are trivially zero because there are no physical states to scatter.

There can be non-trivial scattering amplitudes for Chern–Simons theory provided matter fields are introduced. The **Chern–Simons matter Lagrangian** is typically written

$$\mathcal{L} = \mathcal{L}_{\mathrm{CS}} + \mathcal{L}_{\phi\psi} \,, \tag{11.31}$$

where the matter Lagrangian $\mathcal{L}_{\phi\psi}$ encodes the interactions of scalars ϕ and fermions ψ with the Chern–Simons gauge field as well as their mutual interactions. In 3d, the (complex) scalar- and fermion-interactions with dimensionless couplings are of the form $\phi^3 \bar\phi^3$ and $\bar\psi \psi \bar\phi \phi$. Thus for superconformal theories, $\mathcal{L}_{\phi\psi}$ takes the form

$$\mathcal{L}_{\phi\psi} = -D^\mu \bar\phi D_\mu \phi + i \bar\psi \slashed{D} \psi + V_{\psi \bar\psi \phi \bar\phi} + V_{\phi^3 \bar\phi^3} \,, \tag{11.32}$$

where $V_{\psi\bar{\psi}\phi\bar{\phi}}$ and $V_{\phi^3\bar{\phi}^3}$ are quartic and sextic interaction terms. The explicit form of these terms depends on the theory; we will show you two examples, namely the $\mathcal{N} = 8$ and $\mathcal{N} = 6$ superconformal 3d theories (Sections 11.3.5 and 11.3.6). But let us first explore the properties of amplitudes in 3d a little further.

11.3.3 Special kinematics and poles in amplitudes

The are 3-particle interaction terms in the Lagrangians discussed in Section 11.3.2, but we have learned in Section 11.3.1 that all on-shell 3-point amplitudes vanish in 3d. Nonetheless, the 3-particle vertices still make their presence felt by hiding in special kinematic limits of higher-point amplitudes. As an example of this, consider the limit $s_{12} = \langle 12 \rangle^2 \to 0$ of a 4-point amplitude. In this limit, $|1\rangle$ becomes proportional to $|2\rangle$, so $|1\rangle = \alpha|2\rangle$ for some α. Further, we must have $(1 + \alpha^2)s_{23} = 0$, since

$$0 = p_4^2 = (p_1 + p_2 + p_3)^2 \xrightarrow{s_{12}\to 0} 0 = s_{13} + s_{23} = (1 + \alpha^2)s_{23} . \qquad (11.33)$$

There are two types of solutions to this constraint. For generic α, s_{23} must be zero and one can conclude that all Lorentz invariants vanish. This is in line with our previous discussion that there are no Lorentz invariants for on-shell 3-point kinematics. However, the constraint $(1 + \alpha^2)s_{23} = 0$ also admits a solution that allows non-trivial Lorentz invariants: $\alpha = \pm i$. For $\alpha = \pm i$, we have $p_1 = -p_2$ and similarly $p_3 = -p_4$. Thus this corresponds to the kinematic configuration where two particles are traveling in straight lines:

$$k_1 = k_2 \qquad k_3 = k_4 . \qquad (11.34)$$

From momentum conservation, any exchange between the two particle lines must have zero momentum, so when we approach the $s_{12} \to 0$ limit the amplitude should develop a singularity associated with the propagator of an exchanged soft particle.

To see this in an explicit example, consider the 3d YM gluon amplitude (11.27)

$$A_4[1234] = -\frac{\langle 13 \rangle^4}{\langle 12 \rangle^2 \langle 23 \rangle^2} . \qquad (11.35)$$

Taking the limit $|1\rangle \to i\,|2\rangle$, the amplitude indeed develops a non-trivial $1/s_{12}$ singularity with a non-vanishing residue:

$$A_4[1^-2^+3^-4^+]\Big|_{|1\rangle \,\to\, i\,|2\rangle} = -\frac{s_{23}}{s_{12}} . \qquad (11.36)$$

The $1/s_{12}$ singularity reflects the $1/p^2$ behavior of the gluon propagator.

Thus, although there are no massless 3-point amplitudes in 3d, the 4-point amplitude still develops a non-trivial "soft" pole. The origin of this singularity comes from the exchange of a soft particle between two particles going in straight lines. Note that the exchanged particle has momentum $p^\mu \to 0$, so it is not strictly going on-shell. This is also reflected in the observation that amplitude (11.36) does not factorize into two 3-point amplitudes.

Importantly, the precise behavior of the singularity is dictated by the propagator of the intermediate particle. If the exchanged particle is an ordinary Yang–Mills gluon, then we should observe a $1/p^2$ singularity. That is what happened in the example (11.36). However, if it is a fermion or a Chern–Simons gauge boson, one should find a $1/\sqrt{p^2}$ singularity.[6] In the case of a Chern–Simons boson, it comes from the propagator of the gauge field in the Lagrangian $\mathcal{L}_{\mathrm{CS}}$ of (11.29); in Landau gauge it is

$$\langle A^\mu(p) A^\nu(-p) \rangle = \frac{\epsilon^{\mu\rho\nu} p_\rho}{p^2} . \tag{11.37}$$

We are going to use information about poles in this special kinematic limit to constrain the possible 4-point amplitudes in Section 11.3.5.

11.3.4 $D = 3$ superconformal algebra

We stated in Section 11.1 that the minimal spinors in 3d are 2-component Majorana spinors. They satisfy the Majorana reality condition, and so do the supersymmetry charges. Thus for $\mathcal{N}$-fold supersymmetry in 3d, we have $\mathcal{N}$ real supercharges and the R-symmetry group is $SO(\mathcal{N})$. Since the 3d theories discussed in this chapter have $\mathcal{N} = 8$ or $\mathcal{N} = 6$ supersymmetry, we focus on $\mathcal{N}$=even in the following. When $\mathcal{N} = 2M$, the real supercharges can be grouped into $M = \mathcal{N}/2$ complex spinors Q^{aA} and their complex conjugates $\widetilde{Q}^a_A$. Here $A = 1, \ldots, M$ is the index of the reduced $SU(M)$ R-symmetry.

We introduce M on-shell superspace coordinates η_i^A for each external leg. The supercharge can now be written as

$$\widetilde{Q}^a_A = \sum_i |i\rangle^a \eta_{iA} , \qquad Q^{aA} = |i\rangle^a \frac{\partial}{\partial \eta_{iA}} . \tag{11.38}$$

You can quickly see that $\{\widetilde{Q}^a_A, Q^{bB}\} = \delta_A{}^B P^{ab}$.

The generators (11.38) are part of a larger symmetry group: the $OSp(\mathcal{N}|4)$ superconformal group. The notation $OSp(\mathcal{N}|4)$ means that the bosonic generators include the $SO(\mathcal{N})$

---

[6] Our reasoning here is valid only for $n = 4$. For example, the n-point Parke–Taylor amplitude with $n > 4$ has only $1/\langle i, i+1\rangle$ poles that do not exhibit the $1/p^2$ of the Yang–Mills gluon propagator. A careful inspection of the 3-point gluon vertex reveals that the only non-vanishing term is proportional to $k_2^\mu(\epsilon_1 \cdot \epsilon_2)$ in the limit where legs 1, 2 are the two gluons propagating in a straight line. This is dotted into the remaining Feynman diagram which for $n = 4$ is simply another 3-point vertex that in this limit contributes just one term proportional to $k_3^\mu(\epsilon_3 \cdot \epsilon_4)$. Hence, on this soft pole, the residue is simply given by the product of the two 3-point vertices. For $n > 4$, the remaining Feynman diagrams give multiple contributions, and thus the residue of this soft pole contains several terms that can cancel to leave behind a milder singularity. Therefore, our discussion of soft-pole structure is only valid for $n = 4$.

R-symmetry as well as the $Sp(4)$ conformal symmetry generators. More precisely, the generators are:

$$P^{ab} = \sum_i |i\rangle^a |i\rangle^b$$

$$\tilde{Q}_A^a = \sum_i |i\rangle^a \eta_{iA} \qquad\qquad Q^{aA} = \sum_i |i\rangle^a \partial_{\eta_{iA}}$$

$$M_{ab} = \sum_i \langle i|_{(a} \partial_{|i\rangle^{b)}} \qquad\qquad D = \sum_i \left(\tfrac{1}{2}|i\rangle^a \partial_{|i\rangle^a} + \tfrac{1}{2}\right)$$

$$R_{AB} = \sum_i \eta_{iA}\eta_{iB} \quad R_A{}^B = \sum_i \left(\eta_{iA}\partial_{\eta_{iB}} - \tfrac{1}{2}\delta^A{}_B\right) \quad R^{AB} = \sum_i \partial_{\eta_{iA}}\partial_{\eta_{iB}} \qquad (11.39)$$

$$\tilde{S}_{aA} = \sum_i \partial_{|i\rangle^a} \eta_{iA} \qquad\qquad S_a^A = \sum_i \partial_{|i\rangle^a} \partial_{\eta_{iA}}$$

$$K_{ab} = \sum_i \partial_{|i\rangle^a} \partial_{|i\rangle^b} \, .$$

The $SO(\mathcal{N})$ R-symmetry generators are separated into $U(\mathcal{N}/2)$ generators $R_A{}^B$ and coset generators R_{AB} and R^{AB} of $SO(\mathcal{N})/U(\mathcal{N}/2)$.

As an important application for these generators, let us explore what kind of constraint the $U(1)$ piece of the $U(\mathcal{N}/2)$ R-symmetry imposes on the superamplitudes in a 3d $\mathcal{N} = 2M$ superconformal theory. The $U(1)$ piece is given by

$$R_C{}^C = \sum_i \left(\eta_{iC}\partial_{\eta_{iC}} - \frac{M}{2}\right). \qquad (11.40)$$

The $R_C{}^C$ generator annihilates the superamplitude, $R_C{}^C \mathcal{A}_n = 0$, if

$$\sum_i \eta_{iC}\partial_{\eta_{iC}} \mathcal{A}_n = n\frac{M}{2}\mathcal{A}_n . \qquad (11.41)$$

The LHS simply counts the Grassmann degree of η_As in $\mathcal{A}_n$. Since one cannot have fractional degrees of η in $\mathcal{A}_n$, the equation (11.41) can only hold for odd M if $n = $ even. So we learn that *only even-multiplicity scattering amplitudes can be non-vanishing for a superconformal theory with M odd.*

The *same is also true for M even.* This is because superconformal theories generally require the presence of a gauge field whose self-interaction is described by the Chern–Simons Lagrangian (11.29). The form of the Chern–Simons matter Lagrangian (11.31), (11.32) implies that any odd-multiplicity Feynman diagram has at least one external leg associated with a gauge field. As discussed previously, the Chern–Simons gauge field does not carry any physical degrees of freedom, so this means that the odd-multiplicity scattering amplitudes must vanish.

So we learn that in a 3d superconformal theory, the Grassmann degree of the superamplitudes, i.e. the N^KMHV-level, is rigidly tied to the number of external particles, contrary to its freer life in 4d. For example for $\mathcal{N} = 8$, an MHV superamplitude has Grassmann degree 8 and by (11.41) it exists only for $n = 4$ external particles in a 3d superconformal theory; a 6-point superamplitude on the other hand must have Grassmann degree 12, so it has to be NMHV. Thus in a 3d $\mathcal{N} = 8$ superconformal theory, there is no tower of MHV superamplitudes, no equivalent of the n-gluon Parke–Taylor amplitude. Similarly, in a 3d $\mathcal{N} = 6$ superconformal theory, the 4-point superamplitude must have Grassmann degree 6.

11.3.5 $\mathcal{N} = 8$ superconformal theory: BLG

A 3d superconformal theory with $\mathcal{N} = 8$-fold supersymmetry has an on-shell spectrum with eight scalars (ϕ, ϕ^{AB}, $\bar{\phi}$) and eight fermions (ψ^A, $\bar{\psi}_A$). Just as in $\mathcal{N} = 4$ SYM in 4d, it is convenient to encode the degrees of freedom in an on-shell superfield,

$$\Phi = \phi + \eta_A \psi^A - \frac{1}{2}\eta_A\eta_B \phi^{AB} - \frac{1}{3!}\epsilon^{ABCD}\eta_A\eta_B\eta_C \bar{\psi}_D + \eta_1\eta_2\eta_3\eta_4 \bar{\phi}. \quad (11.42)$$

We have $\eta_A \to -\eta_A$ under little group transformations, so the superfield Φ is inert. This means that the superamplitudes are also invariant under little group transformations.

Since there are no massless 3-point amplitudes, let us consider the most general 4-point tree superamplitude that enjoys $\mathcal{N} = 8$ 3d superconformal symmetry. To start with, invariance under $\mathcal{N} = 8$ supersymmetry implies that an n-point superamplitude takes the form

$$\mathcal{A}_n = \delta^3(P)\, \delta^{(8)}(\widetilde{Q})\, f_n(|i\rangle, \eta_i), \quad (11.43)$$

where $\delta^{(8)}(\widetilde{Q}) = \prod_{A=1}^4 (\frac{1}{2}\widetilde{Q}_A^a \widetilde{Q}_{aA})$. The function f_n is constrained further by the superconformal generators (11.39).

As noted at the end of Section 11.3.4, the $U(1)$ generator (11.41) requires the 4-point superamplitude to have degree 8. Since the supermomentum delta function is already degree 8 in the η_is, we infer that f_4 can only depend on the bosonic variables $|i\rangle$.

Next, annihilation of the superamplitude by the dilatation operator D in (11.39) implies

$$D\mathcal{A}_n = 0 \quad \to \quad \sum_i \left(\frac{1}{2}|i\rangle^a \partial_{|i\rangle^a}\right)\mathcal{A}_n = -\frac{n}{2}\mathcal{A}_n. \quad (11.44)$$

As in 4d (see Exercise 5.2), the operator $\sum_i(\frac{1}{2}|i\rangle^a\partial_{|i\rangle^a})$ counts the mass dimension when acting on a function of spinor brackets. It also acts on the delta functions in $\mathcal{A}_n$, giving a factor of -3 on $\delta^3(P)$ and 4 on $\delta^{(8)}(\widetilde{Q})$. Thus, by (11.44), dilatation invariance requires the mass-dimension of f_4 to be $-\frac{4}{2} - (-3 + 4) = -3$.

▶ **Exercise 11.5**

Use the example around equation (5.4) to show that $\sum_i(\frac{1}{2}|i\rangle^a\partial_{|i\rangle^a})\,\delta^3(P) = -3\delta^3(P)$.

To summarize, from invariance under $U(1)$ R-symmetry and dilatation, we conclude that f_4 is a purely bosonic function of mass-dimension -3. Finally, taking into account that the superamplitude must be little group invariant, we can write the 4-point superamplitude [148] as

$$\mathcal{A}_4(\Phi_1\Phi_2\Phi_3\Phi_4) = \delta^3(P)\, \delta^{(8)}(\widetilde{Q})\, \frac{1}{\langle 12\rangle\langle 23\rangle\langle 31\rangle}. \quad (11.45)$$

We can now project out the 4-scalar amplitude $A_4(\phi\phi\bar{\phi}\bar{\phi})$ using (11.42): the Grassmann delta function produces a factor of $\langle 34 \rangle^4$, so we get (with the help of momentum conservation)

$$A_4(\phi\phi\bar{\phi}\bar{\phi}) = \frac{\langle 34 \rangle^4}{\langle 12 \rangle \langle 23 \rangle \langle 13 \rangle} = -\frac{\langle 34 \rangle^3}{\langle 24 \rangle \langle 23 \rangle} , \qquad (11.46)$$

The astute reader should object: multiplying the solution (11.45) by an arbitrary function of $\frac{\langle 13 \rangle \langle 24 \rangle}{\langle 14 \rangle \langle 23 \rangle}$ still satisfies all previous criteria. This is indeed a valid objection; however, such a function would change the pole structure of component amplitudes, such as (11.46), generated by $\mathcal{A}_4$. We have imposed in (11.45) that the amplitudes only have $1/\sqrt{p^2}$ poles. Why? Well, since the only scalar-fermion interactions are of the form $\phi^3\bar{\phi}^3$ and $\bar{\psi}\psi\bar{\phi}\phi$, poles in the tree-level amplitude $A_4(\phi\phi\bar{\phi}\bar{\phi})$ cannot arise from scalar or fermion propagators. Hence the only option is that they come from gauge boson exchanges. Since we are considering a 3d superconformal theory, the gauge boson self-coupling must be dimensionless; this rules out 3d Yang–Mills theory and rules in Chern–Simons gauge theory. Hence all poles in $A_4(\phi\phi\bar{\phi}\bar{\phi})$ must be $1/\sqrt{p^2}$ and this fixes the 4-point tree superamplitude in an $\mathcal{N} = 8$ superconformal 3d theory to be (11.45).

The result (11.45) for the superamplitude has a very important property: it is antisymmetric under the exchange of any two external particles. This property is inherited by the component amplitude $A_4(\phi\phi\bar{\phi}\bar{\phi})$ in (11.46) and that contradicts the expected Bose symmetry. We encountered something similar in Section 2.6 when we wrote down the 3-point gluon amplitudes in 4d from just little group scaling and dimensional analysis. The resolution was to include the antisymmetric structure constants f^{abc} of the Yang–Mills gauge group.

At the superamplitude level, the same issue arises: the physical degrees of freedom are contained in the bosonic superfield Φ, so $\mathcal{A}_4(\Phi_1 \Phi_2 \Phi_3 \Phi_4)$ should be Bose symmetric under the exchange of any two external legs. But – as you see from (11.45) – it is fully antisymmetric. We could avoid this contradiction if the amplitudes were color-ordered. However, the presence of the $1/\langle 24 \rangle$ pole in for example (11.46) invalidates this interpretation. Instead, the contradiction can be resolved if we include more than one supermultiplet, each one with a label a_i. Then we can introduce a new 4-index "coupling constant" $f^{a_1 a_2 a_3 a_4}$ that is completely antisymmetric in all four indices. Using this we write

$$\mathcal{A}_4\left(\Phi_1^{a_1} \Phi_2^{a_2} \Phi_3^{a_3} \Phi_4^{a_4}\right) = \delta^3(P)\, \delta^{(8)}(\widetilde{Q})\, \frac{f^{a_1 a_2 a_3 a_4}}{\langle 12 \rangle \langle 23 \rangle \langle 31 \rangle} . \qquad (11.47)$$

Now Bose symmetry is respected. Thus *by requiring $\mathcal{N} = 8$ superconformal symmetry in 3d, the 4-point superamplitude forces us to introduce a completely **antisymmetric 4-index coupling constant***.

This new coupling constant looks similar to the totally antisymmetric 3-index structure constant of Yang–Mills theory $f^{a_1 a_2 a_3}$. This resemblance is not a coincidence. In the search for an $\mathcal{N} = 8$ super Chern–Simons matter theory, Bagger, Lambert, and Gustavsson (BLG) [151, 152] found a Lagrangian whose gauge symmetry is built on a *Lie 3-algebra*.

This algebra is defined through a triple product

$$[T^a, T^b, T^c] = f^{abc}{}_d \, T^d \,. \tag{11.48}$$

The gauge indices are raised/lowered with $h^{ab} = \mathrm{Tr}\, T^a T^b$ and its inverse. The structure constants $f^{abcd} = f^{abc}{}_e h^{ed}$ are totally antisymmetric. Just as the structure constants of the usual gauge Lie 2-algebra satisfy the Jacobi identity (2.69), the 3-algebra structure constants are required to satisfy a four-term "fundamental identity":

$$f^{fgd}{}_e f^{abce} - f^{fga}{}_e f^{bcde} + f^{fgb}{}_e f^{cdae} - f^{fgc}{}_e f^{dabe} = 0 \,. \tag{11.49}$$

The fields in **BLG theory** [151, 152] consist of eight scalars $X_a^{I_v}$ with $I_v = 1, \ldots, 8$ transforming as a vector of $SO(8)$, eight real spinors $\Psi_a^{I_c}$ with $I_c = 1, \ldots, 8$ transforming as a chiral spinor of $SO(8)$, and a Chern–Simons gauge field A_μ^{ab}. The BLG Lagrangian is [153]

$$\frac{1}{\kappa_{\mathrm{CS}}} \mathcal{L}_{\mathrm{BLG}} = \frac{1}{48} \epsilon^{\mu\nu\rho} \left(\frac{1}{2} f^{abcd} A_{\mu ab} \partial_\nu A_{\rho cd} + \frac{1}{3} f^{cda}{}_g f^{efgb} A_{\mu ab} A_{\nu cd} A_{\rho ef} \right)$$
$$- \frac{1}{2} D^\mu X_a^{I_v} D_\mu X_a^{I_v} + \frac{i}{2} \bar{\Psi}_a^{I_c} \slashed{D} \Psi_a^{I_c} + i3 f^{abcd} \bar{\Psi}_a \Gamma^{I_v J_v} \Psi_b X_c^{I_v} X_d^{J_v}$$
$$- 12 f^{abcd} f_a{}^{efg} (X_b^{I_v} X_c^{J_v} X_d^{K_v})(X_e^{I_v} X_f^{J_v} X_g^{K_v}) \,. \tag{11.50}$$

In the Lagrangian construction [151, 152], the need for the antisymmetric 4-index structure constant comes from the requirement that the supersymmetry transformations on the fields close into the correct algebra. Linear combinations of the eight scalars and fermions can be identified as the $(\phi, \phi_{AB}, \bar{\phi})$ and $(\psi_A, \bar{\psi}^A)$ components of our superfield (11.42). Indeed, the 4-point amplitudes computed from the Lagrangian (11.50) match [148] the component amplitudes of the 4-point superamplitude (11.47).

It is quite non-trivial for an antisymmetric f^{abcd} to satisfy (11.49) and currently the only known example is if a is an index of $SO(4)$ and $f^{abcd} \sim \epsilon^{abcd}$. In search for other examples, there were many attempts to relax the symmetry properties of the 4-index structure constant [154–156]. However, as we have shown from the on-shell analysis, $\mathcal{N} = 8$ superconformal symmetry only allows for a totally antisymmetric structure constant. Indeed, all known examples of Lie 3-algebras with f^{abcd} not totally antisymmetric correspond to Chern–Simons matter theories with $\mathcal{N} < 8$ supersymmetries.

11.3.6 $\mathcal{N} = 6$ superconformal theory: ABJM

Let us now consider a 3d superconformal theory with $\mathcal{N} = 6$ supersymmetry. The R-symmetry is $SO(6) = SU(4)$ and the physical degrees of freedom are four complex scalars X_A and four complex fermions ψ^{Aa} as well as their complex conjugates $\bar{X}^A$ and $\bar{\psi}_{Aa}$. They transform in the fundamental or anti-fundamental of $SU(4)$ and $A = 1, 2, 3, 4$. To arrange these states in on-shell superspace, we introduce three anticommuting variables η_A and write

$$
\begin{aligned}
\Phi &= X_4 + \eta_A \psi^A - \frac{1}{2} \epsilon^{ABC} \eta_A \eta_B X_C - \eta_1 \eta_2 \eta_3 \psi^4, \\
\bar{\Psi} &= \bar{\psi}_4 + \eta_A \bar{X}^A - \frac{1}{2} \epsilon^{ABC} \eta_A \eta_B \bar{\psi}_C - \eta_1 \eta_2 \eta_3 \bar{X}^4.
\end{aligned}
\tag{11.51}
$$

We have split the fields as $X_A \to (X_4, X_A)$ and $\psi^A \to (\psi^4, \psi^A)$, and similarly for $\bar{X}^A$ and $\bar{\psi}_A$. So only an $SU(3)$ subgroup of the $SU(4)$ is manifest in this on-shell superspace formalism.

The on-shell superspace representation (11.51) involves a bosonic superfield Φ and a fermionic superfield $\bar{\Psi}$. Having two superfields is standard for superamplitudes in theories with less-than-maximal supersymmetry. For example in 4d $\mathcal{N} < 4$ SYM, the spectrum is not CPT self-conjugate and therefore a superfield is needed for each of the CPT conjugate supermultiplets; details and applications of the formalism can be found in [87]. In 3d, the need for two superfields comes from R-symmetry. Just as in 4d $\mathcal{N} < 4$ SYM, where the two superfields have states that are parity-conjugate with respect to each other, in 3d the two superfields contain states that are conjugate to each other under R-symmetry.

Since fermions transform with a minus under 3d little group transformations, the super-amplitude must be odd under $|i\rangle \to -|i\rangle$ and $\eta_i \to -\eta_i$ if i is a $\bar{\Psi}$ state. Following the same steps as for the $\mathcal{N} = 8$ BLG theory in Section 11.3.5, we then find that the 4-point su-peramplitude in a 3d $\mathcal{N} = 6$ superconformal theory is fixed up to a multiplicative constant to be [157]

$$
\mathcal{A}_4 [\bar{\Psi}_1 \Phi_2 \bar{\Psi}_3 \Phi_4] = \delta^3(P) \, \delta^{(6)}(\tilde{Q}) \, \frac{1}{\langle 14 \rangle \langle 43 \rangle}.
\tag{11.52}
$$

The 4-point superamplitude (11.52) precisely encodes the color-ordered 4-point amplitudes of an $\mathcal{N} = 6$ Chern–Simons matter theory that was constructed by Aharony, Bergman, Jafferis, and Maldacena (ABJM) [158]. The theory, known as the ***ABJM theory***, contains two gauge fields $A^a{}_b$ and $\hat{A}^{\dot{a}}{}_{\dot{b}}$ with gauge group $U(N) \times U(N)$. The matter fields are in the bi-fundamental, meaning that they transform in the fundamental of one $U(N)$ gauge group and the anti-fundamental of the other $U(N)$. Hence, the index structures of the matter fields are $(X_A)^{\dot{a}}{}_a$, $(\bar{X}^A)^a{}_{\dot{a}}$, $(\psi^A)^{\dot{a}}{}_a$, and $(\bar{\psi}_A)^a{}_{\dot{a}}$. The Lagrangian is [159, 160]

$$
\mathcal{L}_{\text{ABJM}} = \frac{k}{2\pi} \left[\frac{1}{2} \epsilon^{\mu\nu\rho} \, \text{Tr} \left(A_\mu \partial_\nu A_\rho + \frac{2i}{3} A_\mu A_\nu A_\rho - \hat{A}_\mu \partial_\nu \hat{A}_\rho - \frac{2i}{3} \hat{A}_\mu \hat{A}_\nu \hat{A}_\rho \right) \right.
$$
$$
\left. - (D^\mu X_A)^\dagger D_\mu X_A + i \bar{\psi}_A \not{D} \psi^A + \mathcal{L}_4 + \mathcal{L}_6 \right],
\tag{11.53}
$$

where the covariant derivatives for the bi-fundamental fields are

$$
\begin{aligned}
D^\mu X_A &\equiv \partial_\mu X_A + i \hat{A}_\mu X_A - i X_A A_\mu, \\
(D^\mu X_A)^\dagger &\equiv \partial_\mu \bar{X}^A + i A_\mu \bar{X}^A - i \bar{X}^A \hat{A}_\mu,
\end{aligned}
\tag{11.54}
$$

with the same definitions for ψ^A and $\bar{\psi}_A$.

The quartic and sextic interaction terms in (11.53) are

$$\mathcal{L}_4 = i\, \mathrm{Tr}\left(\bar{X}^B X_B \bar{\psi}_A \psi^A - X_B \bar{X}^B \psi^A \bar{\psi}_A + 2 X_A \bar{X}^B \psi^A \bar{\psi}_B - 2 \bar{X}^A X_B \bar{\psi}_A \psi^B \right.$$
$$\left. - \epsilon_{ABCD} \bar{X}^A \psi^B \bar{X}^C \psi^D + \epsilon^{ABCD} X_A \bar{\psi}_B X_C \bar{\psi}_D \right), \tag{11.55}$$

$$\mathcal{L}_6 = \frac{1}{3}\, \mathrm{Tr}\left(X_A \bar{X}^A X_B \bar{X}^B X_C \bar{X}^C + \bar{X}^A X_A \bar{X}^B X_B \bar{X}^C X_C + 4 \bar{X}^A X_B \bar{X}^C X_A \bar{X}^B X_C \right.$$
$$\left. - 6 X_A \bar{X}^B X_B \bar{X}^A X_C \bar{X}^C \right). \tag{11.56}$$

For theories whose external states are bi-fundamental matter fields, the color structure of the amplitude is given in terms of a product of Kronecker deltas. In particular, with $n = 2m$ the full color-dressed amplitude is [157]

$$\sum_{\sigma \in S_m,\ \bar{\sigma} \in \bar{S}_{m-1}} \mathcal{A}_n(\bar{1}, \sigma_1, \bar{\sigma}_1, \ldots, \bar{\sigma}_{m-1}, \sigma_m)\, \delta^{\dot{a}_{\sigma_1}}_{\dot{a}_{\bar{1}}} \cdots \delta^{\dot{a}_{\sigma_m}}_{\dot{a}_{\bar{\sigma}_{m-1}}}\, \delta^{a_{\bar{\sigma}_1}}_{a_{\sigma_1}} \cdots \delta^{a_{\bar{1}}}_{a_{\sigma_m}}, \tag{11.57}$$

where the sums are over all distinct permutations of m even sites and $m-1$ odd sites. Each partial amplitude $\mathcal{A}_n$ is multiplied by a product of Kronecker deltas, and this naturally defines an ordering, very similar to Yang–Mills amplitudes. However, since the on-shell degrees of freedom are contained in two distinct supermultiplets, the color-ordered superamplitude is not cyclically invariant, but invariant up to a sign under cyclic rotation of *two* sites:

$$\mathcal{A}_{n=2m}\left[\bar{\Psi}_1 \Phi_2 \ldots \Phi_{2m} \right] = (-1)^{m-1} \mathcal{A}_{n=2m}\left[\bar{\Psi}_3 \Phi_4 \ldots \Phi_{2m} \bar{\Psi}_1 \Phi_2 \right]; \tag{11.58}$$

the minus signs come from the exchanges of $\bar{\Psi}$s. For the superamplitude (11.52), the 2-site cyclic property (11.58) is ensured by momentum conservation.

After having seen a Lie 3-algebra appear in the $\mathcal{N} = 8$ superconformal BLG theory in Section 11.3.5, you may wonder if the above Lagrangian can also be rewritten in terms of a 3-algebra. Indeed it can! In fact, we can read off the properties of the 4-index structure constants from the 4-point superamplitude (11.52). It is symmetric under the exchange of the legs that correspond to the fermionic supermultiplet $\bar{\Psi}$, while it is antisymmetric under the exchange of the bosonic multiplets Φ. This is opposite from the expected symmetry properties of $\mathcal{A}_4(\bar{\Psi}_1 \Phi_2 \bar{\Psi}_3 \Phi_4)$, and therefore one can consider dressing the superamplitude with a 4-index structure constant $f^{a_2 a_4 \bar{a}_1 \bar{a}_3}$ that is antisymmetric with respect to the exchange of barred or unbarred indices, respectively. The color-dressed superamplitude is then[7]

$$\mathcal{A}_4\left(\bar{\Psi}_1^{\bar{a}_1} \Phi_2^{a_2} \bar{\Psi}_3^{\bar{a}_3} \Phi_4^{a_4} \right) = \delta^3(P)\, \delta^{(6)}\big(\tilde{Q}\big)\, \frac{f^{a_2 a_4 \bar{a}_1 \bar{a}_3}}{\langle 14 \rangle \langle 43 \rangle}. \tag{11.59}$$

---

[7] You might wonder why this issue did not come up when we stated that the amplitude in (11.52) matched that derived from the Lagrangian (11.53). The reason is that it matched in the context of a color-ordered amplitude where the exchange of external lines is not a symmetry. In contrast, here we are considering a fully color-dressed amplitude. In other words, we are asking what properties the color factor should have such that the amplitude can be considered as a color-dressed amplitude.

It has been shown [161] that the Lagrangian (11.53) is completely equivalent to an alternative one where the matter fields carry the 3-algebra indices indicated in (11.59).

11.3.7 BCFW recursion in 3d

We argued in Section 11.3.4 that only even-point amplitudes are non-vanishing in 3d superconformal theories. This means that the 4-point superamplitudes are the building blocks of higher-point amplitudes in these theories. Conveniently, we found that the 4-point tree-level superamplitudes in $\mathcal{N} = 8$ and $\mathcal{N} = 6$ theories in Sections 11.3.5 and 11.3.6 are completely determined by the requirements of symmetries and pole structure. Now is the time to go to higher-point, and of course our favorite tool is BCFW recursion.

To get started, we have to set up a BCFW recursion relation in 3d. And 3d is different from all other $D > 3$ in terms of defining a BCFW deformation. To see this, recall from Chapter 3 that we shift two external momenta i and j linearly,

$$p_i \rightarrow p_i + zq, \quad p_j \rightarrow p_j - zq, \tag{11.60}$$

with a vector q that satisfies

$$q \cdot p_i = q \cdot p_j = q^2 = 0. \tag{11.61}$$

This ensures that the shifted momenta remain on-shell and that invariants $\hat{P}^2_{ij\ldots k}$ are linear in z, so that each propagator going on-shell corresponds to a unique simple pole in the z-plane.

Unfortunately (or, very interestingly, if that is how you like it), in 3d the only q that satisfies these constraints is $q = 0$. The reason is this. A 3d vector q with $q^2 = 0$ can be written as a bi-spinor $q = |q\rangle^a |q\rangle^b$. The $|q\rangle$ is a 2-component spinor so it cannot be linearly independent from the $|i\rangle$ and $|j\rangle$ of the two light-like momenta we are shifting. Hence

$$|q\rangle = \alpha|i\rangle + \beta|j\rangle \tag{11.62}$$

for some numbers α and β. Solving for α and β subject to the constraints $q \cdot p_i = q \cdot p_j = 0$ in (11.61) gives $\alpha = \beta = 0$ and hence $q = 0$.

So in order to make progress, we need to relax some of the constraints imposed on the shifted momenta. We cannot give up on momentum conservation and on-shellness for the shifted momenta. Instead, we can either shift three or more external momenta or give up on the property that the momenta shift linearly in z. The former is similar to the shift associated with CSW (Section 3.4) and comes at the price of involving many diagrams for the superamplitudes. Opting for the solution with fewer diagrams, we choose to give up on linearity in z and consider the following general 2-line "deformation" [162]

$$\begin{pmatrix} |\hat{i}\rangle \\ |\hat{j}\rangle \end{pmatrix} = R(z) \begin{pmatrix} |i\rangle \\ |j\rangle \end{pmatrix}. \tag{11.63}$$

$R(z)$ is a 2×2 matrix that depends on z. Since we want the shift to respect momentum conservation, the matrix R must satisfy:

$$R(z)^T R(z) = I. \tag{11.64}$$

Thus $R(z)$ is an orthogonal matrix and we can parameterize it as

$$R(z) = \begin{pmatrix} \cos\theta & -\sin\theta \\ \sin\theta & \cos\theta \end{pmatrix} = \begin{pmatrix} \frac{z+z^{-1}}{2} & -\frac{z-z^{-1}}{2i} \\ \frac{z-z^{-1}}{2i} & \frac{z+z^{-1}}{2} \end{pmatrix}. \tag{11.65}$$

If we define the deformation on the fermionic variables η_i and η_j in the same fashion, supermomentum conservation is also preserved by the shift:

$$\hat{\tilde{q}}_{iA} + \hat{\tilde{q}}_{jA} = \left(|\hat{i}\rangle, |\hat{j}\rangle\right) \begin{pmatrix} \hat{\eta}_{iA} \\ \hat{\eta}_{jA} \end{pmatrix} = \left(|i\rangle, |j\rangle\right) R^T(z) R(z) \begin{pmatrix} \eta_{iA} \\ \eta_{jA} \end{pmatrix} = \tilde{q}_{iA} + \tilde{q}_{jA}. \tag{11.66}$$

The deformation matrix (11.65) becomes the identity when $z = 1$, so unshifted kinematics corresponds to $z = 1$ and not $z = 0$. This leads to the following contour integral representation of the unshifted tree-level amplitude:

$$\mathcal{A}_n = \frac{1}{2\pi i} \oint_{z=1} \frac{\hat{\mathcal{A}}_n(z)}{z-1}, \tag{11.67}$$

where the contour wraps just the pole at $z = 1$. If the deformed superamplitude $\hat{\mathcal{A}}_n(z)$ vanishes as $z \to \infty$,[8] one can perform a contour-deformation and evaluate the amplitude as a sum of the residues at finite $z \neq 0, 1$.

Just as in 4d, the poles at finite $z \neq 0, 1$ correspond to propagators going on-shell. Let us take a closer look at what the singularities look like. Without loss of generality, we choose 1 and n as the deformed momenta:

$$\hat{p}_1^{ab} = \frac{1}{2}(p_1^{ab} + p_n^{ab}) + z^2 q^{ab} + z^{-2}\tilde{q}^{ab},$$
$$\hat{p}_n^{ab} = \frac{1}{2}(p_1^{ab} + p_n^{ab}) - z^2 q^{ab} - z^{-2}\tilde{q}^{ab}. \tag{11.68}$$

Here q and $\tilde{q}$ are given by

$$q^{ab} = \frac{1}{4}(|1\rangle + i|n\rangle)^a (|1\rangle + i|n\rangle)^b, \qquad \tilde{q}^{ab} = \frac{1}{4}(|1\rangle - i|n\rangle)^a (|1\rangle - i|n\rangle)^b. \tag{11.69}$$

Defining $P_{12...i}^{ab} = p_1^{ab} + p_2^{ab} + \cdots + p_i^{ab}$, the on-shell condition for the shifted propagator $\hat{P}_{12...i}^2$ takes the form

$$\hat{P}_{12...i}^2 = \langle \tilde{q} | P_{23...i} | \tilde{q} \rangle z^{-2} + \langle q | P_{23...i} | q \rangle z^2 - (P_{23...i} \cdot P_{i+1...n-1}) = 0, \tag{11.70}$$

where $(p_i \cdot p_j) = p_i^\mu p_{j\mu}$ and $\langle i | P | j \rangle \equiv \lambda_i^a P_a{}^b \lambda_{jb}$. One can explicitly write down the values of z that correspond to the propagator going on-shell

$$\left\{(z_{1,i}^*)^2, (z_{2,i}^*)^2\right\} = \frac{(P_{2...i} \cdot P_{i+1...n-1}) \pm \sqrt{(P_{2...i})^2 (P_{i+1...n-1})^2}}{2\langle q | P_{2...i} | q \rangle}. \tag{11.71}$$

---

[8] One should also make sure that there are no poles at $z = 0$. Exchanging $1/z \leftrightarrow z$ in (11.65) can be compensated by extra sign factors in the kinematics of the shifted legs, so if $\mathcal{A}_n(z)$ vanishes as $z \to \infty$ for generic kinematics, then it also vanishes at $z = 0$.

▶ **Exercise 11.6**

Prove the following useful identity

$$\left((z_{1,i}^*)^2 - 1\right)\left((z_{2,i}^*)^2 - 1\right) = \frac{P_{12...i}^2}{\langle q | P_{2...i} | q \rangle}\,. \tag{11.72}$$

As the propagator goes on-shell, the amplitude factorizes into two lower-point amplitudes. This allows us to write the sum of residues at $z \neq 1$ as a sum over distinct single propagator diagrams with legs 1 and n on opposite sides of the propagator. For Chern–Simons matter theories, we also require that only even multiplicity subamplitudes appear on each side of the propagator. For each propagator, one needs to sum over the four solutions, $(z_{1,i}^*, -z_{1,i}^*, z_{2,i}^*, -z_{2,i}^*)$ to the on-shell constraint (11.70). The final result is then [162]

$$\mathcal{A}_n = \sum_i \int d^3\eta_I \left(\hat{\mathcal{A}}_L(z_{1,f}^*; \eta_I) \frac{H(z_{1,f}^*, z_{2,f}^*)}{P_{12...i}^2} \hat{\mathcal{A}}_R(z_{1,f}^*; i\eta_I) + (z_{1,f}^* \leftrightarrow z_{2,f}^*) \right), \tag{11.73}$$

where the function $H(a, b)$ is

$$H(a, b) \equiv \frac{a(b^2 - 1)}{a^2 - b^2} \tag{11.74}$$

and the Grassmann integral takes care of the intermediate state sum. Did you notice the i in $\hat{\mathcal{A}}_R$? That comes from the analytic continuation of the incoming $\rightarrow$ outgoing internal line. In 3d massless kinematics, we only have one type of spinor, namely $|p\rangle$, so with $p^{ab} = -|p\rangle^a |p\rangle^b$ we must have

$$|-p\rangle = i\,|p\rangle\,. \tag{11.75}$$

Hence we must also have $\eta_{-p} = i\,\eta_p$, since – as you can check – this ensures that the arguments of the L and R Grassmann delta functions add up to the overall supermomentum $\widetilde{Q}$.

▶ **Exercise 11.7**

Show that the contour deformation of (11.67) gives the representation (11.73).
[Hints: The identity in (11.72) will be useful. Furthermore, since one of the shifted legs, 1 or n, necessarily corresponds to the fermionic multiplet, we have $\mathcal{A}_L(-z)\mathcal{A}_R(-z) = -\mathcal{A}_L(z)\mathcal{A}_R(z)$.]

The validity of (11.73) relies on whether or not the super-shifted superamplitude vanishes as $z \rightarrow \infty$. It was shown in [162] that this criterion is satisfied for the ABJM and BLG theories.

▶ **Exercise 11.8**

Recall that in Section 3.3 we discussed when a BCFW recursion is valid: we showed that the presence of contact terms, for example ϕ^4, in the action tends to spoil the recursion since such terms go to a constant as $z \to \infty$. This issue can be avoided in supersymmetric theories since amplitudes where such terms are present are related via supersymmetry to those from which it is absent. This is accomplished via the super-BCFW shifts. One can illustrate the idea by carefully choosing the external states such that contact terms do not contribute to a particular component amplitude; then (loosely speaking) supersymmetry ensures that the superamplitude, which contains this well-behaved component amplitude, also goes to zero for $z \to \infty$. Let us test whether such a component amplitude can be found for ABJM theory at 6-point. Consider the 6-point contact term in (11.56). Show that if we choose all R-symmetry indices to be the same, say 1, then the sextic interaction terms vanish. Thus the 6-point scalar amplitude with all scalars having the same $SU(4)$ indices has good large-z behavior.

One thing is deriving the recursion relations, another thing is using them! So let us now apply the 3d recursion relations to compute the 6-point amplitude $A_6(\bar{X}^4 X_4 \bar{X}^4 X_4 \bar{\psi}_4 \psi^4)$ in ABJM theory. For simplicity, we drop the $SU(4)$ indices on the component-fields, i.e. $\bar{X}^4 \to \bar{X}$. Choosing lines 1 and 6 for the shift, the only factorization channel is (123|456), so there is only one diagram, namely

$$ \tag{11.76} $$

The recursion relation (11.73) then reads

$$
\mathcal{A}_6\big(\bar{\Psi}\Phi\bar{\Psi}\Phi\bar{\Psi}\Phi\big)
$$
$$
= \int d^3\eta \left[\hat{\mathcal{A}}_4\big(\bar{\Psi}_{\hat{1}} \Phi_2 \bar{\Psi}_3 \Phi_{\hat{P}_{123}}\big)\Big|_{z=z_1^*} \frac{H(z_1^*, z_2^*)}{P_{123}^2} \mathcal{A}_4\big(\bar{\Psi}_{-\hat{P}_{123}} \Phi_4 \bar{\Psi}_5 \Phi_{\hat{6}}\big)\Big|_{z=z_1^*} + (z_1^* \leftrightarrow z_2^*) \right].
$$
$$ \tag{11.77} $$

To project out the amplitude $A_6(\bar{X}X\bar{X}X\bar{\psi}\psi)$ from the superamplitude (11.77) we need the coefficient of the $(\eta_1)^3 (\eta_3)^3 (\eta_6)^3$ monomial, where $(\eta_i)^3 = \eta_{i1}\eta_{i2}\eta_{i3}$. This follows from (11.51). After manipulation of the Grassmann delta functions and using $\hat{\eta}_1(z)\hat{\eta}_6(z) = \eta_1\eta_6$, we find

$$
A_6\big(\bar{X}X\bar{X}X\bar{\psi}\psi\big) = \hat{A}_4\big(\bar{X}_{\hat{1}} X_2 \bar{X}_3 X_{\hat{P}_{123}}\big)\Big|_{z=z_1^*} \frac{H(z_1^*, z_2^*)}{P_{123}^2} \hat{A}_4\big(\bar{X}_{\hat{P}_{123}} X_4 \bar{\psi}_5 \psi_{\hat{6}}\big)\Big|_{z=z_1^*} + (z_1^* \leftrightarrow z_2^*),
$$
$$ \tag{11.78} $$

where the 4-point amplitudes, obtained from the superamplitude (11.52), are

$$
\hat{A}_4\big(\bar{X}_{\hat{1}} X_2 \bar{X}_3 X_{\hat{P}_{123}}\big) = -\frac{\langle \hat{1}3\rangle^3}{\langle \hat{1}\hat{P}_{123}\rangle \langle \hat{P}_{123}3\rangle} \quad \text{and} \quad \hat{A}_4\big(\bar{X}_{\hat{P}_{123}} X_4 \bar{\psi}_5 \psi_{\hat{6}}\big) = \frac{\langle \hat{P}_{123}\hat{6}\rangle^2}{\langle 65\rangle} . \tag{11.79}
$$

By (11.71), the poles in the z-plane are located at

$$
z_1^{*2} = \frac{\langle 16\rangle^2 - \big(\langle 23\rangle - \langle 45\rangle\big)^2}{\big(\langle 1| + i\langle 6|\big) P_{45}\big(|1\rangle + i|6\rangle\big)}, \qquad z_2^{*2} = \frac{\langle 16\rangle^2 - \big(\langle 23\rangle + \langle 45\rangle\big)^2}{\big(\langle 1| + i\langle 6|\big) P_{45}\big(|1\rangle + i|6\rangle\big)} . \tag{11.80}
$$

After repeated use of momentum conservation and the Schouten identity, we find that the 6-point amplitude is

$$A_6(\bar{X}X\bar{X}X\bar{\psi}\psi) = -\frac{1}{2P_{123}^2}\left[\frac{(\langle 2|P_{123}|6\rangle + i\langle 31\rangle\langle 45\rangle)^3}{(\langle 1|P_{123}|4\rangle + i\langle 23\rangle\langle 56\rangle)(\langle 3|P_{123}|6\rangle + i\langle 12\rangle\langle 45\rangle)}\right.$$
$$\left.-\frac{(\langle 2|P_{123}|6\rangle - i\langle 31\rangle\langle 45\rangle)^3}{(\langle 1|P_{123}|4\rangle - i\langle 23\rangle\langle 56\rangle)(\langle 3|P_{123}|6\rangle - i\langle 12\rangle\langle 45\rangle)}\right].$$
$$(11.81)$$

Here the first term is the result of evaluating the first term in (11.77) while the second term above is the $(z_1^* \leftrightarrow z_2^*)$ contribution.

▶ **Exercise 11.9**

Let us derive (11.81) from (11.78). First prove

$$\left((z_1^*)^2 + 1\right)\left((z_2^*)^2 - 1\right) = \frac{-i\langle 1|P_{23}|6\rangle + \langle 23\rangle\langle 45\rangle}{\langle q|P_{2\dots i}|q\rangle}. \tag{11.82}$$

Next, use (11.82) to show that

$$\langle \hat{1}3\rangle z_1^* \left((z_2^*)^2 - 1\right) = \frac{i\langle 23\rangle\left(\langle 2|P_{123}|6\rangle + i\langle 31\rangle\langle 45\rangle\right)}{2\langle q|P_{23}|q\rangle}. \tag{11.83}$$

Now continue to manipulate the tree amplitudes (11.79) to derive the first line in (11.81).

You may worry about the apparently spurious poles in the expression (11.81), since each only appears in one term and not the other and thus cannot cancel. But have no fear, these are really local poles in disguise! To see this, we rewrite them as (see Exercise 11.10)

$$\frac{1}{\langle 1|P_{123}|4\rangle - i\langle 23\rangle\langle 56\rangle} = \frac{\langle 1|P_{123}|4\rangle + i\langle 23\rangle\langle 56\rangle}{\langle 1|P_{123}|4\rangle^2 + \langle 23\rangle^2\langle 56\rangle^2} = \frac{\langle 1|P_{123}|4\rangle + i\langle 23\rangle\langle 56\rangle}{P_{123}^2 P_{234}^2}. \tag{11.84}$$

Thus each spurious-looking pole in (11.81) is really a product of local poles. Note that this tells us that the two terms in the BCFW result (11.81) are individually local and free of spurious poles! The reason behind this will be discussed further in Sections 11.3.8 and 11.3.10.

▶ **Exercise 11.10**

The final manipulation in (11.84) made use of the identity

$$\langle i|p_j + p_k|l\rangle^2 - (p_i + p_j + p_k + p_l)^2\langle jk\rangle^2 = (p_i + p_j + p_k)^2(p_j + p_k + p_l)^2,$$

which holds for any four massless vectors p_i, p_j, p_k, p_l in 3d. Prove it.

▶ **Exercise 11.11**

Although the two terms in (11.81) are individually local, they actually need to come in the combination in (11.81): show that the relative minus sign is necessary for the amplitude to have the correct little group properties.

11.3.8 ABJM and dual conformal symmetry

Let us dive straight into the deep end and define 3d dual variables y_i^{ab} and θ_{iA}^a such that $y_i^{ab} - y_{i+1}^{ab} = p_i^{ab}$ and $\theta_{iA}^a - \theta_{i+1,A}^a = \tilde{q}_{iA}^a$. Momentum and supermomentum delta functions for a 4-point superamplitude are then

$$\delta^3(P)\delta^{(\mathcal{N})}(\tilde{Q}) \to \delta^3(y_1 - y_5)\delta^{(\mathcal{N})}(\theta_1^a - \theta_5^a). \tag{11.85}$$

We define **_dual conformal inversion_** on the variables y_i and θ_i the same way in any spacetime dimension, namely as in (5.37). It then follows from (11.85) that the inversion weights of the momentum and supermomentum delta function exactly cancel for $\mathcal{N} = 6$ supersymmetry. Using

$$y_{i,i+2}^2 = s_{i,i+1} = \langle i, i+1 \rangle^2, \tag{11.86}$$

we deduce (as in Exercise 5.6) the dual inversion rule for a 3d angle bracket

$$I[\langle i, i+1 \rangle] = \frac{\langle i, i+1 \rangle}{\sqrt{y_i^2 y_{i+2}^2}}. \tag{11.87}$$

For the 4-point superamplitude (11.52) of $\mathcal{N} = 6$ ABJM theory, this then implies

$$I\left[\mathcal{A}_4(\bar{\Psi}_1 \Phi_2 \bar{\Psi}_3 \Phi_4)\right] = \sqrt{y_1^2 y_2^2 y_3^2 y_4^2}\, \mathcal{A}_4(\bar{\Psi}_1 \Phi_2 \bar{\Psi}_3 \Phi_4). \tag{11.88}$$

It can be shown [162], using the $\mathcal{N} = 6$ super-BCFW recursion relations, that dual inversion on the n-point tree-level superamplitude gives

$$I[\mathcal{A}_n] = \left(\prod_{i=1}^n \sqrt{y_i^2}\right) \mathcal{A}_n. \tag{11.89}$$

Thus the 3d ABJM tree-level superamplitudes are dual conformal *covariant* with uniform inversion weight $\frac{1}{2}$ on each leg.

Under dual conformal inversion, the superamplitudes of 4d $\mathcal{N} = 4$ SYM transform covariantly with uniform inversion weight 1 on each leg. In Section 5.3 we argued that, as a result, the dual conformal boosts $\mathcal{K}^\mu$ annihilate the superamplitudes only after the non-trivial weights have been compensated by a shift of $\mathcal{K}^\mu$, as below (5.42). This shift is crucial for defining the dual superconformal symmetry and extending it together with the ordinary superconformal symmetry to the $SU(2, 2|4)$ Yangian of the 4d planar $\mathcal{N} = 4$ SYM superamplitudes.

In 3d, the dual conformal symmetry can be enlarged into the dual superconformal symmetry group $OSp(6|4)$ [163]. The symmetry group acts on the dual space that consists of coordinates $(y_i^{ab}, \theta_{iA}^a, r_{iAB})$, where the extra R-symmetry coordinate r_{iAB} is defined by:

$$r_{iAB} - r_{i+1,AB} = \eta_{iA}\eta_{iB}. \tag{11.90}$$

The group $OSp(6|4)$ is also the supergroup for the ordinary superconformal symmetry of the ABJM Lagrangian in (11.53). The combination of the dual and ordinary superconformal symmetries forms an infinite dimensional $OSp(6|4)$ Yangian algebra [157], very similar in nature to the $SU(2, 2|4)$ Yangian symmetry of 4d planar $\mathcal{N} = 4$ SYM.

The super-BCFW construction (11.77) actually gives the tree-level 6-point superamplitude in terms of two Yangian invariants,

$$\mathcal{A}_6^{\text{tree}} \equiv \mathcal{A}_6^{\text{tree}}(\bar{\Psi}_1\Phi_2\bar{\Psi}_3\Phi_4\bar{\Psi}_5\Phi_6) = Y_1 + Y_2. \tag{11.91}$$

The two Yangian invariants Y_1 and Y_2 arise precisely from the two BCFW-terms in (11.77). We will not need their explicit form; they can be found in [162].

It will be relevant for us to also consider the tree superamplitude with shifted sites,

$$\mathcal{A}_{6,\text{shifted}}^{\text{tree}} \equiv \mathcal{A}_6^{\text{tree}}(\Phi_1\bar{\Psi}_2\Phi_3\bar{\Psi}_4\Phi_5\bar{\Psi}_6) = \mathcal{A}_6^{\text{tree}}(\bar{\Psi}_2\Phi_3\bar{\Psi}_4\Phi_5\bar{\Psi}_6\Phi_1). \tag{11.92}$$

It has a super-BCFW representation that can be written as

$$\mathcal{A}_{6,\text{shifted}}^{\text{tree}} = \mathcal{A}_6^{\text{tree}}(\Phi_1\bar{\Psi}_2\Phi_3\bar{\Psi}_4\Phi_5\bar{\Psi}_6) = Y_1 - Y_2. \tag{11.93}$$

Now the important point is that two physical objects, $\mathcal{A}_6^{\text{tree}}$ and $\mathcal{A}_{6,\text{shifted}}^{\text{tree}}$, are written as distinct linear combinations of the same two Yangian invariants: this is only possible if each of the two Yangian invariants is local, i.e. free of spurious poles. We have already noted the locality for the particular component amplitude (11.81). Now you see why it was needed. Note that this distinguishes 3d ABJM theory from 4d planar $\mathcal{N} = 4$ SYM in which the dual superconformal invariant 5-brackets had spurious poles.

11.3.9 Loops and on-shell diagrams in ABJM

The loop-level superamplitudes can be explored using unitarity methods (Chapter 6). Using the dual inversion property of the tree-level superamplitudes, it can be shown that the planar loop superamplitudes of ABJM, prior to integration, are dual conformal covariant, i.e. they satisfy (11.89). Thus perturbatively, planar ABJM has a structure very similar to planar $\mathcal{N} = 4$ SYM, they are almost baby brothers/sisters. This is rather surprising given that the two theories have very distinct Lagrangians and live in different spacetime dimensions. Moreover, in quantum field theory textbooks, one learns that $D < 4$ theories generically have more severe IR divergences compared to $D = 4$. Thus one might expect that although planar ABJM is very similar to $\mathcal{N} = 4$ SYM at the pre-integrated level, the similarity would be completely scrambled by the potentially more severe IR-divergence in $D = 3$.

To see if this is the case, let us take a look at the planar loop amplitudes in detail. The 1-loop amplitudes in ABJM are purely rational functions [164–166]. This can be understood as a consequence of dual conformal symmetry, since the only dual conformal covariant scalar integral is the massive triangle integral, and it integrates to

$$I_3(K_1, K_2, K_3) = -\frac{i\pi}{2}\frac{1}{\sqrt{-K_1^2}\sqrt{-K_2^2}\sqrt{-K_3^2}}, \qquad (11.94)$$

where K_1, K_2, K_3 are the sums of the external momenta going out of each of the three corners and $K_i^2 \neq 0$. There are no triangle diagrams with massless corners $K_i^2 = 0$; this follows from generalized unitarity methods using the fact that 3-point loop amplitudes vanish. The integrated result (11.94) has transcendentality 1 thanks to the factor of π.

Since 6-point is the lowest multiplicity at which the triangle integral (11.94) contributes after integration, we conclude that the **4-point 1-loop amplitude vanishes** up to $O(\epsilon)$ in dimensional regularization $D = 3 - 2\epsilon$.[9]

The **6-point 1-loop superamplitude** is [164, 165, 168][10]

$$\mathcal{A}_6^{\text{1-loop}} = -i\left(\frac{N}{k}\right)\mathcal{A}_{6,\text{shifted}}^{\text{tree}}$$

$$\times \left(\langle 12 \rangle\langle 34 \rangle\langle 56 \rangle \, I_3(P_{12}, P_{34}, P_{56}) + \langle 23 \rangle\langle 45 \rangle\langle 61 \rangle \, I_3(P_{23}, P_{45}, P_{61})\right), \quad (11.95)$$

where N comes from the gauge group $U(N) \times U(N)$, and k is the Chern–Simons level in (11.53). The tree superamplitude $\mathcal{A}_{6,\text{shifted}}^{\text{tree}}$ was defined in (11.92). Using the integrated result (11.94) for the scalar triangle integrals I_3, we find that the 1-loop 6-point superamplitude is

$$\mathcal{A}_6^{\text{1-loop}} = -\frac{\pi}{2}\left(\frac{N}{k}\right)\mathcal{A}_{6,\text{shifted}}^{\text{tree}}$$

$$\times \left(\text{sgn}(\langle 12 \rangle)\,\text{sgn}(\langle 34 \rangle)\,\text{sgn}(\langle 56 \rangle) + \text{sgn}(\langle 23 \rangle)\,\text{sgn}(\langle 45 \rangle)\,\text{sgn}(\langle 61 \rangle)\right). \quad (11.96)$$

Here we have introduced

$$\text{sgn}(\langle ij \rangle) \equiv \frac{\langle ij \rangle}{\sqrt{-K_{ij}^2}} = \frac{\langle ij \rangle}{|\langle ij \rangle|}, \qquad (11.97)$$

which equals ± 1 depending on the kinematics. Thus, remarkably, the 1-loop 6-point superamplitude can be either zero or non-vanishing depending on the kinematics! This peculiar behavior has to do with an interesting topological feature of light-like momenta in 3-dimensions. In 3d Minkowski space, a light-like vector can be written as $p_i^\mu = E_i(1, \cos\theta_i, \sin\theta_i)$. This means that light-like vectors can be projected to points on

---

[9] The 4-point 1-loop integrand is non-trivial. It is given by a loop-momentum dependent integrand that integrates to zero up to $O(\epsilon)$ [167].

[10] This result is only valid up to $O(\epsilon)$. There are additional integrands, whose coefficients are proportional to the tree amplitude, that integrate to zero up to $O(\epsilon)$.

a circle S^1. From

$$\langle ij \rangle = \sqrt{-2p_i \cdot p_j} = i\sqrt{E_i E_j} \, \sin\left(\frac{\theta_i - \theta_j}{2}\right) \qquad (11.98)$$

we see that the sign of $\langle ij \rangle$ changes whenever the two points that represent p_i and p_j cross each other on the S^1. Thus the 1-loop amplitude encounters a sudden jump, from zero to non-vanishing or vice versa, whenever two points on the S^1 cross each other:

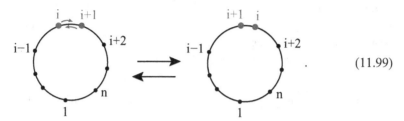

$$(11.99)$$

The two configurations (11.99) are topologically inequivalent.[11] Thus this sudden jolt is the amplitude's way of telling us that we are changing the topology of momentum space!

▶ Exercise 11.12

From 4d, we are familiar with loop amplitudes being proportional to the tree amplitudes, so it may seem odd that here in 3d the 1-loop 6-point amplitude (11.96) is proportional to the shifted tree amplitude. Verify that the LHS and RHS of (11.96) have the same little group scaling thanks to the sign-functions.

The first non-trivial loop contribution to the ***4-point amplitude*** enters at ***2-loop order*** and it is given by [167, 169]

$$\mathcal{A}_4^{\text{2-loop}} = \left(\frac{N}{k}\right)^2 \mathcal{A}_4^{\text{tree}} \left[-\frac{\left(-\mu^{-2} y_{13}^2\right)^{-\epsilon} + \left(-\mu^{-2} y_{24}^2\right)^{-\epsilon}}{(2\epsilon)^2} \right.$$
$$\left. + \frac{1}{2} \ln^2\left(\frac{y_{13}^2}{y_{24}^2}\right) + 4\zeta_2 - 3\ln^2 2 + \mathcal{O}(\epsilon) \right]. \quad (11.100)$$

Notice that the IR-divergent part is equivalent to that of the 1-loop 4-point superamplitude (6.42) of planar $\mathcal{N} = 4$ SYM, with $\epsilon \to 2\epsilon$ because this is 2-loops.[12] Not only is the IR-structure of this amplitude identical to that of planar $\mathcal{N} = 4$ SYM, but so is the finite $\ln^2\left(y_{13}^2/y_{24}^2\right)$ piece!

---

[11] This can be made more precise. By judiciously adding 2π to the angles θ_i, one can arrange the angles such that a given kinematics configuration has all angles strictly increasing according to their color ordering, i.e. $0 < \theta_{i+1} - \theta_i < 2\pi$. This gives a well defined "winding number" $w = (\theta_n - \theta_1)/(2\pi)$. Now as two points cross each other, the winding number changes by one, indicating a distinct topological sector.

[12] The overall scaling is not fixed because we have not identified the 4d $\mathcal{N} = 4$ SYM gauge coupling g with a power of N/k in ABJM.

Moving on to the **6-point 2-loop amplitude**, one finds [170]

$$\mathcal{A}_6^{\text{2-loop}} = \left(\frac{N}{k}\right)^2 \left\{ \frac{\mathcal{A}_6^{\text{tree}}}{2}\left[\text{BDS}_6 + R_6\right] + \frac{\mathcal{A}_{6,\text{shifted}}^{\text{tree}}}{4i}\left[\ln\frac{u_2}{u_3}\ln\chi_1 + \text{cyclic}\times 2\right] \right\}. \tag{11.101}$$

Here BDS_6 is the 1-loop MHV amplitude (6.57) of $\mathcal{N}=4$ SYM, again with proper rescaling of the regulator $\epsilon \to 2\epsilon$ to account for being at 2-loops. As the remaining pieces are finite, the BDS Ansatz captures the IR-divergent as well as the resulting non-dual-conformal part of the amplitude. So once again, we observe that the IR structure of planar ABJM theory is identical to that of planar $\mathcal{N}=4$ SYM! The "remainder" function R_6 in (11.101) is

$$R_6 = -2\pi^2 + \sum_{i=1}^{3}\left[\text{Li}_2(1-u_i) + \frac{1}{2}\ln u_i \ln u_{i+1} + (\arccos\sqrt{u_i})^2\right], \tag{11.102}$$

where the u_is are the dual conformal cross-ratios defined in (6.53). The shifted tree $\mathcal{A}_{6,\text{shifted}}^{\text{tree}}$ was encountered in (11.95). Finally, the function χ_1 in (11.101) is

$$\chi_1 = \frac{\langle 12\rangle\langle 45\rangle + i\langle 3|P_{123}|6]}{\langle 12\rangle\langle 45\rangle - i\langle 3|P_{123}|6]}, \tag{11.103}$$

and "cyclic$\times 2$" means we sum over all cyclic rotations by two sites, $i \to i+2$.

▶ **Exercise 11.13**

Seeing both $\mathcal{A}_6^{\text{tree}}$ and $\mathcal{A}_{6,\text{shifted}}^{\text{tree}}$ in the same amplitude, you should check that the other factors in (11.101) indeed compensate for the little group weight difference.

Now that we have seen explicit examples of planar loop amplitudes in ABJM theory, let us turn to the subject of **Leading Singularities** and **on-shell diagrams**. We studied these for planar 4d $\mathcal{N}=4$ SYM in Chapter 8. Because of the dual superconformal symmetry of the loop-integrands, multi-loop amplitudes of ABJM theory can be calculated with Leading Singularity methods. In 3d, a maximal cut takes three propagators on-shell for each loop-momentum. At 1-loop order, the only dual conformal scalar integral is the massive triangle (11.94), so this plays the equivalent role of the box-diagram in 4d. In 4d, we build the Leading Singularity on-shell diagrams from vertices that are the fundamental 3-point MHV and anti-MHV superamplitudes. These vanish in 3d, so here we use the 4-point superamplitudes instead. In ABJM theory, the first non-trivial 1-loop Leading Singularity is the 6-point diagram

$$\tag{11.104}$$

As noted above, each vertex represents an on-shell 4-point superamplitude of ABJM theory. There is no distinction of "black" and "white" vertices because there is only one type of 4-point superamplitude in ABJM.

In 4d $\mathcal{N} = 4$ SYM, we found that the 4-point Leading Singularity box diagram represents the 4-point tree amplitudes (see Section 8.3). Similarly, in ABJM, it turns out that the 6-point Leading Singularity triangle diagram (11.104) reproduces the tree-level 6-point superamplitude. To see this, we isolate the 3rd vertex in (11.104) and label the legs as

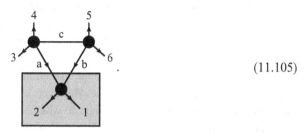

$$(11.105)$$

The internal momenta a and b are on-shell, so they each have two degrees of freedom. Of the combined $4 = 2 + 2$ degrees of freedom in a and b, three are fixed in terms of momenta 1 and 2 by the momentum conservation delta function at the bottom vertex in (11.105). Thus the spinor variables of a and b can be parameterized in terms of $|1\rangle$ and $|2\rangle$ using a single free variable. With a little thought – or, even better, a little calculation – one finds that the following parameterization solves the momentum conservation constraints

$$|a\rangle = \cos\theta \,|1\rangle - \sin\theta \,|2\rangle \,, \qquad |b\rangle = \sin\theta \,|1\rangle + \cos\theta \,|2\rangle \,. \qquad (11.106)$$

This is exactly the BCFW deformation (11.65) of legs 1 and 2. Indeed, the final on-shell condition $p_c^2 = 0$ becomes the factorization condition that the parameter θ (i.e. z) must satisfy.

▶ Exercise 11.14

Verify that the supermomentum delta function on the bottom vertex enforces the following identification $\eta_a = \cos\theta \,\eta_1 - \sin\theta \,\eta_2$ and $\eta_b = \sin\theta \,\eta_1 + \cos\theta \,\eta_2$.

Now it is very tempting to conclude that the on-shell diagram (11.105) can be also understood as a BCFW diagram for the 6-point tree superamplitude in ABJM. This is true, but we have to make sure that we produce the BCFW recursion formula (11.73), including the weight-factor $H(a, b)$ defined in (11.74) and the propagator of the factorization channel. Taking into account the Jacobian factors associated with the triple cut and the bottom vertex, it has been shown [168] that $H(a, b)$ and the factorization propagator are indeed produced. Thus we have found the 3d analogue of the "BCFW bridge," namely

$$= \int d^3\eta \left(\mathcal{A}_4\left(\Phi\bar{\Psi}\Phi\bar{\Psi}\right) \frac{H(z_1^*, z_2^*)}{(P_{234})^2} \mathcal{A}_4\left(\bar{\Psi}\Phi\bar{\Psi}\Phi\right) + (z_1^* \leftrightarrow z_2^*) \right) = \mathcal{A}_6^{\text{tree}} \,.$$

$$(11.107)$$

Recall that in 4d, the Leading Singularity is closely related to the integral coefficients in expressions like (6.26). Previously we have seen that the 1-loop 6-point amplitude is proportional to $\mathcal{A}_{6,\text{shifted}}^{\text{tree}}$, so it is perhaps puzzling that the 6-point Leading Singularity (11.107) is just $\mathcal{A}_6^{\text{tree}}$. This has to do with a subtlety of the Jacobian factors. Recall that the integral coefficients can be determined by unitarity cuts. When we apply unitarity cuts, we are substituting the propagators with delta functions, as discussed in Section 6.2. As we solve the delta function constraints, we generate a Jacobian factor with an absolute value. On the other hand, when we are computing the Leading Singularity, we treat the delta functions as contour integrals, thus while localizing on a pole, the Jacobian factor does not come with an absolute value. In the 1-loop cases that we encountered in 4d, the Jacobian factors for the two loop-momentum solutions are identical, so the presence of an absolute value did not make a difference. However, in 3d, the Jacobian factors for the two loop-momentum solutions differ by a sign, so whether or not there is an absolute value on the Jacobian makes a difference [170]. The result of this is that the 1-loop 6-point amplitude is proportional to $\mathcal{A}_{6,\text{shifted}}^{\text{tree}}$ while the 6-point Leading Singularity is just $\mathcal{A}_6^{\text{tree}}$.

▶ **Exercise 11.15**

The above discussion indicates that if we had a relative plus sign for the two BCFW terms on the RHS of (11.81), the result would be $\mathcal{A}_{6,\text{shifted}}^{\text{tree}}$ instead of $\mathcal{A}_6^{\text{tree}}$. Verify that with a relative plus sign, (11.81) has the correct little group property of $\mathcal{A}_{6,\text{shifted}}^{\text{tree}}$. [Hint: you need to take into account that the coefficient for the η-polynomial corresponds to a different component amplitude in the shifted amplitude.]

Instead of (11.105), we could have computed the on-shell diagram

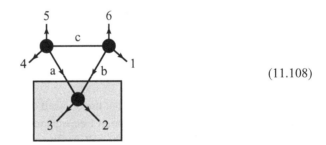

$$(11.108)$$

and the result would have been exactly the same, namely $\mathcal{A}_6^{\text{tree}}$. This gives us the ABJM equivalent of the "square move" (8.48) in 4d $\mathcal{N} = 4$ SYM. The ABJM "triangle move" is

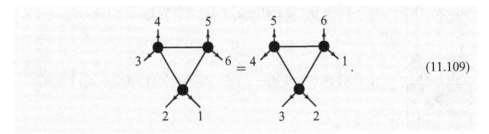

$$(11.109)$$

This is dubbed the ***Yang–Baxter move***, because it is precisely the graphical representation of the Yang–Baxter equation that plays an important role in integrable theories. It is usually represented as

$$(11.110)$$

Just as in $\mathcal{N} = 4$ SYM, the on-shell diagrams at higher multiplicity have an interesting structure. A description of this can be found in [120].

11.3.10 The orthogonal Grassmannian

Let us see if we can reason our way to a proper Grassmannian formula for the 3d superamplitudes of ABJM theory. We begin with

$$\prod_{a=1}^{k} \delta^{2|3} \left(\sum_i C_{ai} \Lambda_i \right) \equiv \prod_{a=1}^{k} \delta^2 \left(\sum_i C_{ai} |i\rangle \right) \delta^{(3)} \left(\sum_i C_{ai} \eta_i \right), \qquad (11.111)$$

where $\Lambda_i = (|i\rangle, \eta_i)$. As in the 4d case, treated in Chapter 9, the function (11.111) is invariant under generators in (11.39) which are linear in derivatives. If we consider the generators that are quadratic in derivatives, for example the conformal boost generator, we find

$$\left(\sum_{i=1}^{n} \frac{\partial}{\partial |i\rangle^a} \frac{\partial}{\partial |i\rangle^b} \right) \prod_{a=1}^{k} \delta^2 \left(\sum_j C_{aj} |j\rangle \right) = \left(\sum_{i=1}^{n} C_{a'i} C_{b'i} \right) f_{ab} \prod_{a \neq a', b'} \delta^2 \left(\sum_j C_{aj} |i\rangle \right). \qquad (11.112)$$

Here f_{ab} is a function that includes either a single derivative or double derivatives of the delta functions, depending on whether $a = b$ or not. The important piece in (11.112) is the prefactor: it tells us that to ensure invariance under conformal boosts, we need to dress (11.111) with an extra term that enforces $CC^T = 0$, i.e.

$$\delta^{\frac{k(k+1)}{2}} \left(CC^T \right) \prod_{a=1}^{k} \delta^{2|3} \left(\sum_i C_{ai} \Lambda_i \right). \qquad (11.113)$$

The product CC^T is a symmetric $k \times k$ matrix, so setting it to zero takes $k(k+1)/2$ constraints, as indicated in the delta function.

▶ Exercise 11.16

Show that the combination in (11.113) is also invariant under the multiplicative generators such as P^{ab}.

Now we can readily write down an Ansatz for an $\mathcal{N} = 6$ superconformal invariant integral over a Grassmannian $\mathrm{Gr}(k, n)$ subject to the orthogonal constraint $CC^T = 0$:

$$\int d^{n \times k} C \; f(M) \, \delta^{\frac{k(k+1)}{2}} \left(CC^T \right) \prod_{a=1}^{k} \delta^{2|3} \left(\sum_i C_{ai} \Lambda_i \right), \qquad (11.114)$$

where $f(M)$ is a function that only depends on the minors of the Grassmannian, so that it preserves $SL(k)$ invariance. In order to interpret (11.114) as an integral over a Grassmannian manifold, it has to be $GL(k)$ invariant. All terms in (11.114) are $SL(k)$ invariant, and the $GL(1)$ weight count of the delta functions plus the measure gives $nk - k(k + 1) - 2k + 3k = k(n - k)$. This tells us that the function $f(M)$ needs to have $GL(1)$ weight $-k(n - k)$.

We need more input to fix $f(M)$ and the extra information comes from little group analysis. Under the $\mathbb{Z}_2$ little group, we have $|i\rangle \to -|i\rangle$ and $\eta_i \to -\eta_i$, so invariance of the delta functions in (11.114) requires $C_{ai} \to -C_{ai}$. For an amplitude with a $\bar{\Psi}$-supermultiplet on the odd-sites, the superamplitude should pick up a minus sign whenever we perform a $\mathbb{Z}_2$ transformation on the odd-numbered legs, while it should be inert for even legs with their Φ-supermultiplet. Take $n = 2k$, and $k = $ even: then the product of k consecutive minors,

$$f(M) = \prod_{i=1}^{k} \frac{1}{M_i}, \qquad (11.115)$$

indeed satisfies the little group criteria. (Exercise 11.17 helps you see this.) Furthermore, since $n = 2k$, the function (11.115) has $GL(1)$ weight $-k^2$, precisely as needed for overall $GL(1)$ invariance.

▶ Exercise 11.17

For $k = 3$ (and hence $n = 6$), verify that (11.115) indeed picks up a minus sign under little group scaling for odd legs and is invariant for even legs. Show that for $k = $ odd, the function $f(M) = \prod_{i=2}^{k+1} \frac{1}{M_i}$ does the right job.

We conclude that the 3d Grassmannian formula for ABJM theory is given by an orthogonal Grassmannian integral [171] which for $k = $ even is

$$\mathcal{L}_{2k,k}^{\mathrm{O}} = \int \frac{d^{2k^2} C}{GL(k)} \left(\prod_{i=1}^{k} \frac{1}{M_i} \right) \delta^{\frac{k(k+1)}{2}} \left(CC^T \right) \prod_{a=1}^{k} \delta^{2|3} \left(\sum_i C_{ai} \Lambda_i \right). \qquad (11.116)$$

The superscript "O" indicates it is an ***orthogonal Grassmannian***. When $k = $ odd, the product of minors is replaced by $\prod_{i=2}^{k+1} \frac{1}{M_i}$, as shown in Exercise 11.17. Some comments are in order:

- *Momentum conservation* is enforced in (11.116) in a slightly differently manner than in the 4d version (9.17) of the Grassmannian integral because we only have the $|i\rangle$-spinors in 3d. Here is how it goes. The orthogonality condition forces the Grassmannian to be a collection of null vectors in an n-dimensional space. The bosonic delta function $\delta^2(C \cdot |i\rangle)$ requires the two n-dimensional vectors $\{|i\rangle\}$ to be orthogonal to C. This means that $\{|i\rangle\}$ lies in the complement of C, which is nothing but C^T, and thus $\{|i\rangle\}$ must also be null: $\sum_i |i\rangle^a |i\rangle^b = 0$.

- *Two-site cyclicity?* The integral $\mathcal{L}^O_{2k,k}$ does not appear to have the correct cyclic invariance by two sites as discussed (11.58). However, thanks to the orthogonality condition it can be shown that

$$M_i M_{i+1} = (-1)^k M_{i+k} M_{i+1+k} . \tag{11.117}$$

Therefore the formula (11.116) is indeed invariant under cyclic rotation by two sites up to a factor of $(-1)^{k-1}$, as required.

▶ **Exercise 11.18**

Show that (11.117) holds true at 4-points: using $GL(2)$ invariance and the orthogonality condition, we can choose to fix the 2×4 matrices C to take the form

$$C = \begin{pmatrix} 1 & 0 & i \sin\theta & -i \cos\theta \\ 0 & 1 & i \cos\theta & i \sin\theta \end{pmatrix}. \tag{11.118}$$

Verify that $CC^T = 0$ and that (11.117) holds.

- *The dimension of the integral* (11.116) is found by counting how many free variables are left after localization by the delta functions. To start with, there is a total of $2k^2$ integration variables. The bosonic delta functions fix $k(k+1)/2 + 2k - 3$ constraints, with the -3 coming from the removal of the constraints that enforce momentum conservation. Subtracting the k^2 redundancy of $GL(k)$, the dimension of the integral is then $\frac{(k-2)(k-3)}{2}$. Thus for 4- and 6-point amplitudes ($k = 2, 3$), the delta functions completely localize the Grassmannian integral.

Let us now take a closer look at (11.116) for $n = 4$. Gauge fix the $GL(2)$ by taking

$$C = \begin{pmatrix} c_{21} & 1 & c_{23} & 0 \\ c_{41} & 0 & c_{43} & 1 \end{pmatrix}. \tag{11.119}$$

This leaves four parameters that can be fixed by the four delta functions in $\delta(C \cdot |i\rangle)$. Denote the solutions to $C \cdot |i\rangle = 0$ by $c^*_{\bar{r}s}$, with barred labels indicating even legs and un-barred odd legs. Then the delta functions can be rewritten as

$$\delta^4(C \cdot |i\rangle) = \frac{1}{\langle 13 \rangle^2} \prod_{\bar{r},s} \delta^4(c_{\bar{r}s} - c^*_{\bar{r}s}), \qquad \begin{pmatrix} c^*_{21} & c^*_{23} \\ c^*_{41} & c^*_{43} \end{pmatrix} = -\frac{1}{\langle 13 \rangle} \begin{pmatrix} \langle 23 \rangle & \langle 12 \rangle \\ \langle 43 \rangle & \langle 14 \rangle \end{pmatrix}. \tag{11.120}$$

▶ **Exercise 11.19**

Which property of the external momenta does it take for the above solution $c^*_{\bar{r}s}$ to solve the orthogonality constraint? (Show it!)

Localizing the Grassmannian integral on to $c^*_{\bar{r}s}$, we then find

$$\delta^3(CC^T) = \frac{\langle 13 \rangle^6}{\langle 24 \rangle^3} \delta^3(P), \qquad \delta^6(C \cdot \eta) = \frac{\langle 24 \rangle^3}{\langle 13 \rangle^6} \delta^{(6)}(\tilde{Q}), \qquad \frac{1}{M_1 M_2} = \frac{\langle 13 \rangle^2}{\langle 14 \rangle \langle 34 \rangle}. \tag{11.121}$$

Combining (11.120) and (11.121), we recover the superamplitude (11.52) of ABJM theory.

For $n = 6$, the integral (11.116) is again completely localized by the bosonic delta functions onto two solutions, each corresponding to one of the BCFW terms in (11.77). Recall that these two terms are individually local (see (11.81)). We can now understand why. At 6-point these are the only possible invariants produced by the Grassmannian integral, so this means that *if the orthogonal Grassmannian integral produces all possible dual conformal invariants*[13] the Leading Singularity of the 6-point superamplitude must be some linear combination of them. However, we already know that there are two distinct local rational functions for $n = 6$, namely $\mathcal{A}_n^{\text{tree}}$ and $\mathcal{A}_n^{\text{1-loop}} \propto \mathcal{A}_{n,\text{shifted}}^{\text{tree}}$. As we noted at the end of Section 11.3.8, since they are distinct, this can only mean one thing, namely that the two terms in (11.77) are individually local and free of spurious poles.

We conclude this section with a comparison of the 3d and 4d Grassmannians. In the 4d Grassmannian, the choice of contour that gives the tree superamplitude forces the Grassmannian $\text{Gr}(k, n)$ to be localized to a $\text{Gr}(2, n)$ submanifold. For $n = 6$ we saw how this is intimately related to locality, since the tree contour selected three residues whose sum was free of spurious poles. In 3d, on the other hand, the orthogonal Grassmannian integral localizes completely for $n = 6$ and gives us two local objects without any need for us to pick any contour. Does this mean that the localization to the $\text{Gr}(2, n)$ submanifold is not really related to locality? The answer turns out to be "no" in an interesting way. It was found in [173] that for $n = 6$, the orthogonality constraint indeed forces the Grassmanian to localize to a $\text{Gr}(2, n)$ submanifold. Thus, the Grassmannian for $\mathcal{N} = 4$ SYM achieves locality for the 6-point NMHV superamplitude by choosing a particular "tree-contour," while for 6-point ABJM superamplitudes, the Grassmannian achieves locality by subjecting itself to the orthogonal constraint. The invariant between the two cases is the $\text{Gr}(2, n)$ submanifold, which was previously [125, 126] linked to Witten's twistor string formulation [54]. So is there perhaps a 3d twistor string theory? A twistor-like string theory with target space $SU(2, 3|5)$ was constructed in [174] and it reproduces the $\text{Gr}(2, n)$ formula of the ABJM amplitudes. It is quite fascinating how the study of scattering amplitudes reveals the existence of a new twistor string theory!

There is also a momentum twistor formulation of the Grassmannian integral for ABJM amplitudes (see Elvang *et al.* [60]). It could be the first step towards an ABJM amplituhedron.

---

[13] Using on-shell diagrams, one can show that the Leading Singularities obtained from the result of loop-level recursion can always be identified with residues of the orthogonal Grassmannian integral [172].

Supergravity amplitudes

We have seen that on-shell methods are particularly powerful for theories with gauge redundancy. Gravity has in a sense even more redundancy than Yang–Mills theory because of the diffeomorphism invariance. So perhaps there are hidden structures waiting to be discovered in gravity amplitudes? In this chapter, we discuss what is currently known about the scattering amplitudes in perturbative supergravity theories, including their UV behavior, and we review the interesting connections between gauge theory amplitudes and gravity amplitudes, relations that are often phrased loosely as "gravity $=$ (gauge theory)^2." These relations are explored further in Chapter 13.

12.1 Perturbative gravity

In a typical course on General Relativity you learn about Einstein's equation and its solutions, for example the Schwarzchild black hole and Friedmann–Robertson–Walker cosmology. (And if you have a hot course, you will also learn about black rings.) These are solutions to the classical equations of motion of gravity, just as the point-particle Coulomb potential, electromagnetic waves, or Dirac monopoles are solutions to the Maxwell equations in electromagnetism. Here we are interested in the scattering of perturbative states at weak coupling. From the point of view of perturbation theory, monopoles and black holes are considered non-perturbative states that are typically suppressed by powers e^{-1/g^2} in the weak-coupling $g \ll 1$ scattering processes.

As you know well from your QFT courses, scattering amplitudes are obtained after quantization of the field theory: start with the Lagrangian, extract the Feynman rules, and off we go to calculate scattering perturbatively. Of course, in the previous 200-something pages, we have tried to convince you that recursion relations and other on-shell methods offer much more insight and efficiency than the good old Feynman rules, but to understand what we mean by **perturbative gravity**, let us start with the Lagrangian approach and Feynman rules. This will also give us a greater appreciation for the powers of the modern on-shell methods.

The Einstein equation, $G_{\mu\nu} = 8\pi T_{\mu\nu}$, is the classical equation of motion that follows from the variational principle applied to the **Einstein–Hilbert action**

$$S_{\text{EH}} = \frac{1}{2\kappa^2} \int d^D x \, \sqrt{-g} \, R \, + \, S_{\text{matter}} \,, \tag{12.1}$$

where R is the Ricci scalar and $2\kappa^2 = 16\pi G_N$. We have written the action in D spacetime dimensions with a D-dimensional Newton's constant G_N. The metric $g_{\mu\nu}(x)$ is a field in the field theory described by (12.1). The variation $\delta g_{\mu\nu}$ of $\sqrt{-g}\, R$ gives (after partial integration and a little work [175–177]) the Einstein tensor part, $G_{\mu\nu} = R_{\mu\nu} - \frac{1}{2}g_{\mu\nu}R$, of Einstein's equation, while the metric variation of the "matter" action S_{matter} in (12.1) gives the stress-tensor part, $T_{\mu\nu} = \frac{2}{\sqrt{-g}}\frac{\delta S_{\text{matter}}}{\delta g^{\mu\nu}}$. In the following, we use the term **pure gravity** to describe the field theory (12.1) without matter fields, $S_{\text{matter}} = 0$.

Quantum field theory in curved spacetime is a highly non-trivial and interesting subject which has important consequences such as Hawking radiation of black holes. But this is not what we are going to discuss here. Our focus is the application of standard quantum field theory in flat spacetime to scattering of the particles associated with the quantization of the gravitational field $g_{\mu\nu}$. More precisely, we expand the gravitational field around flat space $g_{\mu\nu} = \eta_{\mu\nu} + \kappa h_{\mu\nu}$ and regard the fluctuating field $h_{\mu\nu}$ as the **graviton field**. To start with, let us just consider pure gravity without matter and expand the Einstein–Hilbert action in powers of $\kappa h_{\mu\nu}$. Since the Ricci-scalar R involves two derivatives, every term in the expansion has two derivatives. Suppressing the increasingly intricate index-structure, we write these terms schematically as $h^{n-1}\partial^2 h$ for $n = 2, 3, 4, \ldots$, so that the action becomes

$$S_{\text{EH}} = \frac{1}{2\kappa^2}\int d^D x\, \sqrt{-g}\, R = \int d^D x \left[h\partial^2 h + \kappa\, h^2 \partial^2 h + \kappa^2\, h^3 \partial^2 h + \kappa^3\, h^4 \partial^2 h + \cdots \right].$$

(12.2)

There are infinitely many terms. There are two reasons for this: (a) in R, the series expansion of the inverse metric generates an infinite series and (b) the expansion of the determinant $g = \det g_{\mu\nu}$ is finite, but the square root in $\sqrt{-g}$ generates an infinite series.[1] There are no mass terms in (12.2), so the particles associated with quantization of the gravitational field $h_{\mu\nu}$ are massless: they have spin-2 and are called **gravitons**.

In order to extract Feynman rules from (12.2) we first have to gauge fix the action. A typical choice is the **de Donder gauge**, $\partial^\mu h_{\mu\nu} = \frac{1}{2}\partial_\nu h_\mu{}^\mu$, which brings the quadratic terms in the action to the form

$$h\partial^2 h \;\to\; -\frac{1}{2}h_{\mu\nu}\Box h^{\mu\nu} + \frac{1}{4}h_\mu{}^\mu \Box h_\nu{}^\nu.$$

(12.3)

The propagator resulting from these quadratic terms is

$$P_{\mu_1\nu_1,\mu_2\nu_2} = -\frac{i}{2}\left(\eta_{\mu_1\mu_2}\,\eta_{\nu_1\nu_2} + \eta_{\mu_1\nu_2}\,\eta_{\nu_1\mu_2} - \frac{2}{D-2}\,\eta_{\mu_1\nu_1}\,\eta_{\mu_2\nu_2} \right)\frac{1}{k^2}.$$

(12.4)

Each graviton leg is labeled by two Lorentz-indices. The external line rule is to dot in graviton polarization vectors. In 4d, the polarizations encode the two helicity $h = \pm 2$ physical graviton states. They can be constructed as products of the spin-1 polarization vectors (2.50):

$$e_-^{\mu\nu}(p_i) = \epsilon_-^\mu(p_i)\epsilon_-^\nu(p_i), \qquad e_+^{\mu\nu}(p_i) = \epsilon_+^\mu(p_i)\epsilon_+^\nu(p_i).$$

(12.5)

---

[1] By a field redefinition, we can use $g_{\mu\nu} = e^{-h_{\mu\nu}}$ instead; this brings the metric and its inverse on an equal footing and therefore offers a simpler expansion [178].

Note that this ensures the correct little group scaling t^{-2h_i} of the on-shell graviton scattering amplitude.

The infinite set of 2-derivative interaction terms $h^{n-1}\partial^2 h$ yields Feynman rules for n-graviton vertices for *any* $n = 3, 4, 5, \ldots$ For example, the de Donder gauge 3-vertex takes the form

$$V_3(p_1, p_2, p_3) = p_1^{\mu_3} p_2^{\nu_3} \eta^{\mu_1 \nu_2} \eta^{\mu_2 \nu_1} + \text{(many other terms with various index-structures)}. \tag{12.6}$$

You can look up the full expression for the 3-vertex in [179].

The 3-term de Donder propagator (12.4) and the infinite set of complicated interaction terms should make it clear that calculation of even tree-level graviton scattering amplitudes from Feynman diagrams is not a business for babies. The 4-point graviton tree amplitude was calculated brute force with Feynman diagrams in [22] where each of the four contributing Feynman diagrams fill about a page or so of elaborate index-delight. Nonetheless, the final result can be brought to a very simple form: in 4d, it can be written in spinor helicity formalism as

$$M_4^{\text{tree}}(1^- 2^- 3^+ 4^+) = \frac{\langle 12 \rangle^7 [12]}{\langle 13 \rangle \langle 14 \rangle \langle 23 \rangle \langle 24 \rangle \langle 34 \rangle^2} = \frac{\langle 12 \rangle^4 [34]^4}{stu}. \tag{12.7}$$

We have already encountered this expression in Exercise 2.33. Note that we will be using M_n to denote (super)gravity amplitudes to distinguish them from (super) Yang–Mills amplitudes A_n.

Of course, you already know where we are headed: on-shell methods and recursion relations make the calculation of tree-level graviton scattering amplitudes much more fun and efficient – and they have the power to reveal structures in the amplitudes that were not visible at the level of the Lagrangian. The short version of the story is that little group scaling fixes the possible 3-graviton amplitudes and recursion then allows you to compute all other tree-level graviton processes. Loop-level amplitudes can be constructed with unitarity techniques (Chapter 6). Thus the infinite set of interaction terms in the Lagrangian is not needed from the point of view of the on-shell scattering amplitudes: the terms' role in life is to ensure off-shell diffeomorphism invariance of the gravitational action. It is an interesting aspect of on-shell recursion relations that they eliminate the need for infinitely many interaction terms.

Let us specialize to $D = 4$ and be more explicit about the graviton scattering amplitudes. Dimensional analysis and little group scaling fix the 3-point graviton amplitudes to be

$$M_3(1^- 2^- 3^+) = \frac{\langle 12 \rangle^6}{\langle 23 \rangle^2 \langle 31 \rangle^2} = A_3[1^- 2^- 3^+]^2,$$

$$M_3(1^+ 2^+ 3^-) = \frac{[12]^6}{[23]^2 [31]^2} = A_3[1^+ 2^+ 3^-]^2. \tag{12.8}$$

At tree-level, the graviton amplitudes with all-plus or all-minus helicity arrangements vanish in pure gravity as do those with just one $\pm$-helicity:

$$M_n^{\text{tree}}(1^+2^+\ldots n^+) = M_n^{\text{tree}}(1^-2^+\ldots n^+) = M_n^{\text{tree}}(1^+2^-\ldots n^-)$$
$$= M_n^{\text{tree}}(1^-2^-\ldots n^-) = 0. \tag{12.9}$$

This is most easily proven using the supersymmetry Ward identities, just as we did in (4.26), (4.27) for gluon amplitudes. The tree gravity amplitudes have to obey these same Ward identities as in a supergravity theory because the supersymmetric partners couple quadratically; hence it is only at loop-level the pure graviton amplitudes can distinguish themselves from the supergravity amplitudes. In particular, (12.9) has to hold at tree-level.

▶ **Exercise 12.1**

For simplicity, we dropped the explicit powers of the gravitational coupling κ in (12.7) and (12.8), and we continue to do so henceforth. What is the mass-dimension of κ in 4d? Show that the 4-graviton amplitude (12.7) has the correct mass-dimension (cf. (2.102)).

We categorize graviton amplitudes the same way as gluon amplitudes with designation N^KMHV. An important difference is that the graviton scattering amplitudes are not color-ordered. Using BCFW recursion relations,[2] relatively compact graviton amplitudes can be found for the MHV sector. One of the earliest formulas is BGK (Berends, Giele, and Kuijf) [180] written here in the form presented in [84] valid for $n > 4$:

$$M_n^{\text{tree}}(1^-2^-3^+\ldots n^+) = \sum_{P(3,4,\ldots,n-1)} \frac{\langle 12\rangle^8 \prod_{l=3}^{n-1}\langle n|2+3+\cdots+(l-1)|l]}{\left(\prod_{i=1}^{n-2}\langle i,i+1\rangle\right)\langle 1,n-1\rangle\langle 1n\rangle^2\langle 2n\rangle^2\left(\prod_{l=3}^{n-1}\langle ln\rangle\right)} \cdot \tag{12.10}$$

The sum is over all permutations of the labels $(3, 4, \ldots, n-1)$.

Another form of the same MHV graviton tree amplitude expresses M_n^{tree} in terms of color-ordered gluon amplitudes A_n^{tree} of Yang–Mills theory:

$$M_n^{\text{tree}}(1^-2^-3^+\ldots n^+) = \sum_{P(i_3,i_4,\ldots,i_n)} s_{1i_n}\left(\prod_{k=4}^{n-1}\beta_k\right) A_n^{\text{tree}}\left[1^-2^-i_3^+ i_4^+\ldots i_n^+\right]^2, \tag{12.11}$$

where $n \geq 4$ and

$$\beta_k = -\frac{\langle i_k i_{k+1}\rangle}{\langle 2 i_{k+1}\rangle}\langle 2|i_3 + i_4 + \cdots + i_{k-1}|i_k]. \tag{12.12}$$

The result (12.11) can be derived [181, 182] using a $[-,-\rangle$ BCFW shift.

---

[2] BCFW for tree graviton amplitudes is valid when based on shifts $[-,-\rangle$, $[-,+\rangle$, and $[+,+\rangle$.

There are also other graviton MHV formulas available in the literature, for example the "soft-factor" formula [183]. You may find that these MHV expressions are terribly complicated compared with Parke–Taylor; however, they are remarkably simple when compared with the impenetrable mess a Feynman diagram calculation would produce.

Beyond the MHV level, one can readily use BCFW to calculate explicit results. You might be curious if there is also a CSW-like expansion for gravity amplitudes. The MHV vertex expansion [35] based on the Risager-shift (see discussion below Exercise 3.10) works for NMHV graviton amplitudes with $n < 12$ particles. It fails [36] for $n \geq 12$ because the large-z falloff of the n-point amplitude under Risager-shift is $1/z^{12-n}$ and the Cauchy contour deformation argument needed to derive the recursion relations therefore picks up a term at infinity for $n > 11$. The all-line shift discussed in Section 3.4 also fails (for interesting reasons [26]). For further discussion of CSW for gravity, see [17, 20, 26, 36].

The relation between gravity and gauge theory amplitudes is clearly visible in the 4d MHV expressions (12.8) and (12.11), but there are more general relations available. The first such example is the **_KLT relations_**, derived in string theory by Kawai, Lewellen and Tye [21]: the KLT relations give the n-point tree-level closed string scattering amplitude as a sum over products of pairs of n-point open string partial amplitudes, with coefficients that depend on the kinematic variables as well as the string tension $1/(2\pi\alpha')$. This is natural, albeit non-trivial, since the closed string vertex operators are products of open string vertex operators. The non-triviality of the KLT relations is that the factorization into open string amplitudes survives the integrals over the insertion points of the vertex operators. In the limit of infinite tension, $\alpha' \to 0$, the closed string amplitudes with massless spin-2 string external states become the regular graviton scattering amplitudes M_n we have discussed above. And in this limit, the open-string partial amplitudes with external massless spin-1 states become the color-ordered gluon amplitudes A_n. Thus, in the limit $\alpha' \to 0$, KLT offers a relationship between tree-level M_n and A_n for each n. For $n = 4$, 5, and 6 the field theory KLT relations are

$$M_4^{\text{tree}}(1234) = -s_{12}\, A_4^{\text{tree}}[1234]\, A_4^{\text{tree}}[1243],$$

$$M_5^{\text{tree}}(12345) = s_{23}s_{45}\, A_5^{\text{tree}}[12345]\, A_5^{\text{tree}}[13254] + (3 \leftrightarrow 4),$$

$$M_6^{\text{tree}}(123456) = -s_{12}s_{45}A_6^{\text{tree}}[123456]\left(s_{35}A_6^{\text{tree}}[153462] + (s_{34} + s_{35})A_6^{\text{tree}}[154362]\right)$$
$$+ \mathcal{P}(2, 3, 4). \tag{12.13}$$

In the 6-point case, $\mathcal{P}(2, 3, 4)$ stands for the sum of all permutations of legs $2, 3, 4$. At 7-point and higher, the KLT relations are more complicated; they can be found in Appendix A of [184]. The relation between gravity and gluon scattering is not at all visible in the Lagrangian (12.2), although field redefinitions and clever gauge choices can bring the first few terms in the gravitational action into a more KLT-like form; see [185–187] and the review [188].

Note that there is no specification of helicities of the external states in (12.13): this is because the KLT relations are valid in D-dimensions. In 4d, the KLT relations work for any helicity assignments of the gravitons on the LHS; if the ith graviton has helicity $h_i = +2$, then the gluons labeled i in the amplitudes on the RHS have helicity $h_i = +1$; similarly for negative helicity. This ensures that the little group scaling works out on both sides of the KLT relations. We may then say that KLT in 4d uses

$$\text{graviton}^{\pm 2}(p_i) = \text{gluon}^{\pm 1}(p_i) \otimes \text{gluon}^{\pm 1}(p_i). \tag{12.14}$$

This is also encoded in the graviton polarizations (12.5).

Inspecting the relationship (12.14) between gravitons and gluons, we could also ask what happens if we combine gluons of *opposite* helicity in the KLT relations. The result is something that has the little group scaling of a scalar on the gravity side. In fact, what you get is the **dilaton and axion**:

$$\left.\begin{array}{c}\text{dilaton}\\\text{axion}\end{array}\right\} = \text{gluon}^{\pm 1}(p_i) \otimes \text{gluon}^{\mp 1}(p_i). \tag{12.15}$$

This is completely natural from the string theory point of view where the graviton state comes together with an antisymmetric tensor $B_{\mu\nu}$ and a scalar "trace" mode. The latter is the dilaton and the former has a 3-form field strength $H = dB$ which means that in 4d it is dual to a scalar with shift-symmetry, i.e. an axion. Therefore we can write the relation between the spectra

$$\text{4d axion-dilaton gravity} = (\text{YM theory}) \otimes (\text{YM theory}). \tag{12.16}$$

Given the attention we have poured into the study of amplitudes in (planar) $\mathcal{N} = 4$ SYM in 4d, you may also be curious about what we would get if we tensored the 2^4 states of $\mathcal{N} = 4$ SYM *à la* (12.16). The answer is a very good one: we get the 2^8 states of $\mathcal{N} = 8$ supergravity, which is the 4d supergravity theory with maximal supersymmetry. Supergravity amplitudes, especially those in $\mathcal{N} = 8$ supergravity, are the main focus in the following. We return to the study of "gravity $=(\text{gauge theory})^2$" in Chapter 13, though you will see more of it in the following sections too.

12.2 Supergravity

Supergravity is the beautiful union of gravity and supersymmetry. It is the result of making the supersymmetry transformations local in the sense that the SUSY parameter ϵ is space-time dependent. If you have not previously studied supergravity, you should immediately read the essay [189] and then textbooks such as [37, 39, 190].

The supersymmetry partner of the graviton is called the **gravitino**. It has spin-$\frac{3}{2}$ and (when supersymmetry is unbroken) it is massless. In 4d we characterize a gravitino by its two helicity states $h = \pm\frac{3}{2}$; its Feynman rule for the external line simply combines a $\pm$-helicity spin-1 polarization vector with the $\pm$-helicity spin-$\frac{1}{2}$ fermion wavefunction.

In a 4d supergravity theory with $\mathcal{N}$ supercharges, Q^A and $\widetilde{Q}_A$, the graviton has $\mathcal{N}$ gravitino-partners. We can construct the spectrum by starting with the negative helicity graviton h^- as the highest-weight state and apply the supercharges $\widetilde{Q}_A$. At each step, $\widetilde{Q}_A$ raises the helicity by $\frac{1}{2}$, so when $\widetilde{Q}_A$ is applied once to h^- it produces a negative helicity gravitino ψ_A^-. If $\mathcal{N} = 1$, then the process terminates because of the Grassmann nature of the supercharge. So the $\mathcal{N} = 1$ *pure supergravity* multiplet consists of the two CPT conjugate pairs of graviton and gravitino:

$$\mathcal{N} = 1 \text{ supergravity:} \quad (h^-, \psi^-) \quad \text{and} \quad (\psi^+, h^+). \tag{12.17}$$

When we say *pure supergravity* we mean that there are no other matter-supermultiplets included; we only have the states that are related to the graviton via supersymmetry.

Pure $\mathcal{N} = 2$ *supergravity* has 2×2^2 states

$$\mathcal{N} = 2 \text{ supergravity:} \quad (h^-, \psi_A^-, v^-) \quad \text{and} \quad (v^+, \psi^{A+}, h^+), \tag{12.18}$$

where the two gravitinos ψ_A^- and ψ^{A+} are labeled by $A = 1, 2$ and $v^\pm$ denotes the two helicity states of the spin-1 *gravi-photon*.

Fast-forward to pure $\mathcal{N} = 4$ *supergravity*. Its 2×2^4 states can be characterized as

$$\mathcal{N} = 4 \text{ supergravity} = \left(\mathcal{N} = 4 \text{ SYM}\right) \otimes \left(\mathcal{N} = 0 \text{ SYM}\right). \tag{12.19}$$

By $\mathcal{N} = 0$ SYM we just mean pure Yang–Mills theory. The spectrum (12.19) should be read as follows: the two graviton states are given in terms of the gluon states as in (12.14). Using the spectrum (4.30) of $\mathcal{N} = 4$ SYM, we find:

$$
\begin{aligned}
\text{gravitons:} \quad & h^\pm = g^\pm \otimes g^\pm, \\
\text{gravitinos:} \quad & \psi^{A+} = \lambda^{A+} \otimes g^+ \text{ and } \psi_A^- = \bar{\lambda}_A^- \otimes g^-, \\
\text{gravi-photons:} \quad & v_{AB}^\pm = S_{AB} \otimes g^\pm, \\
\text{gravi-photinos:} \quad & \psi^{A-} = \lambda^{A+} \otimes g^- \text{ and } \psi_A^+ = \bar{\lambda}_A^- \otimes g^+, \\
\text{scalars (dilaton–axion):} \quad & g^\pm \otimes g^\mp,
\end{aligned}
\tag{12.20}
$$

where $g^\pm$ are gluons, $\bar{\lambda}_A^-$ and λ^{A+} are gluinos, and S_{AB} are the six scalars of $\mathcal{N} = 4$ SYM. Totaling the states, we get $2 \times 1 + 2 \times 4 + 2 \times 6 + 2 \times 4 + 2 = 32 = 2 \times 2^4$.

▶ **Exercise 12.2**

Identify the supermultiplets in the theory whose spectrum is $\left(\mathcal{N} = 2 \text{ SYM}\right) \otimes \left(\mathcal{N} = 0 \text{ YM}\right)$.

What is the difference between the two $\mathcal{N} = 4$ supergravity theories whose spectra are $\left(\mathcal{N} = 4 \text{ SYM}\right) \otimes \left(\mathcal{N} = 0 \text{ SYM}\right)$ and $\left(\mathcal{N} = 2 \text{ SYM}\right) \otimes \left(\mathcal{N} = 2 \text{ SYM}\right)$?

Applying the $\mathcal{N}$ supersymmetry generators $\widetilde{Q}_A$ to the graviton top state h^-, we cannot avoid states with spin greater than 2 if $\mathcal{N} > 8$. There are no consistent interactions in flat space for particles with spin greater than 2, so that tells us that maximal supersymmetry in 4d is $\mathcal{N} = 8$. The $\mathcal{N} = 8$ **supergravity** theory is unique: the ungauged theory, which is our focus here, was first written down in [191, 192].[3] Its spectrum of 2^8 states forms a CPT-self-conjugate supermultiplet (just as in $\mathcal{N} = 4$ SYM). As noted at the end of Section 12.1, the spectrum can be characterized as

$$\mathcal{N} = 8 \text{ supergravity} = (\mathcal{N} = 4 \text{ SYM}) \otimes (\mathcal{N} = 4 \text{ SYM}). \tag{12.21}$$

In any supergravity theory, there are **supersymmetry Ward identities** that restrict the amplitudes, just as discussed for gauge theories in Sections 4.2 and 4.4. In particular, the *graviton amplitudes in supergravity* satisfy

$$M_n(1^+2^+\ldots n^+) = M_n(1^-2^+\ldots n^+) = M_n(1^+2^-\ldots n^-) = M_n(1^-2^-\ldots n^-) = 0 \tag{12.22}$$

at all orders in perturbation theory. There are also simple Ward identities among graviton and gravitino MHV amplitudes that give

$$\mathcal{N} \geq 1 \text{ supergravity:} \qquad M_n(1^-\psi^-\psi^+4^+\ldots n^+) = \frac{\langle 13 \rangle}{\langle 12 \rangle} M_n(1^-2^-3^+4^+\ldots n^+), \tag{12.23}$$

just as for gluons and gluinos. In extended ($\mathcal{N} > 1$) supergravity there are further relations, as you will see shortly from the superamplitudes in $\mathcal{N} = 8$ supergravity. The example

$$\mathcal{N} = 8 \text{ supergravity:} \qquad M_n(1^+\ldots i^-\ldots j^-\ldots n^+) = \frac{\langle ij \rangle^8}{\langle 12 \rangle^8} M_n(1^-2^-3^+\ldots n^+) \tag{12.24}$$

is the $\mathcal{N} = 8$ supergravity analogue of the supersymmetry Ward identity (4.35) in $\mathcal{N} = 4$ SYM.

12.3 Superamplitudes in $\mathcal{N} = 8$ supergravity

The spectrum (12.21) of $\mathcal{N} = 8$ supergravity consists of 128 bosons and 128 fermions. Organized by helicity $h = 2, \frac{3}{2}, 1, \frac{1}{2}, 0, -\frac{1}{2}, -1, -\frac{3}{2}, -2$, we can write it out as

$$\begin{gathered} 1 \text{ graviton } h^+, \quad 8 \text{ gravitinos } \psi^A, \quad 28 \text{ gravi-photons } v^{AB}, \\ 56 \text{ gravi-photinos } \chi^{ABC}, \quad 70 \text{ scalars } S^{ABCD}, \quad 56 \text{ gravi-photinos } \chi^{ABCCDE}, \\ 28 \text{ gravi-photons } v^{ABCDEF}, \quad 8 \text{ gravitinos } \psi^{ABCDEFG}, \quad 1 \text{ graviton } h^- = h^{12345678}. \end{gathered} \tag{12.25}$$

---

[3] The gauged $\mathcal{N} = 8$ supergravity theory was presented in [193].

Here $A, B, \ldots = 1, 2, \ldots, 8$ are $SU(8)$ R-symmetry indices and each state above is fully antisymmetric in these labels; this simply reflects that the helicity-h state transforms in the rank $r = 4 - 2h$ fully antisymmetric irreducible representation (irrep) of $SU(8)$ and the multiplicity given in (12.25) is the dimension of the irrep. The 70 scalars are self-dual and satisfy $\overline{S}_{ABCD} = \frac{1}{4!}\epsilon_{ABCDEFGH}S^{EFGH}$. Supersymmetry generators Q^A and $\widetilde{Q}_A$ act on the states in an obvious generalization of (4.31).

Just as in $\mathcal{N} = 4$ SYM, it is highly convenient to combine the states into a superfield, or super-wavefunction, with the help of an on-shell superspace with Grassmann variables η_{iA} whose index $i = 1, \ldots, n$ is a particle label and $A = 1, 2, \ldots, 8$ is a fundamental $SU(8)$ R-symmetry index. The $\mathcal{N} = 8$ superfield is then

$$\Phi_i = h^+ + \eta_{iA}\,\psi^A - \frac{1}{2}\eta_{iA}\eta_{iB}\,v^{AB} + \cdots + \eta_{i1}\eta_{i2}\eta_{i3}\eta_{i4}\eta_{i5}\eta_{i6}\eta_{i7}\eta_{i8}\,h^-. \qquad (12.26)$$

The $SU(8)$ R-symmetry requires that the superamplitudes are degree $8k$ polynomials in the Grassmann variables. This directly gives us the $\mathcal{N} = 8$ supergravity version of the $\mathrm{N}^K\mathrm{MHV}$ classification: the Kth sector contains the superamplitudes of degree $8(K + 2)$ polynomials in the η_{iA}s. It should be clear from (12.26) that the MHV sector ($K = 0$) includes the graviton component amplitude $M_n(1^- 2^- 3^+ \ldots n^+)$.

The super-Poincaré generators – momentum $P^{\dot{a}b}$, rotations/boosts, and the supercharges Q^A and $\widetilde{Q}_A$ are given in (5.1) and (4.46), with the only difference that now $A = 1, 2, \ldots, 8$. Momentum- and supermomentum-conservation require that the general superamplitudes in $\mathcal{N} = 8$ supergravity are of the form

$$\mathcal{M}_n^{\mathrm{N}^K\mathrm{MHV}} = \delta^4(P)\,\delta^{(16)}(\widetilde{Q})\,P_{8K}, \qquad (12.27)$$

where P_{8K} is annihilated by Q^A which acts by differentiation: $Q^A P_{8K} = 0$.

At MHV level, we are already home safe. The Grassmann delta function (12.27) eats up all 16 fermionic variables, so P_0 is η-independent. It can be fixed by requiring that $\mathcal{M}_n$ projects out the correct pure graviton MHV amplitude $M_n(1^- 2^- 3^+ \ldots n^+)$. This is easily accomplished:

$$\mathcal{M}_n^{\mathrm{MHV}} = \delta^4(P)\,\delta^{(16)}(\widetilde{Q})\,\frac{M_n(1^- 2^- 3^+ \ldots n^+)}{\langle 12 \rangle^8}. \qquad (12.28)$$

In particular, the supersymmetry Ward identity (12.24) for graviton amplitudes in $\mathcal{N} = 8$ supergravity follows from (12.28).

Beyond the MHV level, one can solve the supersymmetric Ward identities $Q^A \mathcal{M}_n = \widetilde{Q}_A \mathcal{M}_n = 0$ (just as in the $\mathcal{N} = 4$ SYM case) to find a basis of input-amplitudes that completely determine the full superamplitude. The basis can be labeled by the $K \times 8$ rectangular Young tableaux of $SU(n - 4)$ irreps [42]. Another approach is to use the super-BCFW recursion relations; they are valid for super-shift of any two lines [19, 47].[4]

---

[4] A super-shift version of CSW was discussed in [44] and while it works for all tree superamplitudes in $\mathcal{N} = 4$ SYM, it has limited validity in $\mathcal{N} = 8$ supergravity.

The set-up for the MHV superamplitude (12.28) is perhaps a bit "cheap" because the component amplitude $M_n(1^-2^-3^+\ldots n^+)$, as we have seen in Section 12.1, does not take a particularly compact form and it does not clearly reflect symmetries such as full permutation symmetry of identical external states. So there has been quite a lot of effort towards building an MHV superamplitude that more clearly encodes the symmetries. One representation [194] of the superamplitude builds on a super-BCFW shift in the $\mathcal{N}=7$ formulation of $\mathcal{N}=8$ supergravity.[5] Other MHV formulas use the Grassmannian representations [195, 196] or the twistor string [197, 198]. More recently, new compact formulas for both Yang–Mills and gravity amplitudes have been proposed to be valid in any spacetime dimensions [199, 200].

$\mathcal{N}=8$ supergravity has, as we have noted above, a **global $SU(8)$ R-symmetry**. This symmetry is realized linearly, as you can see on the spectrum and on the amplitudes which vanish unless the external states form an $SU(8)$ singlet. However, the theory also has a "hidden" symmetry: the equations of motion of $\mathcal{N}=8$ supergravity have a **continuous global $E_{7(7)}(\mathbb{R})$ symmetry**. The group $E_{7(7)}$ is a non-compact version of the exceptional group E_7; its maximal compact subgroup is $SU(8)$. $E_{7(7)}$ has rank 7 and is 133 dimensional. It is *not* a symmetry of the action of $\mathcal{N}=8$ supergravity. The best way to think of this is that the $E_{7(7)}$ is spontaneously broken to $SU(8)$. There are $133-63=70$ broken generators, giving 70 Goldstone bosons. Those are exactly the 70 scalars S^{ABCD} in the spectrum (12.25) and they "live" in the coset space $E_{7(7)}/SU(8)$.

As a spontaneously broken symmetry, $E_{7(7)}$ is not linearly realized on the on-shell scattering amplitudes, but instead it manifests itself via **low-energy theorems**.[6] If the momentum of an external scalar S^{ABCD} is taken soft, then the amplitude must vanish because the Goldstone scalars are derivatively-coupled. Basically this says that the moduli space $E_{7(7)}/SU(8)$ is homogeneous: it does not matter what the vevs of the scalars are, all points on moduli space are equivalent. The soft scalar limit probes the neighborhood of a point in moduli space and since the moduli space is homogeneous, the soft scalar limit vanishes. There are also double-soft limits that involve the commutator of two coset generators and these therefore directly reveal, from the on-shell point of view, the coset structure $E_{7(7)}/SU(8)$.

▶ **Exercise 12.3**

Project the amplitude $M_4(S^{1234}S^{5678}h^-h^+)$ out from the MHV superamplitude (12.28). Show that

$$\lim_{p_1\to 0} M_4\big(S^{1234}S^{5678}h^-h^+\big)=0\,. \tag{12.29}$$

In contrast, the scalars in $\mathcal{N}=4$ SYM are not Goldstone bosons, so the soft-scalar limits do not have to vanish. For example, show

$$\lim_{p_1\to 0} A_4\big[S^{12}g^-S^{34}g^+\big]\neq 0\,. \tag{12.30}$$

---

[5] Just as $\mathcal{N}=3$ SYM is identical to $\mathcal{N}=4$ SYM, so is $\mathcal{N}=7$ supergravity identical to $\mathcal{N}=8$ supergravity. The validity of the super-BCFW shifts in $\mathcal{N}=7$ supergravity is proven in [87].

[6] Low-energy theorems were originally developed in pion-physics [201]. For a review, see [202].

The soft limit explores the points of moduli space in the neighborhood of the origin: away from the origin, the $\mathcal{N} = 4$ SYM theory is on the **Coulomb branch**, part of the gauge group is broken, and some of the $\mathcal{N} = 4$ supermultiplets become massive. The non-vanishing limit (12.30) has a nice interpretation. Set $p_1 = \epsilon\, q$ for some light-like $q = -|q\rangle[q|$ and take the soft limit as $\epsilon \to 0$. The limit (12.30) then depends on $|q\rangle$. The soft limit $p_1 \to 0$ leaves an object with momentum conservation on three particles: the result can be interpreted as the small-mass limit of the Coulomb branch amplitude $A_3[W^- S^{34} W^+]$, where $W^\pm$ are the longitudinal modes of the massive spin-1 W-bosons of a massive $\mathcal{N} = 4$ supermultiplet and S^{34} is a massless scalar. From this point of view, q is a reference vector that allows us to project the massive momenta of $W^\pm$ such that the corresponding angle spinors are well-defined. This is actually also needed to define the helicity basis because helicity is not a Lorentz-invariant concept for massive particles; but q breaks Lorentz-invariance and allows us to define a suitable q-helicity basis [26, 140].

The 3-point amplitude $A_3[W^- S^{34} W^+]$ violates the $SU(4)$ R-symmetry of $\mathcal{N} = 4$ SYM at the origin of moduli space. This is fine, because the Coulomb branch breaks the R-symmetry. Minimally, one has $SU(4) \sim SO(6) \to SO(5) \sim Sp(4)$.

The moral of the story is that single-scalar soft limits for $\mathcal{N} = 8$ supergravity amplitudes *vanish* because the 70 scalars are Goldstone bosons of $E_{7(7)} \to SU(8)$. And single-scalar soft limits for $\mathcal{N} = 4$ SYM are *non-vanishing* and reproduce the small-mass limit of the Coulomb branch amplitudes [140]. One can in fact re-sum the entire small-mass expansion from multiple-soft-scalar limits and recover the general-mass Coulomb branch amplitudes [140, 203].

The single-soft-scalar limits of tree amplitudes in $\mathcal{N} = 8$ supergravity were first studied in [36]. Single and double soft limits were discussed extensively and clarified in [47]. The soft-scalar limits play a key role for us in Section 12.5.

12.4 Loop amplitudes in supergravity

It is taught in all good kindergartens that a point-particle theory of gravity is badly UV divergent and non-renormalizable. This means that it is not a good quantum theory. So what is perturbative gravity all about?

Naive power-counting gives a clear indication that gravity with its 2-derivative inter-actions generically has worse UV behavior than for example Yang–Mills theory with its 1- and 0-derivative interactions. Consider for example a generic 1-loop m-gon diagram. In gravity, the numerator of the loop-integrand can have up to $2m$ powers of momenta, while in Yang–Mills theory it is at most m. Both have m propagators, so in gravity this gives

$$\text{gravity 1-loop } m\text{-gon diagram} \sim \int^{\Lambda} d^4\ell \, \frac{(\ell^2)^m}{(\ell^2)^m} \sim \Lambda^4 \,. \tag{12.31}$$

This is power-divergent as the UV cutoff Λ is taken to ∞ for all m. On the other hand, for Yang–Mills theory the m-gon integral has at most ℓ^m in the numerator, so it is manifestly UV finite for $m > 4$.

Now, the power-counting is too naive. There can be cancellations within each diagram. Moreover, we have learned that we should not take individual Feynman diagrams seriously if they are not gauge invariant. So cancellations of UV divergences can take place in the sum of diagrams, rendering the on-shell amplitude better behaved than naive power-counting indicates.

In fact, pure gravity in 4d is finite at 1-loop order [204]: all the 1-loop UV divergences cancel! This is difficult to see by direct Feynman diagram calculations, but it follows rather trivially from the absence of any valid *counterterms*. We review this approach in detail in Section 12.5.

At 2-loop order, it has been demonstrated by Feynman diagram calculations that pure gravity indeed has a UV divergence [205, 206]. In Yang–Mills theory we are not too scared of UV divergences because we know how to treat them with the procedure of renormalization. However, in gravity, it would take an infinite set of local counterterms to absorb the divergences and hence the result is unpredictable: pure gravity is a non-renormalizable theory.

So what is the theory described by the Einstein–Hilbert action? Because it is non-renormalizable, it is not by itself a well-defined *quantum theory* of gravity. Instead, we should regard the field theory defined by the Einstein–Hilbert action as an ***effective field theory***, valid at scales much smaller than the Planck scale $M_{\text{Planck}} \sim 10^{19}$ GeV. To see this, recall that the 4d gravitational coupling κ has mass-dimension -1. So when we do perturbation theory, we should really use the dimensionless coupling $E\kappa$ where E is the characteristic energy of the process. At high enough energies, this dimensionless coupling is no longer small and we cannot trust perturbation theory. So we should not extrapolate to such high energies. In energy units, we have $\kappa^{-1} \sim G_N^{-1/2} = M_{\text{Planck}}$, so this tells us to use gravity, as described by the Einstein–Hilbert action, for energies $E \ll M_{\text{Planck}}$. As a classical effective field theory, General Relativity is enormously successful and captures classical gravitational phenomena stunningly as shown by experimental tests.

Regarding gravity as an effective theory, we can study the low-energy perturbative amplitudes: the tree amplitudes capture the classical physics and there are no divergences to worry about. At 1-loop level, we have mentioned that pure gravity is finite. Could we imagine adding matter to gravity in such a way that its higher-loop amplitudes were also finite? Gravity with generic matter is 1-loop divergent [204, 207], but we know from gauge theories that supersymmetry improves the UV behavior of loop amplitudes, even to such an extreme extent that the maximally supersymmetric Yang–Mills theory, $\mathcal{N} = 4$ SYM, is UV finite: the UV divergences cancel completely at each order in the loop expansion. Could something like that also happen in supergravity? If it did, it would eliminate the need for renormalization and the problems of non-renormalizability would be obsolete. There would still be important questions unresolved about non-perturbative aspects of supergravity; perturbative finiteness would not mean that the theory is UV complete. The question of perturbative UV finiteness of (maximal) supergravity in 4d has received increased attention in the past few years and the on-shell amplitude techniques have facilitated multiple explicit calculations of supergravity loop amplitudes. It should be emphasized that whether or not the perturbative calculations eventually encounter a divergence, one should appreciate that the study of loop amplitudes in supergravity has resulted in a number of new insights, of independent value, about gravity scattering amplitudes. An example is the connection between gravity and Yang–Mills amplitudes via the so-called BCJ dualities (see Chapter 13).

Pure supergravity in 4d is better behaved in the UV than pure gravity: ***all pure supergravity theories in 4d are finite at 1-loop*** [208] ***and 2-loop order*** [209–211], i.e. the first possible UV divergence can appear only at 3-loop order, improving on the 2-loop UV divergence of pure gravity [205, 206]. In the spirit of "the more supersymmetry, the better," it is natural to focus on maximal supersymmetry, i.e. $\mathcal{N} = 8$ supergravity in 4d. An explicit calculation, using the generalized unitarity method, demonstrated that the 3-loop 4-graviton amplitude is UV finite in $\mathcal{N} = 8$ supergravity in 4d [212–214]. This and related observations of unexpected cancellations motivated Bern, Dixon, and Roiban [212] to ask if $\mathcal{N} = 8$ supergravity in 4d is UV finite? They further proposed that the ***critical dimension*** D_c for the first UV divergence of maximal supergravity in D-dimensions follows the same pattern as for maximal super Yang–Mills theory [184, 215], namely

$$D_c(L) = \frac{6}{L} + 4 \quad \text{for} \quad L > 1.$$
(12.32)

It has been shown [216, 217] that the 4-loop 4-graviton amplitude is UV finite in $\mathcal{N} = 8$ supergravity in 4d and that it follows the pattern (12.32). How about 5-loops? A pure-spinor based argument [218] leads to (12.32) for $L = 2, 3, 4$, but implies that the critical dimension for $L = 5$ is $D = 24/5$ and not $26/5$ as (12.32) predicts. This question can be settled by direct computation: at the time of writing, the 5-loop calculation is still in progress, so you will have to watch the news feeds for a resolution.

For $D = 4$, the symmetries of $\mathcal{N} = 8$ supergravity can be used to establish that *all* amplitudes of the theory are UV finite for $L \leq 6$: this explains the finiteness of the 3- and 4-loop 4-graviton amplitudes and predicts that no UV divergence appears in any higher point amplitudes for $L \leq 6$ in 4d. For $L \geq 7$, the known symmetries do not suffice to rule out UV divergences in 4d. The following section reviews how these results are obtained using an on-shell amplitude-based approach [43, 219–221] to counterterms in $\mathcal{N} = 8$ supergravity. We then provide in Section 12.6 an overview of the current status of the UV behavior of supergravity as a function of dimensions D, supersymmetries $\mathcal{N}$, and loop order L.

12.5 $\mathcal{N} = 8$ supergravity: loops and counterterms

Suppose that a supergravity has its *first* UV divergence in an n-point amplitude at L-loop order. Then the effective action for the theory must have a ***local diffeomorphism invariant counterterm*** constructed from n fields (corresponding to the n external states) and $(2L + 2)$ derivatives. The latter statement follows from dimensional analysis because the gravitational coupling κ has mass-dimension -1: for given n, the ratio of the L-loop supergravity amplitude to the tree amplitude has an overall factor of κ^{2L}, so the corresponding local counterterm has to make up the mass-dimension by having $2L$ more derivatives than the 2-derivative tree-level theory.[7]

▶ **Exercise 12.4**

Show that the n-graviton 1-loop amplitude has an overall factor of κ^2 compared with the n-graviton tree amplitude.

If we consider just pure gravity, the possible local diff-invariant counterterms must be Lorentz scalars formed from contractions of Riemann tensors and possibly covariant derivatives. Each Riemann tensor contributes 2-derivatives, so at 1-loop (4-derivatives), the possible candidates are $\sqrt{-g}R^2$, $\sqrt{-g}R_{\mu\nu}R^{\mu\nu}$, and $\sqrt{-g}R_{\mu\nu\rho\sigma}R^{\mu\nu\rho\sigma}$. If we suppress the index contractions, we can write schematically

$$
S_{\text{eff}} = \frac{1}{2\kappa^2} \int d^4x \, \sqrt{-g} \left(\underbrace{R}_{L=0} + \underbrace{\kappa^2 R^2}_{L=1} + \underbrace{\kappa^4 R^3}_{L=2} + \underbrace{\kappa^6 R^4}_{L=3} + \underbrace{\kappa^6 \left(D^2 R^4 + R^5 \right)}_{L=4} \right.
$$
$$
\left. + \underbrace{\kappa^8 \left(D^4 R^4 + D^2 R^5 + R^6 \right)}_{L=5} + \cdots \right), \tag{12.33}
$$

where R denote Riemann tensors and D are covariant derivatives. This should be viewed as a list of possible candidate counterterms; the operators in (12.33) are not necessarily generated in perturbation theory.

▶ **Exercise 12.5**

Trivia question: why are we not including operators $D^{2k}R$ in (12.33)?
   How about operators $D^{2k}R^2$?

[Hint: use the Bianchi identity for the Riemann tensor.]
   Show that operators of the form $D^{2k}R^3$ have vanishing 3-point matrix elements for $k \geq 1$.

Since we consider on-shell amplitudes, we can enforce the equations of motion on the candidate counterterms. In pure gravity, the Einstein equation gives $R_{\mu\nu} = 0$, so this leaves $\sqrt{-g}R_{\mu\nu\rho\sigma}R^{\mu\nu\rho\sigma}$ as the only possibility at 1-loop. However, we are free to add zero to convert $\sqrt{-g}R_{\mu\nu\rho\sigma}R^{\mu\nu\rho\sigma}$ to the Gauss–Bonnet term $\sqrt{-g}\left(R_{\mu\nu\rho\sigma}R^{\mu\nu\rho\sigma} - 4R_{\mu\nu}R^{\mu\nu} + R^2\right)$ which equals a total derivative. Therefore there is ***no local counterterm for pure gravity at 1-loop order!*** And since there is no counterterm, ***pure gravity is not UV divergent at 1-loop***. A cleaner way to say this is that one can do a field redefinition that changes

---

[7] To be a little more precise, the above statement is true for amplitudes with purely bosonic fields. Since external fermions dress the amplitude with dimensionful wavefunctions, each pair of fermions counts one derivative for the purpose of dimensional analysis.

$\sqrt{-g} R_{\mu\nu\rho\sigma} R^{\mu\nu\rho\sigma}$ to the Gauss–Bonnet term, and since a field redefinition does not change the amplitude there cannot be a 1-loop divergence.

At 2-loop order, the candidate counterterm has to be composed of some index contractions of three Riemann tensors – let us denote it by R^3, and here and henceforth leave the $\sqrt{-g}$-factor implicit in candidate counterterm operators. The R^3 counterterm is present for pure gravity, which (as noted in Section 12.4) is 2-loop divergent.

Counterterms have to respect any non-anomalous symmetries of the theory. Supersymmetry is preserved at loop-level, so in supergravity theories counterterm candidates must be supersymmetrizable. We showed in Exercise 2.35 that a matrix element produced by R^3 is fixed by little group scaling to be

$$M_3(1^-2^-3^-)_{R^3} = \text{constant} \times \langle 12 \rangle^2 \langle 23 \rangle^2 \langle 13 \rangle^2 . \tag{12.34}$$

But we also know from (12.22) that all-minus amplitudes violate the supersymmetry Ward identities. So this means that any operator that produces a non-vanishing matrix element $M_3(1^-2^-3^-)_{R^3}$ violates supersymmetry. Since R^3 produces a supersymmetry-violating amplitude, we conclude that R^3 cannot be supersymmetrized [209, 222]. Therefore R^3 is not a viable counterterm and hence ***any pure supergravity must be 2-loop finite!***

In the above argument, you may object that we may not have to care about the 3-point amplitude (12.34) since it vanishes in real kinematics. But it is easy to show (using for example an all-line shift [26]) that a non-vanishing all-minus 3-point amplitude implies that there is a non-vanishing 4-graviton amplitude $M_4(1^-2^-3^-4^-)$ and clearly this violates supersymmetry. Another objection could be: "what if the constant in (12.34) is zero, then there is no contradiction with supersymmetry?" That is true, but if the matrix element vanishes that means that the 3-field part of R^3 is a total derivative, and then we don't care about it anyway because there are no available gravity diff-invariant 4-field operators at 6-derivative order. So, either way, we conclude that pure supergravity is finite at 2-loops.

Let us now specialize to $\mathcal{N} = 8$ supergravity in 4d. The candidate counterterms have to respect $\mathcal{N} = 8$ supersymmetry and also be $SU(8)$ invariant, since the global R-symmetry is non-anomalous [89, 90]. Moreover, they should be compatible with the "hidden" $E_{7(7)}$ symmetry [223]; we will come back to $E_{7(7)}$ later – for now, we explore the constraints of supersymmetry and R-symmetry on the candidate counterterms in $\mathcal{N} = 8$ supergravity.

It is in general difficult to analyze the candidate counterterm operators directly: a full field theory $\mathcal{N} = 8$ supersymmetrization of the independent contractions of Rs and Ds is complicated in component form; for R^4 it has been done explicitly at the linearized level only [224]. A better approach is to use a superfield formalism. There is no off-shell superfield formalism for $\mathcal{N} = 8$ supergravity, but harmonic superspace techniques have been used to constrain the possible counterterms in supergravity theories in various dimensions. We are going to highlight some of the results of the superspace approach in Section 12.6, but otherwise we do not discuss these methods here: this is a book about amplitudes and that is the path we take.

The supersymmetry and R-symmetry constraints on the candidate counterterm operators translate into Ward-identity constraints on the matrix elements produced by counterterms. Let us list the translation of constraints between an operator, whose lowest interaction-term is an n-vertex, and the corresponding n-point matrix element:

n-field operator		n-point matrix element
local with $2L + 2$ derivatives	$\leftrightarrow$	polynomial in $\langle ij \rangle$ and $[kl]$ of degree $2L + 2$
$\mathcal{N} = 8$ SUSY	$\leftrightarrow$	$\mathcal{N} = 8$ SUSY Ward identities
$SU(8)$ R-symmetry	$\leftrightarrow$	$SU(8)$ Ward identities

$$(12.35)$$

($E_{7(7)}$ constraints will be treated separately, starting on page 267.) In addition, the matrix elements have to respect Bose/Fermi symmetry under exchange of identical external states.

The condition that the matrix element is polynomial follows from the locality of the operator and the insistence that it corresponds to the *leading* (i.e. first) UV divergence in the theory; with other operators present, there could be pole terms. The matrix element we consider here is strictly the amplitude calculated from the n-point vertex of the given n-field operator, and therefore it cannot have any poles, i.e. it must be a polynomial in the kinematic variables $\langle ij \rangle$ and $[kl]$. The degree of the polynomial follows from dimensional analysis.

Let us be clear about what our approach is: for a given L, we ask if the *first* UV divergence could appear in an n-point amplitude. If this is so, then there must be a corresponding n-field $(2L + 2)$-derivative counterterm. For example, for $L = 3$ the lowest-n candidate counterterm would be an $SU(8)$-invariant $\mathcal{N} = 8$ supersymmetrization of R^4. To analyze if such an operator exists, we write down all possible matrix elements satisfying the constraints (12.35). If there are no such matrix elements, we conclude there is no corresponding $\mathcal{N} = 8$ SUSY and $SU(8)$-invariant operator and therefore the first divergence in the theory cannot be in the n-point L-loop amplitude. On the other hand, if one or more such matrix elements exist, then the corresponding operator respects *linearized* $\mathcal{N} = 8$ supersymmetry and $SU(8)$ and we may consider it as a candidate counterterm. That does not mean that perturbation theory actually produces the corresponding UV divergence; that would have to be settled by other means, such as an explicit L-loop computation. Thus, the approach here is to use the matrix elements to *exclude* counterterm operators as well as *characterize* candidate counterterms as operators that respect $\mathcal{N} = 8$ SUSY and $SU(8)$-symmetry at the linearized level.

To illustrate the idea, consider R^4.[8] Its 4-point matrix element $M_4(1^- 2^- 3^+ 4^+)_{R^4}$ has to be a degree 8 polynomial in angle and square brackets. Taking into account the little group scaling, there is only one option: the matrix element has to be $M_4(1^- 2^- 3^+ 4^+)_{R^4} = c_{R^4} \langle 12 \rangle^4 [34]^4$, where c_{R^4} is some undetermined constant. By the same arguments as above, dimensional analysis and little group scaling, we know that $M_4(1^- 2^+ 3^- 4^+)_{R^4} = c_{R^4} \langle 13 \rangle^4 [24]^4$. We can then check the MHV-level $\mathcal{N} = 8$ SUSY Ward identity (12.24) which at n-point reads

$$\mathcal{N} = 8 \text{ supergravity:} \quad M_n(1^+ \ldots i^- \ldots j^- \ldots n^+)_{\mathcal{O}} = \frac{\langle ij \rangle^8}{\langle 12 \rangle^8} M_n(1^- 2^- 3^+ \ldots n^+)_{\mathcal{O}}.$$

$$(12.36)$$

---

[8] The relevant contraction of four Riemann tensors is the square of the Bel–Robinson tensor [225].

For our 4-point R^4 matrix elements, we have

$$M_4(1^-2^+3^-4^+)_{R^4} = \frac{\langle 13 \rangle^8}{\langle 12 \rangle^8} M_4(1^-2^-3^+4^+)_{R^4} = \frac{\langle 13 \rangle^8}{\langle 12 \rangle^8} c_{R^4} \langle 12 \rangle^4 [34]^4 = c_{R^4} \langle 13 \rangle^4 [24]^4 ,$$

(12.37)

thanks to momentum conservation $\langle 13 \rangle[34] = -\langle 12 \rangle[24]$. It is not hard to see that the 4-point super-matrix-element

$$\mathcal{M}_4(1234)_{R^4} = \delta^4(P) \, \delta^{(16)}(\tilde{Q}) \, \frac{[34]^4}{\langle 12 \rangle^4}$$

(12.38)

fulfills all criteria (12.35).[9] This means that linear $\mathcal{N} = 8$ supersymmetry and $SU(8)$ do *not* rule out R^4. But it does not mean that it will occur in perturbation theory: in fact, we know from the explicit 3-loop calculation [212] that the 4-graviton amplitude is finite, so R^4 does not occur. Why not? Well, read on, we will get to that later in this section.

Let us now see an example of how the on-shell matrix elements can be used to rule out a counterterm. At 4-loop order, we can write down two pure-gravity 10-derivative operators, $D^2 R^4$ and R^5. The first one, $D^2 R^4$, stands for the possible scalar contractions of two covariant derivatives acting (in some way) on four Riemann tensors. No matter how the indices are contracted, its matrix element turns out to be proportional to $(s + t + u)\mathcal{M}_4(1234)_{R^4}$ and therefore it vanishes. This means that the 4-point interaction in the operator $D^2 R^4$ is a total derivative when evaluated on the equations of motion. So we can rule out the 4-loop 4-graviton amplitude as the first instance of a UV divergence in $\mathcal{N} = 8$ supergravity. The second 10-derivative operator R^5 is a little more interesting.

The 5-point MHV matrix element of R^5 is fixed uniquely by locality, dimensional analysis, and little group scaling up to an overall constant:

$$M_5(1^-2^-3^+4^+5^+)_{R^5} = a_{R^5} \langle 12 \rangle^4 [34]^2 [45]^2 [53]^2 .$$

(12.39)

Next, we check the supersymmetry Ward identity (12.36) in much the same way as for R^4. But now we find

$$M_5(1^-2^+3^-4^+5^+)_{R^5} \overset{?}{=} \frac{\langle 13 \rangle^8}{\langle 12 \rangle^8} M_5(1^-2^-3^+4^+5^+)_{R^5}$$

$$\implies \quad a_{R^5} \langle 13 \rangle^4 [24]^2 [45]^2 [52]^2 \overset{!}{=} a_{R^5} \frac{\langle 13 \rangle^8 [34]^2 [45]^2 [53]^2}{\langle 12 \rangle^4} .$$

(12.40)

This time momentum conservation does not save us. The LHS and RHS of (12.40) are not equal for generic momenta, in particular the LHS is local (i.e. does not have any poles) while the RHS has a pole $1/\langle 12 \rangle^4$. This is a contradiction that can only be resolved when $a_{R^5} = 0$. So that means that the operator R^5 does not have an $\mathcal{N} = 8$ supersymmetrization. And that in turn rules out the 5-point 4-loop amplitude as the first UV divergence in $\mathcal{N} = 8$ supergravity. One can further argue [219, 226] that there are no other possible 10-derivative operators compatible with $\mathcal{N} = 8$ supersymmetry and $SU(8)$, so this means that $\mathcal{N} = 8$ supergravity cannot have its first UV divergence at 4-loop order.

[9] For details of how the R-symmetry acts on the superamplitudes, see Section 2.2 of [42].

Equation (12.40) illustrates a conflict between supersymmetry and locality, a conflict that can be exploited to rule out potential counterterms. We will describe the method for operators of the form $D^{2k}R^5$, then outline the general results. The strategy is to construct the most general matrix element $M_5(1^-2^-3^+4^+5^+)_{D^{2k}R^5}$ that respects the little group scaling, has mass-dimension $2k + 10$, and is Bose symmetric in exchange of same-helicity gravitons. Then ask if there exists a linear combination that respects the SUSY Ward identities. Practically, this was done in [219] using Mathematica. (For more advanced cases, Gröbner basis techniques are very useful [221], and the results found can be reproduced and extended by an analysis based on the superconformal group $SU(2, 2|8)$ [221].)

▷ *Example.* As an example of the procedure in [219], consider $D^2 R^5$. First we find that there are 40 angle-square bracket monomials of degree 12 that have the correct little group scaling of $M_5(1^-2^-3^+4^+5^+)_{D^2R^5}$. This does not take into account redundancy under Schouten or momentum conservation. Now take linear combinations of the 40 monomials to enforce Bose symmetry: this leaves six polynomials as candidates for $M_5(1^-2^-3^+4^+5^+)_{D^2R^5}$. Then impose Schouten and momentum conservation and one finds that only one polynomial survives: this means that the MHV matrix element of $D^2 R^5$ is unique; it takes the form

$$M_5(1^-2^-3^+4^+5^+)_{D^2R^5} = a_{D^2R^5}\, s_{12}\, \langle 12 \rangle^4 [34]^2 [45]^2 [53]^2 \,. \qquad (12.41)$$

This is actually just s_{12} times $M_5(1^-2^-3^+4^+5^+)_{R^5}$ in (12.39). Now repeat the SUSY test (12.40) for M_{5,D^2R^5} to find that the RHS has a pole $1/\langle 12 \rangle^3$, contradicting the locality of the LHS. So the operator $D^2 R^5$ is excluded as a counterterm for $\mathcal{N} = 8$ supergravity, which means that the first divergence cannot be in the 6-loop 5-graviton amplitude.

Let us summarize the result of the process outlined above for a few more operators in the same class – after each step given in the first column, we list how many polynomials remain:

	R^5	$D^2 R^5$	$D^4 R^5$	$D^6 R^5$	$D^8 R^5$
Little grp	1	40	595	4983	29397
Bose symmetry	1	6	63	454	2562
Schouten, mom-cons	1	1	6	9	24
Weakest pole in SUSY Ward id	$\langle 12 \rangle^{-4}$	$\langle 12 \rangle^{-3}$	$\langle 12 \rangle^{-2}$	$\langle 12 \rangle^{-1}$	no pole

$$(12.42)$$

It follows from the last line that the $D^{2k}R^5$ operators are excluded as counterterm candidates for $k = 0, 1, 2, 3$, but not for $k = 4$ where there is one unique matrix element that solves the supersymmetry Ward identities. Thus there is a unique operator $D^8 R^5$ that passes the tests of linearized SUSY; if present, this would correspond to a first UV divergence at 8-loop order in the 5-graviton amplitude. ◁

▶ **Exercise 12.6**

Show that for the operator R^n with $n \geq 3$ there are no n-point MHV matrix elements compatible with the $\mathcal{N} = 8$ supersymmetry Ward identities.

For operators with $n > 5$ fields, one has to distinguish between the different N^KMHV sectors. Beyond MHV level, this can be done using the solutions to the SUSY Ward identities [42] in $\mathcal{N} = 8$ supergravity. As an example of a non-MHV result, the lowest order permissible NMHV-level operator is $D^4 R^6$ at 7-loops. Actually, there are two independent NMHV matrix elements, so this means that there are two independent linearly-supersymmetrizable operators $D^4 R^6$.

A detailed analysis of possible counterterm operators was carried out in [219] (see also [226]) and it was found that

below 7-loop order, the only operators compatible with linearized $\mathcal{N} = 8$ supersymmetry and $SU(8)$ R-symmetry are

$$\underbrace{R^4}_{\text{3-loop}}, \quad \underbrace{D^4 R^4}_{\text{5-loop}}, \quad \underbrace{D^6 R^4}_{\text{6-loop}}. \tag{12.43}$$

At 7-loop order, an infinite tower of linearized $\mathcal{N} = 8$ supersymmetry and $SU(8)$ R-symmetry permissible operators was found

$$\text{7-loops:} \quad D^8 R^4, \quad D^4 R^6, \quad R^8, \quad \phi^2 R^8, \quad \phi^4 R^8, \quad \dots \tag{12.44}$$

The operators $\phi^{2k} R^8$ should be viewed as representatives for the linearized $\mathcal{N} = 8$ supersymmetrization of some contraction of eight Riemann tensors multiplied by an $SU(8)$-singlet combination of $2k$ scalars S^{ABCD}. These do not have purely gravitational $(8+2k)$-point matrix elements, so they cannot be in the MHV or anti-MHV sector. For example, it was shown [221] that $\phi^2 R^8$ only gives N^3MHV matrix elements (four distinct ones).

At 8-loop order and beyond, there are infinite towers of operators that respect linearized $\mathcal{N} = 8$ supersymmetry and $SU(8)$.

You can find a detailed characterization of the counterterms in Table 1 of [221].

There is one symmetry we did not use to restrict the candidate counterterm operators in the above discussion, and that is the $E_{7(7)}$ **"hidden" symmetry**.[10] As discussed at the end of Section 12.3, it manifests itself in the amplitudes of $\mathcal{N} = 8$ supergravity through *low-energy theorems*. These also have to apply to the matrix elements of any acceptable candidate counterterm operator $\mathcal{O}$. In particular, the single-soft-scalar limit must vanish, for example

$$\lim_{p_1 \to 0} M_6 \left(S^{1234} S^{5678} h^- h^- h^+ h^+ \right)_{\mathcal{O}} = 0. \tag{12.45}$$

If the matrix element of an operator does not pass the single-soft-scalar test, then it is not compatible with $E_{7(7)}$ symmetry. If it does pass the test, then we conclude nothing: it could be $E_{7(7)}$ at play or just a coincidence.

---

[10] See also [227, 228] for related aspects of $E_{7(7)}$ symmetry.

It turns out that the single-soft-scalar test is only non-trivial starting at $n = 6$. So precisely (12.45) can be used to test the operators that survived the $\mathcal{N} = 8$ supersymmetry and $SU(8)$ constraints, for example the $L = 3, 5, 6$ operators in (12.43). But it is not easy to use Feynman rules to calculate the 6-point matrix elements of, say, R^4. However, $M_6(S^{1234}S^{5678}h^-h^-h^+h^+)_{R^4}$ can be extracted from the α'-expansion of the closed superstring theory tree amplitude. This may bother you, because there are no continuous global symmetries in string theory, and here we are interested in testing global continuous $E_{7(7)}$. However, the 4d tree-level superstring amplitudes have an accidental global $SU(4) \times SU(4)$ symmetry. This is a consequence of T-duality when 10-dimensional superstring theory is reduced to 4d by compactifying it on a 6-torus. The easiest way to see the $SU(4) \times SU(4)$ symmetry is through the KLT relations: the two open string tree amplitudes on one side of KLT have $SU(4)$ symmetry, so the closed string tree amplitude on the other side of KLT inherits $SU(4) \times SU(4)$; the $SU(4) \times SU(4)$ is enhanced to $SU(8)$ only in the $\alpha' \to 0$ limit [36].

Now, we are going to use the α' contributions from the superstring tree amplitudes, but their $SU(4) \times SU(4)$ is not good enough, we need $SU(8)$. So we *average* the string amplitude over all 35 independent embeddings of $SU(4) \times SU(4)$ into $SU(8)$ to get an $SU(8)$-invariant answer. For example, the leading α'-correction to the closed superstring amplitude is order α'^3. Dimensional analysis implies that this comes from an 8-derivative effective operator with $\mathcal{N} = 8$ supersymmetry and (after averaging) $SU(8)$-invariance. But we know from the previous analysis that there is only one such operator, namely R^4. So after making it $SU(8)$-invariant, the α'^3 contribution from the $SU(8)$-averaged superstring amplitude must be identical (up to an overall constant) to the matrix element of the operator R^4 in $\mathcal{N} = 8$ supergravity [220]! [11] The open superstring tree amplitudes are known in the literature [229], so pulling them through KLT, the $M_6(S^{1234}S^{5678}h^-h^-h^+h^+)$ matrix element can be extracted, $SU(8)$-averaged, and then subjected to the single-soft-scalar test (12.45). [12] And R^4 fails this test: so R^4 ***is not compatible with*** $E_{7(7)}$. Hence $E_{7(7)}$ excludes R^4 as a candidate counterterm and *this explains why the 3-loop 4-graviton amplitude is finite.* The on-shell matrix element technique made it possible to show that the 3-loop divergence of $\mathcal{N} = 8$ supergravity is excluded without doing any loop-amplitude calculations [220]. Note that this is *not* a string theory argument – string theory amplitudes were only used as a tool to calculate the relevant field theory matrix elements – so the argument in [220] is field theoretical.

The analysis outlined above can be repeated [221] for $D^4 R^4$ and $D^6 R^8$ and both are shown to be excluded by $E_{7(7)}$. (See also [231] for a string-based argument for the absence of R^4, $D^4 R^4$ and $D^6 R^8$.) This means that *the symmetries of $\mathcal{N} = 8$ supergravity exclude the divergences in any amplitudes below 7-loop order* [221]. Explicit calculations of the $L = 5, 6$ amplitudes are not yet available, but as explained here they are expected to yield finite results in 4d.

---

[11] In the Einstein frame, the effective operator in superstring theory at order α'^3 is actually $e^{-6\phi}R^4$, where ϕ is the dilaton. The presence of the dilaton (which can be identified as a certain linear combination of the 70 scalar scalars in the $\mathcal{N} = 8$ supergravity spectrum) operator breaks $SU(8)$ to $SU(4) \times SU(4)$. The average over embeddings $SU(4) \times SU(4) \to SU(8)$ eliminates this breaking and restores $SU(8)$.

[12] An earlier test [230] did not involve the $SU(8)$-average.

L

3   R^4 $^{E_{7(7)}}$

4   D^2R^4 R^5

5   D^4R^4 $^{E_{7(7)}}$ D^2R^5 R^6

6   D^6R^4 $^{E_{7(7)}}$ D^4R^5 D^2R^6 R^7

7   D^8R^4 | D^6R^5 | D^4R^6 $_{E_{7(7)}}$ | D^2R^7 | R^8 $_{E_{7(7)}}$ | $\varphi^2D^2R^7$ | φ^2R^8 $_{E_{7(7)}}$ | $\varphi^4D^2R^7$ | φ^4R^8 | $\varphi^6D^2R^7$

8   $D^{10}R^4$ | D^8R^5 | D^6R^6 | D^4R^7 | D^2R^8 | R^9 | $\varphi^2D^2R^8$ | φ^2R^9 | $\varphi^4D^2R^8$ | φ^4R^9 | $\overrightarrow{\text{No}}$

9   $D^{12}R^4$ | $D^{10}R^5$ | D^8R^6 | D^6R^7 | D^4R^8 | D^2R^9 | R^{10} | $\varphi^2D^2R^9$ | φ^2R^{10} | $\varphi^4D^2R^9$ | $\overrightarrow{\text{No}}$ N^4MHV

10   $D^{14}R^4$ | $D^{12}R^5$ | $D^{10}R^6$ | D^8R^7 | D^6R^8 | D^4R^9 | D^2R^{10} | R^{11} | $\varphi^2D^2R^{10}$ | φ^2R^{11} | $\overrightarrow{\text{No}}$ N^3MHV

$\overrightarrow{\text{No MHV}}$ $\overrightarrow{\text{No NMHV}}$ $\overrightarrow{\text{No N}^2\text{MHV}}$ $\overrightarrow{\text{No}}$ N^3MHV

Overview of exclusion of counterterm operators in $\mathcal{N} = 8$ supergravity in 4d.   **Fig. 12.1**

An overview of possible counterterms in 4d $\mathcal{N} = 8$ supergravity is presented in Figure 12.1. Note that at 7-loop order, all but the D^8R^4 operator are excluded. This means that calculation of the 4-graviton 7-loop amplitude can completely settle the question of finiteness at 7-loop order. But this is not known to be the case at 8-loops or higher.

In the following section, we give a brief survey of the current status of the UV behavior of loop amplitudes in supergravity theories.

12.6 Supergravity divergences for various $\mathcal{N}$, L, and D

Current approaches to examining the possible UV divergences of perturbative (super)-gravity can be categorized as follows:

- **Direct computation.** The explicit computations of loop amplitudes are made possible by increasingly sophisticated applications of the generalized unitarity method; this in itself advances the technical tools for attacking higher-loop computations in general field theories. For the purpose of exploring UV divergences in supergravity theories, most efforts focus on the 4-graviton amplitude; this is because (as we have seen in the previous section) the lowest counterterm for pure supergravity is of the form $D^{2k}R^4$. See [232] for a discussion on the various details of obtaining multi-loop integrands and extracting the UV divergences.

- **Symmetry analysis.** The leading UV divergence of a theory has a corresponding local gauge-invariant operator that must respect all non-anomalous global symmetries of the theory. Analyzing the symmetry properties of operators, one can rule out UV divergences and identify candidate counterterms. In Section 12.5, we took an approach based on the on-shell matrix elements of the candidate operators, and used it to rule out divergences in 4d $\mathcal{N} = 8$ supergravity for $L \leq 6$. Alternatively, one can also use extended-superspace to construct possible invariant operators; for early constructions, see [233, 234]. If the invariant operator can be expressed as a superspace integral over a subset of superspace coordinates, then it is considered a "BPS operator" and it is subject to non-renormalization theorems. If it is given as a full superspace integral, then it is non-BPS and expected to receive quantum corrections, thus serving as a candidate counterterm. The distinction between BPS and non-BPS invariants relies on subtle assumptions about the number of supersymmetries that can be linearly realized off-shell. This lies outside the scope of this presentation, so we refer you to [235] and references therein.

- **_Pure spinor formalism._** The "pure spinor formalism" [236] is a first-quantized approach (in contrast, QFT is a second-quantized approach) to scattering processes in 10d maximal supersymmetric theories. Using the 10d loop-integrand obtained in the pure spinor formalism, one can infer properties of the loop amplitudes in $D \leq 10$ and this can be helpful for assessing potential UV divergences. For more details of this approach to multi-loop amplitudes, see [218].

- **_Role of non-perturbative states?_** The 2^8 massless states of ungauged $\mathcal{N} = 8$ supergravity in 4d match exactly the spectrum of massless states of closed Type IIB superstring theory compactified on a six-torus T^6. At the classical level, $\mathcal{N} = 8$ supergravity in 4d can be viewed as the low-energy ($\alpha' \to 0$) limit of Type IIB superstring theory on T^6. However, it was pointed out in [237] that one cannot obtain *perturbative* $\mathcal{N} = 8$ supergravity in 4d as a consistent truncation of the string spectrum: in the limit $\alpha' \to 0$, keeping the 4d coupling small forces infinite towers of additional states to become light, e.g. Kaluza–Klein states, winding modes, Kaluza–Klein monopoles, and/or wrapped branes. Thus one obtains from string theory not just the spectrum of $\mathcal{N} = 8$ supergravity, but a slew of additional light states. This argument is independent of whether $\mathcal{N} = 8$ supergravity is finite in 4d or not. However, it does mean that even if pure $\mathcal{N} = 8$ supergravity in 4d were to be a well-defined theory of quantum gravity, its UV completion would not be Type IIB string theory.

  A related objection [238][13] to the program of studying finiteness of perturbative supergravity is that even the 4d $\mathcal{N} = 8$ supergravity theory itself contains non-perturbative states, namely BPS black holes, that in certain regions of moduli-space become light enough that they may enter the perturbative expansion [238]. Such contributions would never enter the unitarity method approach to explicit calculation of amplitudes.

As we have noted, there are certainly examples of divergences in various perturbative (non-super)gravity theories in 4d: 1-loop in gravity with generic matter [204, 207], 2-loop in pure gravity [205, 206], and at 1-loop [240] in dilaton–axion gravity whose spectrum was given in (12.16). The first example of a UV divergence in pure supergravity was found at 4-loop order in the 4-graviton amplitude of $\mathcal{N} = 4$ supergravity [241].

We end this chapter by summarizing what explicit computations of supergravity loop amplitudes have revealed so far about the critical dimension D_c of supergravity with various numbers of supersymmetry:

Maximal supergravity (32 supercharges)

Loop-order	1	2	3	4
D_c	8 [82]	7 [242]	6 [213]	$\frac{11}{2}$ [216]

$$\text{(12.46)}$$

---

[13] See also [239].

Half-maximal supergravity (16 supercharges)

Loop-order	1	2	3	4
D_c	8 [243]	6 [244]	> 4 [245]	≤ 4 [241]

$$(12.47)$$

Half-maximal supergravity with matter (both with 16 supercharges)

Loop-order	1	2	3
D_c	4 [246]	≤ 4 [247]	≤ 4 [247]

$$(12.48)$$

In the above, "≤ 4" indicates an upper bound for the critical dimension and we have given a literature reference for each result.

The absence of a UV divergence for half-maximal supergravity at 3-loops in 4d [245] as well as 2-loops in 5d [244], was *not* anticipated by superspace-based analyses. This unexpected result prompted a conjecture of the existence of an off-shell formalism that preserves the full 16 supersymmetries [248]. This would imply finiteness for half-maximal supergravity with matter at 2-loop in 5d, but explicit calculations [247] have shown that a UV divergence is actually present, thus contradicting the conjecture.

The study of the UV structure of perturbative supergravity theories in diverse dimensions has resulted in some interesting insights into the relation between gravity and gauge theory scattering amplitudes: this includes the color-kinematics duality that is the subject of the next chapter. It is relevant to note that no matter what one thinks of the program to study the UV-behavior of supergravity, these new insights would have been difficult to come by without the effort put into explicit calculations and the lessons learned in the process.

A recurring theme in our discussion of perturbative supergravity in Chapter 12 is captured by the abstract expression "gravity = (gauge theory)^2." It enters in the context of the spectrum of states, for example as

$$\mathcal{N} = 8 \text{ supergravity} = \left(\mathcal{N} = 4 \text{ SYM}\right) \otimes \left(\mathcal{N} = 4 \text{ SYM}\right), \tag{13.1}$$

and also carries over to the scattering amplitudes. For instance, the gravity 3-point amplitude equals the square of Yang–Mills 3-gluon amplitude (12.8). That is a special case of the KLT formula (12.13) which expresses the tree-level n-graviton amplitude as sums of products of color-ordered n-gluon Yang–Mills amplitudes. The KLT formula extends to all tree-level amplitudes in $\mathcal{N} = 8$ supergravity following the prescription (13.1) for "squaring" the spectrum.

While the KLT formula follows the "gravity = (gauge theory)^2" storyline, it is unsatisfactory in some respects. First, the formula becomes highly tangled at higher points, as it involves nested permutation sums and rather complicated kinematic invariants. Second, since it involves products of *different* color-ordered amplitudes, it is not really a squaring relation (except at 3-points). Finally, it is only valid at tree-level. You may think that it is asking too much to have gravity amplitudes, arising from the complicated Einstein–Hilbert Lagrangian, closely related to amplitudes of the much simpler Yang–Mills theory. But one lesson we have learned so far is not to let the Lagrangians obscure our view! In this chapter, we explore a form of "gravity = (gauge theory)^2" that makes the amplitude squaring relation more direct and has been proposed to be valid at both tree- and loop-level.

We begin by giving a qualitative answer to the simple question: why is the KLT formula so complicated? In our study of color-ordered amplitudes, we often exploit the fact that the allowed physical poles are those that involve adjacent momenta, e.g. $1/P^2_{i,i+1,\ldots,j-1,j}$. This is a special feature linked to the color-ordered Feynman rules. But for gravity, there is no color-structure and hence no canonical sense of ordering of the external states. Thus the poles that appear in a gravity amplitude can involve any combination of external momenta. This tells us that in order to faithfully reproduce the pole structure of a gravity amplitude from "gravity = (gauge theory)^2," we need color-ordered Yang–Mills amplitudes *with different orderings*. For higher points, the proliferation of physical poles in the gravity amplitude forces us to include more and more Yang–Mills amplitudes with distinct orderings. This is why the KLT formula involves a sum over a growing number of different color-ordered Yang–Mills amplitudes. The complicated kinematic factors in KLT are needed to cancel double poles.

▶ **Exercise 13.1**

Justify the kinematic factors and distinct orderings of the Yang–Mills amplitude in the KLT relations (12.13).

The above discussion suggests that for the comparison with gravity, it may be more useful to consider the fully color-dressed Yang–Mills amplitudes instead of the color-ordered partial amplitudes. The former has the same physical poles as the gravity amplitude. Indeed, this is a productive path, so to get started, we review some useful properties of the color-structure of Yang–Mills amplitudes.

13.1 The color-structure of Yang–Mills theory

The full color-dressed n-point tree amplitude of Yang–Mills theory can be conveniently organized in terms of diagrams with only cubic vertices, such as

$$(13.2)$$

The tree amplitude is then written as a sum over all distinct trivalent diagrams, labeled by i,

$$A_n^{\text{tree}} = \sum_{i \in \text{trivalent}} \frac{c_i n_i}{\prod_{\alpha_i} p_{\alpha_i}^2} . \qquad (13.3)$$

The denominator is given by the product of all propagators (labeled by α_i) of a given diagram. The numerators factorize into a group-theoretic color-part c_i, which is a polynomial of structure constants f^{abc}, and a purely kinematic part n_i, which is a polynomial of Lorentz-invariant contractions of polarization vectors ϵ_i and momenta p_i. As an example, the 4-point amplitude is

$$A_4^{\text{tree}} = \quad + \quad + \quad = \frac{c_s n_s}{s} + \frac{c_u n_u}{u} + \frac{c_t n_t}{t} ,$$

$$(13.4)$$

Three trivalent diagrams whose color factors c_i, c_j, and c_k are related by the Jacobi identity. Note that the diagrams share the same propagators except one, indicated by a thicker gray line. We denote the unshared inverse propagators as s_i, s_j, and s_k.

where the color factors

$$c_s \equiv \tilde{f}^{a_1 a_2 b}\, \tilde{f}^{b\, a_3 a_4}\,, \quad c_t \equiv \tilde{f}^{a_1 a_3 b}\, \tilde{f}^{b\, a_4 a_2}\,, \quad c_u \equiv \tilde{f}^{a_1 a_4 b}\, \tilde{f}^{b\, a_2 a_3}\,, \tag{13.5}$$

as already introduced in (2.68). The normalization of the structure constants $\tilde{f}^{abc}$ was discussed in footnote 5 on page 30 of Chapter 2.

The numerators n_i can be constructed straightforwardly using Feynman rules. Feynman diagrams with only cubic vertices directly contribute terms of the form $\frac{c_i n_i}{\prod_{\alpha_i} p^2_{\alpha_i}}$. The Yang–Mills 4-point contact terms can be "blown up" into s-, t- or u-channel 3-vertex pole diagrams by trivial multiplication by $1 = t/t = s/s = u/u$. Note that since $c_s + c_t + c_u = 0$, this does not give a unique prescription for how to assign a given contact term into the cubic diagrams, so the numerators in (13.3) are not uniquely defined.

We can deform the numerators n_i in several ways without changing the result of the amplitude. For example, one can trivially shift the polarization vectors as $\epsilon_i(p_i) \to \epsilon_i(p_i) + \alpha_i p_i$; this changes the kinematic numerator factors n_i, but not the overall amplitude because it is gauge invariant. A more non-trivial deformation uses the color factor Jacobi identity $c_s + c_t + c_u = 0$: taking $n_s \to n_s + s\Delta$, $n_t \to n_t + t\Delta$, and $n_u \to n_u + u\Delta$, where Δ is an arbitrary function, leaves the amplitude invariant since the net deformation is proportional to $c_s + c_t + c_u$.

In general, for any set of three trivalent diagrams whose color factors are related through a Jacobi identity,

$$c_i + c_j + c_k = 0\,, \tag{13.6}$$

the following numerator-deformation leaves the amplitude invariant:

$$n_i \to n_i + s_i \Delta\,, \quad n_j \to n_j + s_j \Delta\,, \quad n_k \to n_k + s_k \Delta\,. \tag{13.7}$$

Here $1/s_i$, $1/s_j$, and $1/s_k$ are the unique propagators that are *not* shared among the three diagrams, as shown in Figure 13.1. Since Δ can be an arbitrary function, it is similar to a gauge parameter, except that now it is not a transformation of the gauge field, but rather a transformation of the numerator factors n_i. Because of this similarity, the freedom (13.6), (13.7) is often called **generalized gauge transformation** [9]. It plays an important role in linking Yang–Mills and gravity.

The fact that the numerators n_i are not unique nor gauge invariant should not raise any alarm. After all, the individual Feynman diagrams are not physical observables. For practical purposes, it is useful to focus on gauge-invariant quantities. Note that if the color factors are organized in a basis that is independent under Jacobi identities, the coefficient in front of each basis element is necessarily gauge invariant. These coefficients then serve as "partial-amplitudes" that constitute part of the full amplitude, but are fully gauge invariant.

A straightforward way to obtain such partial-amplitudes is to start with the full color-dressed amplitude in (13.3) and use the color Jacobi identity to systematically disentangle the color factors. This can be achieved in a graphical fashion introduced in [8]: start with the color factor of an arbitrary Feynman diagram (with all contact vertices blown up into two cubic vertices as discussed earlier) and convert it by repeated use of the Jacobi identity into a sum of color factors in *multi-peripheral form*

$$\to \quad \tilde{f}^{a_1 a_{\sigma_1} b_1} \tilde{f}^{b_1 a_{\sigma_2} b_2} \ldots \tilde{f}^{b_{n-3} a_{\sigma_{n-2}} a_n},$$
(13.8)

where the positions of legs 1 and n are fixed and σ represents a permutation of the remaining $n-2$ legs. As an example, consider a color diagram that has a Y-fork extending from the baseline. Applying the Jacobi identity on the propagator in the Y-fork, the diagram is converted to a linear combination of two diagrams in multi-peripheral form:

(13.9)

Any trivalent diagram can be cast into a linear combination of diagrams of multi-peripheral form (13.8). The important point is then that the color factors in multi-peripheral form are not related by any Jacobi identities, so there is a total of $(n-2)!$ independent color factors at a given n. The full color-dressed tree amplitude can be expressed in terms of this color basis and then the coefficient of each color factor is a gauge-invariant quantity, denoted for now by $\tilde{A}_n$. We write the full color-dressed amplitude in the multi-peripheral basis as

$$A_n^{\text{tree}} = \sum_{\sigma \in S_{n-2}} \tilde{f}^{a_1 a_{\sigma_1} b_1} \tilde{f}^{b_1 a_{\sigma_2} b_2} \ldots \tilde{f}^{b_{n-3} a_{\sigma_{n-2}} a_n} \, \tilde{A}_n(1, \sigma_1, \sigma_2, \ldots, \sigma_{n-2}, n),$$
(13.10)

where the sum is over all permutations of lines $2, 3, \ldots, n-1$.

Recall from Section 2.5 that we introduced an alternative, manifestly crossing, symmetric representation that uses trace factors of generators as the basis for the color factor. In this trace-basis the color-dressed amplitude is

$$A_n^{\text{tree}} = \sum_{\sigma \in S_{n-1}} \text{Tr}(T^{a_{\sigma_1}} T^{a_{\sigma_2}} \cdots T^{a_{\sigma_{n-1}}} T^{a_n}) A_n[\sigma_1, \sigma_2, \ldots, \sigma_{n-1}, n], \qquad (13.11)$$

where one sums over all permutations of lines $1, 2, \ldots, n-1$ and $A_n[\ldots]$ is our familiar *color-ordered amplitude*. Note that there are $(n-1)!$ distinct traces in (13.11), but since there are only $(n-2)!$ independent color factors, the trace "basis" is over-complete and the color-ordered partial amplitudes satisfy special linear relations. These linear relations are the *Kleiss–Kuijf relations* [7], the simplest of which is the $U(1)$ *decoupling identity* shown in (2.85). These relations were discussed in Section 2.5; they reduce the number of independent color-ordered amplitudes from $(n-1)!$ to $(n-2)!$.

The $(n-2)!$ partial amplitudes $\tilde{A}_n$ are exactly the color-ordered partial amplitudes that are independent under the Kleiss–Kuijf relations. $\tilde{A}_n$ is not unique, since we could have chosen any other two legs to replace 1 and n as reference legs in the multi-peripheral color-basis (13.10). This reflects the fact that there is no unique choice of independent color-ordered partial amplitudes under the Kleiss–Kuijf relations.

In summary, we have reviewed that for a given Yang–Mills n-point tree amplitude, there is a total of $(n-2)!$ color factors that are independent under the Jacobi identities. A convenient choice of independent color factors is those that appear in a multi-peripheral representation, and they can be chosen to be a suitable subset of the $(n-1)!$ color-ordered amplitudes $A_n[\ldots]$.

13.2 Color-kinematics duality: BCJ, the tree-level story

The discussion in Section 13.1 may appear to be a deviation from our path to gravity, but it is a useful detour, as we will see shortly. We begin with the only amplitude we know where gravity is given as a direct square of Yang–Mills, namely the 3-point amplitude. We would like to understand what is so special about the 3-point amplitude that is not shared by its higher-point counterparts. The 3-point superamplitude of $\mathcal{N} = 4$ SYM is

$$\mathcal{A}_3 = \frac{\delta^{(8)}(\tilde{Q})}{\langle 12 \rangle \langle 23 \rangle \langle 31 \rangle}. \qquad (13.12)$$

This amplitude is cyclic invariant, as required for a color-ordered superamplitude. But note that it is also totally antisymmetric, exactly as the 3-point color factor f^{abc}. Hence, the 3-point superamplitude has kinematics that reflect the structure of the color factor of a 3-vertex. Taking a leap of faith, we might wish to generalize this to higher-points, such that the kinematics of each individual trivalent diagram satisfies the same properties as its color factor, including Jacobi identities as in Figure 13.1. But what do we mean by the kinematics of each diagram? As we have seen, this is not a gauge-invariant statement. It certainly cannot include propagators, as you can see from Figure 13.1. Thus we jump to the conclusion that perhaps the numerator factors n_i in (13.3) can be arranged to have the same properties as the corresponding color factors c_i?

The **color-kinematics duality** was first proposed for Yang–Mills theories by Bern, Carrasco, and Johansson (BCJ) [9]. The duality states that scattering amplitudes of Yang–Mills theory, and its supersymmetric extensions, can be given in a representation where the numerators n_i have the *same* algebraic properties as the corresponding color factors c_i. More precisely, using the representation (13.3), the BCJ proposal is that one can always find a representation such that the following parallel relations hold for the color and kinematic factors:

$$c_i = -c_j \quad \Leftrightarrow \quad n_i = -n_j$$
$$c_i + c_j + c_k = 0 \quad \Leftrightarrow \quad n_i + n_j + n_k = 0. \tag{13.13}$$

The duality does *not* state that the numerator factors (13.13) have to be local; they are allowed to have poles.

To illustrate the identity in the first line of (13.13), consider the two diagrams

$$. \tag{13.14}$$

They are related by simply switching two lines on a 3-point vertex, highlighted in gray solid and dashed lines in (13.14). The color factors of the diagrams are related by a minus sign: $c_i = -c_j$, so the color-kinematics duality (13.13) states that there exists a representation where the numerator factors of the two diagrams respect the same antisymmetry property: $n_i = -n_j$.

The second line in (13.13) signifies that the numerator factors must satisfy exactly the same linear relations as their associated color factors. For example, since the color factors of the three diagrams in Figure 13.1 satisfy the Jacobi relation $c_i + c_j + c_k = 0$, the color-kinematics duality states that there is a representation of the numerators of the three diagrams such that $n_i + n_j + n_k = 0$.

At first sight this duality may seem implausible. While the underlying reason for the Jacobi identity to hold for the c_is is the non-abelian gauge algebra defined by two-brackets, $[T^a, T^b] = f^{ab}{}_c T^c$, there appears to be no reason for the kinematic numerators to satisfy the same relations. As we now show, the color-kinematics duality is not as impossible as it seems.

Recall that there are $(n-2)!$ independent color factors under the Jacobi identities. If we require that the numerator factors n_i satisfy the same Jacobi identities, then there will only be $(n-2)!$ independent numerators as well. Since there are also only $(n-2)!$ independent color-ordered partial amplitudes, we can express the set of linearly independent partial amplitudes in terms of the $(n-2)!$ numerators:

$$A^{\text{tree}}_{(i)} = \sum_{j=1}^{(n-2)!} \Theta_{ij}\, \hat{n}_j \,. \tag{13.15}$$

Here $i, j = 1, \ldots, (n-2)!$, and $A^{\text{tree}}_{(i)}$ and $\hat{n}_j$ are the independent color-ordered amplitudes and numerators, respectively. The $(n-2)! \times (n-2)!$ matrix Θ_{ij} consists solely of massless scalar propagators. (This matrix was first introduced in [249] as the "propagator matrix.") As an example, for $n = 4$ we choose $A^{\text{tree}}_4[1, 2, 3, 4]$ and $A^{\text{tree}}_4[1, 3, 2, 4]$ as the two independent color-ordered amplitudes. Expanding the color factors in (13.4) in terms of traces (as in Exercise 2.31), we find

$$A^{\text{tree}}_4[1, 2, 3, 4] = -\frac{n_s}{s} + \frac{n_u}{u}, \qquad A^{\text{tree}}_4[1, 3, 2, 4] = -\frac{n_u}{u} + \frac{n_t}{t}. \tag{13.16}$$

Enforcing the color-kinematics duality (13.13) on the numerators gives $n_t = -n_s - n_u$. Choosing n_s, n_u as $(\hat{n}_1, \hat{n}_2)$ in (13.15), we can now identify the 2×2 matrix Θ_{ij} from:

$$\begin{pmatrix} A^{\text{tree}}_4[1, 2, 3, 4] \\ A^{\text{tree}}_4[1, 3, 2, 4] \end{pmatrix} = \begin{pmatrix} -\frac{1}{s} & \frac{1}{u} \\ -\frac{1}{t} & -\frac{1}{u} - \frac{1}{t} \end{pmatrix} \begin{pmatrix} \hat{n}_1 \\ \hat{n}_2 \end{pmatrix} \longrightarrow \Theta_{ij} = \begin{pmatrix} -\frac{1}{s} & \frac{1}{u} \\ -\frac{1}{t} & -\frac{1}{u} - \frac{1}{t} \end{pmatrix}. \tag{13.17}$$

As advertised, the matrix Θ_{ij} consists of propagators. The construction generalizes to higher-points. The explicit form of the 6×6 matrix Θ_{ij} for $n = 5$ was given in [249]. For related work, see [250].

Inverting the matrix Θ_{ij} would give us numerators $\hat{n}_i$ expressed in terms of the color-independent amplitudes, and from the Jacobis one can generate the rest of the numerator factors, thus trivially obtaining a representation that satisfies the color-kinematics duality. If this were true, the color-kinematic duality would be trivial and this doesn't quite smell right. And it isn't: in 4d Yang–Mills theory, the matrix Θ_{ij} does not have full rank, so it cannot be inverted, and we do not have unique numerators $\hat{n}_i$. Indeed looking back at our 4-point example in (13.17), one can easily verify that the 2×2 matrix Θ_{ij} only has rank 1.

There is also another way to see that the numerator factors cannot be uniquely determined. Suppose we have obtained a set of numerators that satisfy color-kinematic identity. Let us assume that we have achieved this at 5-points and then add the following term [251] to the Yang–Mills action:

$$\mathcal{D}_5 = f^{a_1 a_2 b} f^{b a_3 c} f^{c a_4 a_5}$$

$$\times \left(\partial_{(\mu} A^{a_1}_{\nu)} A^{a_2}_{\rho} \frac{\Box}{\Box} A^{a_3 \mu} + \partial_{(\mu} A^{a_2}_{\nu)} A^{a_1 \mu} \frac{\Box}{\Box} A^{a_3}_{\rho} + A^{a_1}_{\rho} A^{a_2 \mu} \frac{\Box}{\Box} \partial_{(\mu} A^{a_3}_{\nu)} \right) \frac{1}{\Box} (A^{a_4 \nu} A^{a_5 \rho}). \tag{13.18}$$

The presence of the trivial $1 = \frac{\Box}{\Box}$ facilitates the identification of n_is in the cubic diagram expansion. The term (13.18) is identically zero thanks to the Jacobi identity, so it does not change the theory. It does, however, modify the Feynman rules so it changes the numerator factors, but in such a way that they still satisfy the color-kinematic Jacobi identity. We conclude that the duality-satisfying numerators $\hat{n}_i$ cannot be unique; and this is why the matrix Θ_{ij} is not invertible in 4d Yang–Mills theory.

Given that the matrix Θ_{ij} has lower rank, there must be linear relations among the color-independent partial amplitudes. To expose such relations for $n = 4$, use the first row of (13.17) to express $\hat{n}_1$ in terms of the partial amplitude $A_4^{\text{tree}}[1, 2, 3, 4]$ and $\hat{n}_2$,

$$\hat{n}_1 = -s A_4^{\text{tree}}[1, 2, 3, 4] + \frac{s}{u}\hat{n}_2 . \tag{13.19}$$

Substituting this solution into the second row of (13.17), we find that

$$A_4^{\text{tree}}[1, 3, 2, 4] = -\frac{\hat{n}_1}{t} - \frac{\hat{n}_2}{t} - \frac{\hat{n}_2}{u} = \frac{s}{t} A_4^{\text{tree}}[1, 2, 3, 4] - \left(\frac{s}{ut} + \frac{1}{t} + \frac{1}{u}\right)\hat{n}_2 . \tag{13.20}$$

The coefficient of $\hat{n}_2$ is proportional to $s + t + u = 0$. Thus, imposing color-kinematics duality gives the following relation among color-ordered amplitudes:

$$t A_4^{\text{tree}}[1, 3, 2, 4] = s A_4^{\text{tree}}[1, 2, 3, 4] . \tag{13.21}$$

This is an example of the **BCJ relations** that we discussed earlier in Section 2.5. In fact, what we did above is equivalent to your calculation in Exercise 2.31. Since the matrix Θ_{ij} is defined with respect to color-ordered amplitudes that are independent under the Kleiss–Kuijf relations, the BCJ amplitude relations are new relations beyond the consequences of the color-structure. These novel relations reflect that color-kinematics duality exists in Yang–Mills theory. It is known [9] that for general n, the $(n - 2)!$-by-$(n - 2)!$ matrix Θ_{ij} has rank $(n - 3)!$, thus implying $(n - 2)! - (n - 3)! = (n - 3)(n - 3)!$ BCJ relations among n-point color-ordered amplitudes. The simplest type of such relations (sometimes called **fundamental BCJ relations**) can be nicely condensed to the form [9]

$$\sum_{i=3}^{n} \left(\sum_{j=3}^{i} s_{2j}\right) A_n^{\text{tree}}[1, 3, \ldots, i, 2, i + 1, \ldots, n] = 0. \tag{13.22}$$

In our 4-point example, $\hat{n}_2$ dropped out of the final equation (13.20). This means that no consistency conditions can be put on $\hat{n}_2$, so we can take $\hat{n}_2$ to be anything we want without affecting the physical amplitude. Such free numerators are sometimes referred to as "pure-gauge." In practice, it is often convenient to set them to zero.

Now you may say that this is all very interesting, but have we lost sight of our original motivation, to get gravity amplitudes from Yang–Mills theory!? No worries, we are already there. A remarkable proposed consequence of the color-kinematics duality is that once duality-satisfying numerators n_i are obtained, the formula

$$M_n^{\text{tree}} = \sum_{i \in \text{cubic}} \frac{n_i^2}{\prod_{\alpha_i} p_{\alpha_i}^2} \tag{13.23}$$

calculates the n-point tree amplitude in the (super)gravity whose spectrum is given by squaring the (super) Yang–Mills spectrum. That is, we simply take the Yang–Mills amplitude formula in (13.3) and replace each color factor c_i with the corresponding duality-satisfying numerator n_i. And, boom, that is gravity! This relation is called the **BCJ double-copy relation**. The formula (13.23) manifestly reproduces all possible poles that should appear in the gravity amplitude. Furthermore, the mass-dimension matches on both sides of the equation.

▷ *Example.* Let us check (13.23) at 4-points. From (13.23), we find that

$$M_4^{\text{tree}} = \left(\frac{n_s^2}{s} + \frac{n_u^2}{u} + \frac{n_t^2}{t} \right) = \left(\frac{n_s^2}{s} + \frac{n_u^2}{u} + \frac{(n_s + n_u)^2}{t} \right), \tag{13.24}$$

where in the second equality, we use color-kinematics duality to set $n_t = -n_s - n_u$. Now use (13.19) and (13.21), remembering that $(\hat{n}_1, \hat{n}_2)$ is identified with (n_s, n_u) and that we can freely set $\hat{n}_2 = 0$: we then find

$$M_4^{\text{tree}} = -\frac{su}{t} A_4^{\text{tree}}[1,2,3,4]^2 = -u\, A_4^{\text{tree}}[1,2,3,4]\, A_4^{\text{tree}}[1,3,2,4]. \tag{13.25}$$

This is just a different form of the KLT formula we encountered previously in (12.13)! A more involved 5-point example was worked out in detail in [9]. Thus, by reproducing the correct KLT relations, the validity of (13.23) is verified at 4- and 5-points. ◁

▶ **Exercise 13.2**

What if we choose a gauge where $\hat{n}_2$ is not zero? Show that when substituting (13.19) and (13.21) into (13.23), $\hat{n}_2$ drops out in the final result. This shows that the gravity formula (13.23) is gauge invariant in the sense of (13.7).

The BCJ squaring relations (13.23) can be exploited to determine an explicit representation of color-kinematics duality-satisfying numerators of the tree amplitude [252, 253]. Recall that by using the color Jacobi relations, we can convert the fully dressed amplitude in (13.3) into the multi-peripheral form (13.8). Assuming that we have duality-satisfying numerators, the double copy representation of the gravity amplitude in (13.23) can now go through exactly the same steps as those that converted (13.3) into (13.8), and obtain a multi-peripheral form of gravity amplitude

$$M_n^{\text{tree}} = \sum_{\sigma \in S_{n-2}} n_{1|\sigma_1, \sigma_2, \ldots, \sigma_{n-2}|n} A_n^{\text{tree}}(1, \sigma_1, \sigma_2, \ldots, \sigma_{n-2}, n), \tag{13.26}$$

where $n_{1|\sigma_1, \sigma_2, \ldots, \sigma_{n-2}|n}$ are the duality-satisfying numerators for the cubic diagrams in multi-peripheral form with legs 1 and n held fixed. Thus we have an expression for the gravity amplitude in terms of a sum of Yang–Mills color-ordered amplitudes times kinematic factors. Note that it looks a lot like the KLT formula. Indeed, as realized first in [252], by lining up a copy of the KLT formula with the color-ordered amplitudes in (13.26), one can readily read-off duality-satisfying n_is. Take, for example, the $n = 4, 5$ KLT formulas (12.13):

$$M_4^{\text{tree}}(1234) = -s_{12} A_4^{\text{tree}}[1234] A_4^{\text{tree}}[1243],$$
$$M_5^{\text{tree}}(12345) = s_{23}s_{45} A_5^{\text{tree}}[12345] A_5^{\text{tree}}[13254] + (3 \leftrightarrow 4). \tag{13.27}$$

Choosing legs 1 and n to be fixed in our multi-peripheral form, we can readily read off:

$$n = 4: \quad n_{1|2,3|4} = -s_{12}\, A_4^{\text{tree}}[1243], \qquad n_{1|3,2|4} = 0\,;$$

$$n = 5: \quad n_{1|2,3,4|5} = s_{23}s_{45}\, A_5^{\text{tree}}[13254]\,, \quad n_{1|2,4,3|5} = s_{24}s_{35}\, A_5^{\text{tree}}[14253]\,, \quad (13.28)$$

$$n_{1|3,4,2|5} = n_{1|4,2,3|5} = n_{1|4,3,2|5} = n_{1|3,2,4|5} = 0\,.$$

From these $(n-2)!$ independent numerators, all remaining numerators can be obtained by applying the Jacobi identities.

If you think that the BCJ double-copy relation (13.23) is too good to be true, have no fear: things are about to get even better! It turns out that the squaring relation (13.23) can be generalized to

$$M_n^{\text{tree}} = \sum_{i \in \text{cubic}} \frac{n_i \tilde{n}_i}{\prod_{\alpha_i} p_{\alpha_i}^2}\,, \tag{13.29}$$

where the gravity numerators are given as the product of two possibly *distinct* Yang–Mills numerators n_i and $\tilde{n}_i$. Only one set of numerators, say n_i, has to satisfy the color-kinematics duality (13.13), while the other copy, $\tilde{n}_i$, can be an arbitrary representation of the Yang–Mills amplitude. To understand why this is so, let us assume that the n_is respect the duality (13.13) while the $\tilde{n}_i$s do not. Define the difference of the two distinct numerators to be

$$\Delta_i \equiv n_i - \tilde{n}_i\,. \tag{13.30}$$

Since n_i and $\tilde{n}_i$ are both valid representations of the same Yang–Mills amplitude, it follows from (13.3) that

$$\sum_{i \in \text{cubic}} \frac{c_i \Delta_i}{\prod_{\alpha_i} p_{\alpha_i}^2} = 0\,. \tag{13.31}$$

In the discussion so far, we have not specified the gauge group, just that it is non-abelian with structure constants that satisfy the Jacobi identities. Thus the only property of the color factors c_i that can make (13.31) hold is the Jacobi relation. Since the n_is satisfy color-kinematics duality, they satisfy the exact same algebraic properties as the c_is, so we conclude that

$$\sum_{i \in \text{cubic}} \frac{n_i \Delta_i}{\prod_{\alpha_i} p_{\alpha_i}^2} = 0\,. \tag{13.32}$$

This establishes the equivalence of (13.23) and (13.29).

Why bother with the existence of (13.29) vs. (13.23)? – we need one set of duality-satisfying numerators for (13.29) anyway, so why not simply square them and just use (13.23)? Well, recall from Section 12.2 that the spectrum of many supergravity theories can be obtained from tensoring two different Yang–Mills theories. For example, the spectrum of pure $\mathcal{N} = 4$ supergravity is the product of $\mathcal{N} = 4$ SYM and pure Yang–Mills theory (12.19). The point is then that the BCJ double-copy relation (13.29) can be used to construct the supergravity scattering amplitude by using the numerators of two distinct (S)YM theories. And importantly, only one copy of the numerators needs to satisfy the duality, not both. Thus, if we have a set of duality-satisfying numerators for $\mathcal{N} = 4$ SYM, by simply combining them in (13.29) with the numerators of ordinary Yang–Mills, say obtained from explicit Feynman diagram computation, we directly get the scattering amplitudes of $\mathcal{N} = 4$ supergravity. This convenient result has powerful consequences as we move on to loop amplitudes in Section 13.3.

We end this section with some concluding remarks regarding the tree-level BCJ color-kinematics duality (13.13) and the double-copy relations (13.23) and (13.29).

First, the existence of numerators that satisfy (13.13) was exemplified for any n in [252, 253] (see also [254]).

Second, the BCJ relations in (13.22) have been successfully derived from string theory using monodromy relations [255, 256] and in field theory using the improved large-z fall-off of non-adjacent BCFW shifts [257]. Another, quite elegant, derivation was given in [258]. In our discussion, the BCJ relations were a consequence of imposing color-kinematic duality on the numerator. Given that the BCJ relations can be proven via string- and field-theory arguments, one can reverse the argument and show that the existence of BCJ relations and Kleiss–Kuijf identities gives rise to algebraic relations on the kinematics [259].

Third, assuming that there exists a duality-satisfying set of *local* numerators for the Yang–Mills tree amplitude, one can rigorously prove that the doubling-relation (13.23) produces the correct gravity tree amplitude for any n [251]. The proof is established inductively by showing that the difference between (13.23) and the gravity amplitude obtained from BCFW recursion vanishes if one assumes (13.23) to hold for all lower-point amplitudes.

Finally, you may wonder if duality-satisfying numerators can be obtained directly from the Feynman rules of some Lagrangian. In 4d, this can be done for MHV amplitudes [260], but difficulties arise beyond MHV. In general dimensions, a straightforward construction of the cubic diagrams using Feynman rules does not give duality-satisfying numerators, even if the freedom of how to assign contact terms is taken into account. However, modification of the action by non-local terms can give duality-satisfying numerators straight from the Feynman rules of the deformed action [251],

$$\mathcal{L}_{YM} = \mathcal{L} + \mathcal{L}'_5 + \mathcal{L}'_6 + \cdots \tag{13.33}$$

Here $\mathcal{L}$ is the conventional Yang–Mills Lagrangian and $\mathcal{L}'_n$, $n > 4$ are terms that involve n fields and vanish by the Jacobi identity so that the theory is actually not changed. As an example, the quintic terms are

$$\mathcal{L}'_5 \sim \mathrm{Tr}\,[A^\nu, A^\rho]\frac{1}{\square}\Big([[\partial_\mu A_\nu, A_\rho], A^\mu] + [[A_\rho, A^\mu], \partial_\mu A_\nu] + [[A^\mu, \partial_\mu A_\nu], A_\rho]\Big). \tag{13.34}$$

Even though the deformation is non-local, it is completely harmless: $\mathcal{L}'_5$ is simply zero because the terms in the parentheses vanish by the Jacobi identity. Thus by adding a particular zero to the action, one can expose the intricate relation between gravity and Yang–Mills theory. (One may say that this points to a curious deficiency of the action, namely that it treats all zeros in the same way.) A systematic approach to generating explicit higher-order deformations $\mathcal{L}'_n$ is given in [261].

We have seen how the tree-level squaring relation between gravity and Yang–Mills is more straightforward when phrased in terms of the non-gauge-invariant numerators n_i than in terms of the gauge-invariant color-ordered Yang–Mills amplitudes, as in KLT. The true power of this is revealed when it is applied to loop-integrands: the KLT relations hold only at tree-level but, as we will see next, the BCJ squaring relations survive at loop-level.

13.3 Color-kinematics duality: BCJ, the loop-level story

We begin with color-kinematic duality for loop amplitudes of Yang–Mills theory. Any diagram involving the 4-point contact term can be blown up into cubic vertices, as discussed in Section 13.1, so we consider only trivalent loop-diagrams. The full L-loop color-dressed Yang–Mills amplitude can then be written as

$$\mathcal{A}_n^{L\text{-loop}} = \sum_{j \in \text{cubic}} \int \left(\prod_{l=1}^{L} \frac{d^D \ell_l}{(2\pi)^D} \right) \frac{1}{S_j} \frac{n_j \, c_j}{\prod_{\alpha_j} p_{\alpha_j}^2}, \tag{13.35}$$

where the notation follows that defined for (13.3) and S_j is the symmetry factor of the diagram. It was proposed in [262] that there exist representations (13.35) where the kinematic numerators n_i satisfy the same algebraic relation as that of color factors, i.e. (13.13). And once such numerators are found, the gravity amplitude is given by the double-copy formula [262]

$$\mathcal{M}_n^{L\text{-loop}} = \sum_{j \in \text{cubic}} \int \left(\prod_{l=1}^{L} \frac{d^D \ell_l}{(2\pi)^D} \right) \frac{1}{S_j} \frac{n_j \, \tilde{n}_j}{\prod_{\alpha_j} p_{\alpha_j}^2}, \tag{13.36}$$

in which only one of the two copies n_i, $\tilde{n}_i$ is required to satisfy color-kinematics duality.

The validity of (13.36) can be justified through unitarity cuts [67]: assuming that gauge-theory numerators n_i satisfy the duality, the gravity integrand built by taking double copies of numerators has the correct cuts in all channels. To see this, consider a set of generalized unitarity cuts that break the loop amplitude down to products of tree amplitudes. On the cut, the gauge-theory integrand factorizes into products of tree amplitudes whose numerator factors satisfy all color-kinematics dualities relevant for each tree amplitude, because they are merely a subset of the relations required by the loop-level duality. As an example consider the following unitarity cut of the 3-loop cubic diagram:

$$\tag{13.37}$$

We have labeled the *uncut* propagators $1/p_i^2$ by $i = 1, \dots, 5$ and the numerator of the diagram is denoted by n. On the cut, the LHS must be equivalent to the RHS, which is the product of kinematic factors of the factorized tree-diagrams. If the numerators n satisfy all Jacobi identities associated with the parent diagram, the numerators n_a, n_b, and n_c must satisfy the Jacobi identities of the individual tree-diagrams. Now squaring the duality-satisfying numerators in the Yang–Mills tree amplitude, one obtains the gravity tree amplitude. Thus, squaring the Yang–Mills loop-numerators, one is guaranteed to obtain the correct cut for the gravity loop amplitude. In other words, (13.36) is guaranteed to satisfy all unitarity cuts and therefore give the correct answer [262].

If any readers have come all the way with us to this page, then they may question if the above argument only justifies (13.36) as the correct answer for the cut-constructible part of the (super)gravity loop amplitude: what about rational terms that are not cut constructible? Recall that rational terms can be obtained by considering the unitarity cuts in higher-dimensions, where the extra-dimensional momenta can be interpreted as the regulator (see (6.34)). But in the discussion so far, there was no specification of the spacetime dimension: the color-kinematics duality and the double-copy relation are valid in arbitrary dimensions! In other words, (13.36) produces the correct cut in any spacetime dimension, and thus it also faithfully reproduces the rational terms.

Now the validity of (13.36) relies on the existence of duality-satisfying numerators n_i. Do we know that there always exists such a representation – and, if so, how to systematically construct it? Indeed, this is the million dollar question.[1] Unlike at tree-level, there is currently not a formal proof of the existence of duality-satisfying numerators. However, we have explicit examples of such numerators in multiple cases, as summarized in Section 13.3.4. For now, let us see some non-trivial examples.

13.3.1 1-loop 4-point $\mathcal{N} = 4$ SYM

In Section 6.3, we used the unitarity method to compute the *color-ordered* 1-loop 4-point superamplitude of $\mathcal{N} = 4$ SYM and found the result to be

$$\mathcal{A}_4^{\text{1-loop}}[1234] = su\,\mathcal{A}_4^{\text{tree}}[1234]\,I_4(p_1, p_2, p_3, p_4)\,, \qquad (13.38)$$

where $I_4(p_1, p_2, p_3, p_4)$ is the scalar box integral. The fully *color-dressed* 1-loop amplitude can be written in terms of color-ordered amplitudes as [40, 103]

$$\mathcal{A}_4^{\text{1-loop, full}} = c_{1234}^{(1)}\,\mathcal{A}_4^{\text{1-loop}}[1234] + c_{1342}^{(1)}\,\mathcal{A}_4^{\text{1-loop}}[1342] + c_{1423}^{(1)}\,\mathcal{A}_4^{\text{1-loop}}[1423]\,, \quad (13.39)$$

with $c_{ijkl}^{(1)}$ the 1-loop color factor of a box diagram with consecutive external legs (i, j, k, l), e.g.

$$c_{1234}^{(1)} = \tilde{f}^{ea_1b}\,\tilde{f}^{ba_2c}\,\tilde{f}^{ca_3d}\,\tilde{f}^{da_4e}\,. \qquad (13.40)$$

Let us now show that (13.39) does satisfy color-kinematic duality. Take one of the four propagators in the box diagram and apply the Jacobi identity to, say, the propagator between

---

[1] Let's be clear: the authors (H.E. and Y.-t.H.) will not pay anyone a million dollars – or any other amount.

legs 1 and 2:

$$\text{(diagram)} = \text{(diagram)} - \text{(diagram)} . \tag{13.41}$$

The relevant Jacobi identity is

$$c_{1234}^{(1)} - c_{2134}^{(1)} + c_{\mathrm{Tri}:34}^{(1)} = 0 , \tag{13.42}$$

where $c_{\mathrm{Tri}:34}^{(1)} \equiv \tilde{f}^{a_1 b a_2} \tilde{f}^{ebc} \tilde{f}^{ca_3 d} \tilde{f}^{da_4 e}$ is the color factor for a triangle diagram. The duality then states that the numerators of the integrals are related as

$$n_{1234}^{(1)} - n_{2134}^{(1)} + n_{\mathrm{Tri}:34}^{(1)} = 0 . \tag{13.43}$$

Looking back at the 1-loop color-dressed result (13.39), we immediately identify

$$n_{1234}^{(1)} = su \, \mathcal{A}_4^{\mathrm{tree}}[1234], \qquad n_{2134}^{(1)} = st \, \mathcal{A}_4^{\mathrm{tree}}[2134], \qquad n_{\mathrm{Tri}:34}^{(1)} = 0 . \tag{13.44}$$

Since $su \, \mathcal{A}_4^{\mathrm{tree}}[1234]$ is fully permutation invariant, it equals $st \, \mathcal{A}_4^{\mathrm{tree}}[2134]$, and therefore the numerator Jacobi identity (13.43) is trivially satisfied! Applying the Jacobi identity to any other propagator of the box diagram, one arrives at the same result.

In principle we should also check the Jacobi identity

$$\text{(diagram)} = \text{(diagram)} - \text{(diagram)} . \tag{13.45}$$

However, for $\mathcal{N} = 4$ SYM, this is trivially satisfied since the numerator associated with each of the above diagrams is zero.

13.3.2 2-loop 4-point $\mathcal{N} = 4$ SYM

At 2-loop order, the color-dressed Yang–Mills amplitude is [103, 242]

$$\mathcal{A}_4^{\mathrm{2\text{-}loop, full}} = \left(c_{1234}^{\mathrm{P}} \, \mathcal{A}_4^{\mathrm{P}}[1234] + c_{3421}^{\mathrm{P}} \, \mathcal{A}_4^{\mathrm{P}}[3421] + c_{1234}^{\mathrm{NP}} \, \mathcal{A}_4^{\mathrm{NP}}[1234] + c_{3421}^{\mathrm{NP}} \, \mathcal{A}_4^{\mathrm{NP}}[3421] \right)$$
$$+ \; \mathrm{cyclic}(2, 3, 4) , \tag{13.46}$$

where "cyclic(2, 3, 4)" indicates a sum over the remaining two cyclic permutations of legs 2, 3 and 4. The color factors c_{1234}^{P} and c_{1234}^{NP} are obtained by dressing the planar and non-planar double-box diagrams with structure constants $\tilde{f}^{abc}$:

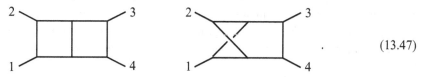

$$\tag{13.47}$$

The explicit amplitudes $\mathcal{A}_4^{\mathrm{P}}[1234]$ and $\mathcal{A}_4^{\mathrm{NP}}[1234]$ are given by the 2-loop scalar-box integrals corresponding to the diagrams (13.47), multiplied by the numerator factors

$$n_{1234}^{\mathrm{P}} = s^2 u \, \mathcal{A}_4^{\mathrm{tree}}[1234], \qquad n_{1234}^{\mathrm{NP}} = s^2 t \, \mathcal{A}_4^{\mathrm{tree}}[1234]. \tag{13.48}$$

Color-kinematic duality now imposes the following linear relations among the numerators of the scalar integrals:

$$\tag{13.49}$$

The above two identities are satisfied by the numerators in (13.48) by virtue of the absence of integrals with triangle sub-loops as well as the permutation invariance of $su \, \mathcal{A}_4^{\mathrm{tree}}[1234]$.

▶ **Exercise 13.3**

What is the identity associated with the Jacobi relation applied to the thick gray propagator in the following diagram:

$$\tag{13.50}$$

Do the numerators in (13.48) satisfy this identity?

13.3.3 3-loop 4-point $\mathcal{N} = 4$ SYM

Thus far in our examples, the duality-satisfying numerators n_i have been independent of the loop-momenta. At 3-loop order, the representation given in [262] for the 4-point amplitude uses duality-satisfying numerators that do depend on the loop-momenta. The scalar integrals that participate in the 3-loop amplitude of $\mathcal{N} = 4$ SYM answer are shown in Figure 13.2. The full amplitude is given as a sum of these integrals (along with permutations of the external legs), their associated color factors and kinematic numerators, suitably normalized

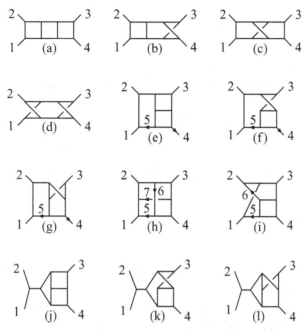

The scalar integrals that are involved in the BCJ calculation of the 3-loop amplitude of $\mathcal{N}=4$ SYM. (Reprinted figure with permission from Z. Bern, J. J. M. Carrasco, and H. Johansson, Phys. Rev. Lett. **105**, 061602 (2010) [262]. Copyright 2010 by the American Physical Society. http://journals.aps.org/prl/abstract/10.1103/PhysRevLett.105.061602.)

Fig. 13.2

by the symmetry factor of the diagram. The kinematic numerators are [262]

Integral	$\mathcal{N}=4$ SYM numerator	
(a)–(d)	s^2	
(e)–(g)	$\left(s\left(-\tau_{35}+\tau_{45}+u\right)-u\left(\tau_{25}+\tau_{45}\right)+t\left(\tau_{25}+\tau_{35}\right)-s^2\right)/3$	
(h)	$\left(s\left(2\tau_{15}-\tau_{16}+2\tau_{26}-\tau_{27}+2\tau_{35}+\tau_{36}+\tau_{37}-t\right)\right.$ $\left.+u\left(\tau_{16}+\tau_{26}-\tau_{37}+2\tau_{36}-2\tau_{15}-2\tau_{27}-2\tau_{35}-3\tau_{17}\right)+s^2\right)/3$	(13.51)
(i)	$\left(s\left(-\tau_{25}-\tau_{26}-\tau_{35}+\tau_{36}+\tau_{45}+2u\right)\right.$ $\left.+u\left(\tau_{26}+\tau_{35}+2\tau_{36}+2\tau_{45}+3\tau_{46}\right)+t\,\tau_{25}+s^2\right)/3$	
(j)–(l)	$s(u-t)/3$	

An overall factor of $su\mathcal{A}_4^{\text{tree}}$ has been removed, and $\tau_{ij}=2k_i\cdot l_j$, where k_i and l_j are the momenta labeled in each diagram in Figure 13.2. The numerators in the table satisfy all Jacobi identities of the corresponding color factors. For example,

$$\includegraphics = \includegraphics + \includegraphics \qquad (13.52)$$

is trivially satisfied because diagram (e) and (f) have the same numerator factor and the third diagram in (13.52) vanishes.

▶ Exercise 13.4

Verify that the numerators (13.51) satisfy the following identity:

$$ (13.53) $$

Note that in this example all three numerators are non-zero.

13.3.4 Summary

There is no formal proof that duality-satisfying numerators can always be found for loop amplitudes in (super) Yang–Mills, however, there is considerable evidence in favor of this property. We present here a list of non-trivial examples for which the BCJ duality-satisfying numerators have been constructed:

- Up to 4-loops for 4-point in $\mathcal{N} = 4$ SYM [232, 262].
- Up to 2-loops for 5-point in $\mathcal{N} = 4$ SYM [263].
- At 1-loop up to 7-points in $\mathcal{N} = 4$ SYM [264].
- Up to 2-loops for 4-point for the all-plus pure Yang–Mills amplitude [262].
- 1-loop 4-point for pure Yang–Mills theory in arbitrary dimensions [240].
- 1-loop n-point all-plus or single-minus helicity amplitudes in pure Yang–Mills theory [265].
- 1-loop 4-point amplitudes in theories with less than maximal supersymmetry [266].
- 1-loop 4-point for an abelian orbifold of $\mathcal{N} = 4$ SYM [267].
- 1-loop 4-point Yang–Mills theory with matter [268].

Although most progress has been made for $\mathcal{N} = 4$ SYM, the examples are not restricted to the maximally supersymmetric theory or to 4d.

13.4 Implications for UV behavior of supergravity

With duality-satisfying numerators for ($\mathcal{N} = 4$) super Yang–Mills amplitudes, we do not need to do much work to compute supergravity amplitudes! We now give several examples of this application of BCJ.

$\mathcal{N} = 8$ supergravity

By squaring the duality-satisfying numerators of the 1-, 2- and 3-loop 4-point amplitudes in Sections 13.3.1–13.3.3, we immediately obtain the integrands of the 1-, 2- and 3-loop 4-point amplitudes in $\mathcal{N} = 8$ supergravity. At 1- and 2-loop order, the $\mathcal{N} = 4$ SYM numerators are independent of the loop-momentum, so since the scalar box and the scalar double box integrals are UV finite in 4d, we immediately see that the 4-point amplitudes in $\mathcal{N} = 8$ supergravity are finite in 4d at 1- and 2-loops. The same conclusion extends to 3-loop order by simple power-counting even though the 3-loop numerators in (13.51) depend on the loop-momenta. Thus, as promised, without further calculations, we have just reproduced the result that the 4-point amplitudes of $\mathcal{N} = 8$ supergravity are finite in 4d up to and including 3-loop order. Of course, this agrees with previous explicit calculations [208–214] and the counterterm analysis discussed in Section 12.5.

Critical dimension for maximal pure supergravity

As mentioned in Section 12.4, the critical dimension for UV divergences of maximal supergravity is proposed [212] to match that of maximal SYM,

$$D_c(L) = \frac{6}{L} + 4 \quad \text{for} \quad L > 1. \tag{13.54}$$

At 1- and 2-loop orders, the duality-satisfying numerators of $\mathcal{N} = 4$ SYM are loop-independent, so the critical dimension is simply determined by the scalar integrals; it is therefore universal between $\mathcal{N} = 4$ SYM and $\mathcal{N} = 8$ supergravity. At 3-loops, one can use power-counting to see that the integrals with the worst UV behavior are the three diagrams $(j), (k), (l)$ in Figure 13.2. These diagrams dictate the critical dimension at 3-loops for $\mathcal{N} = 4$ SYM. But it follows from (13.51) that the numerator factors of these three diagrams are loop-independent, so squaring them to get the $\mathcal{N} = 8$ supergravity amplitude does not change the critical dimension. One can check that none of the other diagrams has worse behavior than $(j), (k), (l)$ after squaring. Hence the relation (13.54) for the critical dimension also holds for $\mathcal{N} = 8$ supergravity at 3-loop order. A similar BCJ argument extends this result to 4-loop order [232].

▶ **Exercise 13.5**

Use the explicit integrands given in Sections 13.3.2 and 13.3.3 to verify that the critical dimension at 2- and 3-loops is (13.54) for both $\mathcal{N} = 4$ SYM and $\mathcal{N} = 8$ supergravity. What is the critical dimension at 1-loop order?

$\mathcal{N} \geq 4$ supergravity

We obtained $\mathcal{N} \leq 8$ supergravity amplitudes by tensoring two sets of numerators from loop amplitudes in $\mathcal{N} \leq 4$ supersymmetric Yang–Mills theories. Only one of the two copies of numerators needs to satisfy the color-kinematic duality, as discussed around (13.31). Since we already have duality-satisfying numerators for the 1-, 2- and 3-loops 4-point $\mathcal{N} - 4$ SYM amplitudes, we can just tensor them with any Yang–Mills or SYM numerators we like, and obtain $\mathcal{N} \geq 4$ supergravity amplitudes. Since only a small number of cubic diagrams have non-vanishing numerators in the $\mathcal{N} = 4$ SYM copy, we only need a few of the numerators of the other copy.

We begin at 1-loop. Suppose we have an explicit representation of the 1-loop integrand of $\mathcal{N} \leq 4$ SYM computed from Feynman rules. Such a representation usually involves triangles and bubbles as well as diagrams that are not 1-particle-irreducible, but it can be converted into a representation that only involves the box integrals. The price one pays is that the numerators will in general be non-local, but that is not a problem for our application. As an example, the following triangle- and bubble-diagrams can be converted to boxes by introducing inverse propagators:

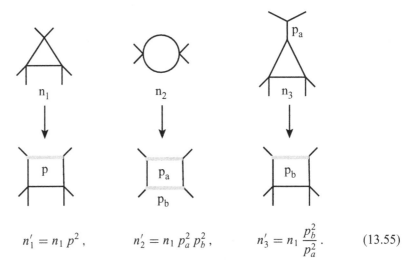

$$n'_1 = n_1 \, p^2 \, , \qquad n'_2 = n_1 \, p_a^2 \, p_b^2 \, , \qquad n'_3 = n_1 \, \frac{p_b^2}{p_a^2} \, . \qquad (13.55)$$

To obtain the $\mathcal{N} \leq 4$ supergravity amplitude, we "tensor" the new numerators n'_i with the duality-satisfying $\mathcal{N} = 4$ SYM numerators n_i in (13.44). Specializing to 4-point, recall that the numerators in (13.44) are not only loop-momentum independent, but also permutation invariant. Independence of the loop-momentum means that the n_is move outside the box-integral. Permutation invariance for the n_is then tells us that the n_i-factor is uniform for each box-integral. In other words, it is just an overall factor, $su \, A_{4,Q=16}^{\text{tree}}[1, 2, 3, 4]$, multiplying the entire 1-loop $\mathcal{N} \leq 4$ SYM amplitude! Thus we have found the following very simple formula for the 4-point 1-loop amplitude in $\mathcal{N} \geq 4$ supergravity [269],

$$M_{4,Q+16}^{(1)} = su \, A_{4,Q=16}^{\text{tree}}[1, 2, 3, 4] \left[A_{4,Q}^{(1)}[1, 2, 3, 4] + A_{4,Q}^{(1)}[1, 3, 4, 2] + A_{4,Q}^{(1)}[1, 4, 2, 3] \right].$$

$$(13.56)$$

The subscript Q indicates the number of supercharges, with $Q = 16$ corresponding to $\mathcal{N} = 4$ supersymmetry. It is remarkable that the 1-loop amplitude of a non-renormalizable gravity theory can be given by a sum of 1-loop amplitudes of a renormalizable one. Note that this relation was exposed only after imposing color-kinematics duality.

Consider now the UV structure in 4d. The 1-loop amplitudes of $\mathcal{N} = 0, 1, 2$ SYM have UV divergences. In 4d SYM, the UV divergence must be proportional to the tree amplitude, otherwise it would imply that a new operator is needed to renormalize the 1-loop divergence.

Thus we conclude that

$$
\left. M_{4,Q+16}^{(1)} \right|_{D=4\,\text{div.}} \sim su\, A_{4,Q=16}^{\text{tree}}[1,2,3,4]
$$

$$
\times \left[A_{4,Q}^{\text{tree}}[1,2,3,4] + A_{4,Q}^{\text{tree}}[1,3,4,2] + A_{4,Q}^{\text{tree}}[1,4,2,3] \right] = 0 .
$$

(13.57)

This result vanishes due to the $U(1)$ decoupling identity (2.85) for the Yang–Mills color-ordered tree amplitudes. Therefore, with the help of BCJ color-kinematics duality, we have shown that pure $\mathcal{N} \geq 4$ supergravity is finite at 1-loop in 4d.

1-loop UV divergence in $\mathcal{N} = 4$ supergravity-matter theory

The simple relation (13.56) can also be used to demonstrate UV divergences. Consider $\mathcal{N} = 4$ supergravity coupled to $\mathcal{N} = 4$ Maxwell theory (i.e. $\mathcal{N} = 4$ SYM with $U(1)$ gauge group). The spectrum of this $\mathcal{N} = 4$ supergravity-matter theory is given by tensoring $\mathcal{N} = 4$ SYM with Yang–Mills minimally coupled to an adjoint scalar. The 1-loop amplitude is exactly the same as (13.56), except that $A_{4,Q}^{(1)}$ is now the 1-loop amplitude of the Yang–Mills-scalar theory. The 4-point 1-loop amplitude in Yang–Mills-scalar theory is UV divergent and is renormalized by a 4-scalar counterterm

$$
\Delta \mathcal{L} = c_{abcd}^{(1)} \, \phi^a \phi^b \phi^c \phi^d ,
$$

(13.58)

where $c_{abcd}^{(1)}$ is the color factor for the box-diagram. Putting this divergence into (13.56) the sum of the three terms is now non-vanishing because the color-structure of (13.58) is not that of a tree amplitude. This then shows that there is a 1-loop UV divergence in the $\mathcal{N} = 4$ supergravity-matter model. This divergence was first computed by standard methods in [246].

▶ **Exercise 13.6**

Use (13.56) to show that the 1-loop UV divergence of the $\mathcal{N} = 4$ gravity-matter system corresponds to an $F^4 = (F_{\mu\nu}F^{\mu\nu})^2$ operator.

Color-kinematics constraints on candidate counterterms

In Section 12.5, we approached UV divergences of supergravity from the viewpoint of local counterterms: we ruled out candidate counterterms based on the known symmetries in $\mathcal{N} = 8$ supergravity. It is reasonable to say that "everything not forbidden is compulsory"[2] and therefore expect that if the known symmetries do not rule out a certain counterterm, then it is likely to appear in the perturbation theory. In this section, we have studied a new structure, BCJ color-kinematics duality, that is very different in nature from the other symmetries imposed on the local counterterms. In examples, we have seen how the BCJ doubling relation reveals the "true" power-counting (at least "truer" than Feynman diagrams) for UV divergences in supergravity amplitudes. So one may now wonder if it is possible that there exist counterterms that respect all known "ordinary" symmetries of the theory and yet are ruled out by color-kinematics duality? We now present such a case.

---

[2] Also known as Gell-Mann's Totalitarian Principle (from T. H. White's *The Once and Future King*).

The 2-loop duality-satisfying numerators (13.48) of $\mathcal{N} = 4$ SYM are momentum independent. Following the same arguments that gave us the 1-loop result (13.56), we find that the 2-loop 4-point amplitude of $\mathcal{N} \geq 4$ supergravity is given as a sum of 2-loop (S)YM amplitudes:

$$M^{(2)}_{4,Q+16}[1,2,3,4] = su\, A^{\text{tree}}_{4,Q=16}[1,2,3,4]\Big[s\Big(A^{\text{P}}_{4,Q}[1,2,3,4] + A^{\text{NP}}_{4,Q}[1,2,3,4]$$

$$+ A^{\text{P}}_{4,Q}[3,4,2,1] + A^{\text{NP}}_{4,Q}[3,4,2,1]\Big) + \text{cyclic}(2,3,4)\Big].$$

$$\tag{13.59}$$

Now let us see what (13.59) says about the UV divergence of supergravity. Consider $\mathcal{N} = 4$ supergravity, hence $Q = 0$ for $A^{\text{P}}_{4,Q}$ and $A^{\text{NP}}_{4,Q}$. In 4d Yang–Mills theory, no counterterm operators are needed to regularize the UV divergence of the 4-point 2-loop amplitude, so the coefficient of the divergence must be generated by the F^2 operator of the classical action. In particular, the 2-loop UV divergence must have tree-level color factors. In 5d, dimension-counting shows that F^3 is the only allowed counterterm; it again only has tree-level color factors.[3] This is because the divergence in 5d is renormalized by a tree diagram with one F^3 counterterm insertion,

$$\tag{13.60}$$

With the intention of using this information, we expand the color-structure c^{P} and c^{NP} in (13.59) into a basis of color factors that are independent under the Jacobi relation. Such a basis is given in Appendix B of [217], and it consists of two tree, one 1-loop, and two 2-loop color factors. So casting (13.59) into the said basis, we know that the coefficients of the 1- and 2-loop color factors have to vanish for the UV-divergent part, since in 4d and 5d only the tree color-structure is generated. The requirement of vanishing 2-loop color factors is [244]

$$0 = \Big[t\Big(A^{\text{P}}_Q[1,3,4,2] + A^{\text{P}}_Q[1,4,2,3] + A^{\text{P}}_Q[3,1,4,2] + A^{\text{P}}_Q[3,2,1,4]$$

$$+ A^{\text{NP}}_Q[1,3,4,2] + A^{\text{NP}}_Q[1,4,2,3] + A^{\text{NP}}_Q[3,1,4,2] + A^{\text{NP}}_Q[3,2,1,4]\Big)$$

$$+ s\Big(A^{\text{P}}_Q[1,3,4,2] + A^{\text{P}}_Q[3,1,4,2] + A^{\text{NP}}_Q[1,3,4,2] + A^{\text{NP}}_Q[3,1,4,2]\Big)\Big]\Big|_{D=4,5\,\text{div.}},$$

$$0 = \Big[s\Big(A^{\text{P}}_Q[1,2,3,4] + A^{\text{P}}_Q[1,3,4,2] + A^{\text{P}}_Q[3,1,4,2] + A^{\text{P}}_Q[3,4,2,1] \tag{13.61}$$

$$+ A^{\text{NP}}_Q[1,2,3,4] + A^{\text{NP}}_Q[1,3,4,2] + A^{\text{NP}}_Q[3,1,4,2] + A^{\text{NP}}_Q[3,4,2,1]\Big)$$

$$+ t\Big(A^{\text{P}}_Q[1,3,4,2] + A^{\text{P}}_Q[3,1,4,2] + A^{\text{NP}}_Q[1,3,4,2] + A^{\text{NP}}_Q[3,1,4,2]\Big)\Big]\Big|_{D=4,5\,\text{div.}}.$$

---

[3] "Tree-level color factor" denotes a contraction of $\tilde{f}^{abc}$s such that no closed loops are formed. For example $\tilde{f}^{abc}\tilde{f}^{cde}$ is a tree color factor for a 4-point amplitude. On the other hand, $\tilde{f}^{abc}\tilde{f}^{cde}\tilde{f}^{efg}\tilde{f}^{gha}$ is a 4-point "1-loop color factor."

Substituting the above into (13.59), one finds that

$$M_{4,16}^{(2)}[1, 2, 3, 4]\Big|_{D=4,5\,\text{div.}} = 0\,. \tag{13.62}$$

So the 2-loop 4-point amplitude in pure $\mathcal{N} \geq 4$ supergravity is finite in both 4d and 5d.

In 4d, the 2-loop R^3 operator can be ruled out by supersymmetry, as we have seen in Section 12.5 from the spinor helicity violating amplitude it generates. This is of course perfectly compatible with the UV finiteness of the 2-loop amplitude (13.62).

However, in 5d the relevant operator at 2-loop order is R^4, and it is compatible with supersymmetry. It has been argued to be duality invariant [248, 270]. So this is an explicit 5d example where a counterterm appears to respect all known symmetries of the theory, yet is not generated because the corresponding 2-loop 4-point amplitude is UV finite.

In 4d, it has been argued [235] that the 3-loop candidate counterterm operator R^4 is allowed by all symmetries of $\mathcal{N} = 4$ supergravity. (It was ruled out in $\mathcal{N} = 8$ supergravity by $E_{7(7)}$.) Yet, by explicit computation, utilizing color-kinematic duality, it has been shown that $\mathcal{N} = 4$ supergravity is actually finite at 3-loop order [245]. So what is going on? The absence of a divergence could be coincidence. Or it could indicate that there is a hidden symmetry that is violated by the would-be-counterterm [271]. The issue is not settled at the time of finishing this book.

13.5 Extensions

We end this section with brief mention of other applications of the BCJ color-kinematics duality. Color-kinematic duality has been extended to scattering amplitudes involving higher-dimension operators [272] and also to form factors [273].

In Section 11.3.5, we encountered the Lie 3-algebra for 3d Chern–Simons matter theory (BLG): it involved 4-index structure constants that, in place of the usual Jacobi identity, satisfy a 4-term fundamental identity (11.49). Surprisingly, color-kinematic duality can also be established for such 3-algebra theories [274] with the basic diagrams built from 4-point vertices only. Which supergravity amplitude is calculated by the BCJ double-copy of duality-satisfying numerators from the 3-algebra Chern–Simons matter theory? At first sight it seems that the answer has to be different from the supergravity amplitude obtained from "squaring" 3d Yang–Mills amplitudes, because diagrams built from quartic vertices must have an even number of external legs n, while trivalent diagrams can have even or odd n. But one can use the double-copy based on 3d maximal ($\mathcal{N} = 8$) super Yang–Mills theory (or the KLT formula) to show that the odd-n supergravity amplitudes vanish in 3d, even though the odd-n Yang–Mills amplitudes are non-vanishing. It has in fact been shown that applying the double-copy trick to 3d $\mathcal{N} = 8$ super Yang–Mills and 3d BLG theory remarkably results in the same supergravity tree amplitudes up to 12 points [274, 275].

By dimensional analysis, 3d gravity is non-renormalizable. It is curious that 3d supergravity amplitudes can be constructed from two distinct color kinematic dualities. Perhaps this puts constraints on the UV behavior of supergravity in 3d.

In this chapter, we list references to other reviews and we highlight a few subjects that were not covered in the main text.

Reviews on on-shell methods for scattering amplitudes

- *Introduction to on-shell methods*
    The QFT textbooks by Srednicki [2], Zee [276], and Schwartz [277] provide brief introductions to the spinor helicity formalism and on-shell recursion relations. In addition, the book [278] offers a comprehensive introduction to on-shell recursion and to loop-integrals and also covers some aspects of scattering amplitudes in $\mathcal{N} = 4$ SYM.

    The following reviews cover various aspects of on-shell methods (the most recent reviews are listed first):
    "A brief introduction to modern amplitude methods" [11].
    "Scattering amplitudes: the most perfect microscopic structures in the universe" [4].
    "Hidden simplicity of gauge theory amplitudes" [52].
    "A first course on twistors, integrability and gluon scattering amplitudes" [279].
    "On-shell methods in perturbative QCD" [280].
    "Calculating scattering amplitudes efficiently" [3].
    "Multiparton amplitudes in gauge theories" [1].
- *Numeric methods and applications in phenomenology*
    "Susy theories and QCD: Numerical approaches" [70].
    "One-loop calculations in quantum field theory: from Feynman diagrams to unitarity cuts" [77].
    "Simplifying multi-jet QCD computation" [281].
    "Loop amplitudes in gauge theories: Modern analytic approaches" [71].
- *Gravity*
    "Introduction to the effective field theory description of gravity" [282].
    "Perturbative quantum gravity and its relation to gauge theory" [188].
    "Ultraviolet behavior of $\mathcal{N} = 8$ supergravity" [283].

Less supersymmetry: $1 \leq \mathcal{N} < 4$

An obvious arena for on-shell methods is theories with non-maximal supersymmetry. Scattering amplitudes in $\mathcal{N} = 1$ supersymmetric theories have been of particular interest in particle phenomenology and on-shell methods were used in such studies [48, 67, 284–286]. The superamplitude and on-shell superspace formalism generalizes to $1 \leq \mathcal{N} < 4$ SYM [87] and has also been used for SYM coupled with matter [85, 287]. It is expected that the on-shell diagram approach to planar amplitudes of $\mathcal{N} = 4$ SYM has a natural extension to $\mathcal{N} < 4$ theories [120].

Amplitudes with massive particles

We have focused on amplitudes with massless particles, but there are also efficient on-shell methods available for amplitudes involving massive particles. The spinor helicity formalism for massless particles, introduced in Section 2, can be generalized to massive particles. There are two approaches to this. In the first, one studies the eigenvectors of the momentum matrix $p_{\alpha\dot\beta} = p_\mu \sigma^\mu_{\alpha\dot\beta}$ to directly get solutions to the massive Dirac equation. In the second approach, the time-like momentum p_i is decomposed along two light-like directions by introducing a null reference vector q_i for each state: $p_i^\mu = p_{i\perp}^\mu - \frac{m_i^2}{2q_i \cdot p_i} q^\mu$. The familiar spinor helicity formalism can then be used for $q_i = -|q_i\rangle[q_i|$ and $p_{i\perp} = -|i_\perp\rangle[i_\perp|$. Helicity is only a Lorentz-invariant quantity for massless particles, but q_i breaks Lorentz-invariance and can therefore be used to define a helicity basis in which we can calculate helicity amplitudes. For an introduction to both approaches, see [288]. The papers [135, 289, 290] studied applications of BCFW recursion relations with massive particles. A recent discussion of the latter approach, as well as applications to CSW-like recursion relations, was given in [26]. Finally, let us mention that there are simple amplitudes with a pair of massive particles. Examples of such "towers" of amplitudes – in a sense massive versions of Parke–Taylor – were presented for scattering processes on the Coulomb branch of $\mathcal{N} = 4$ SYM in [140, 203] (see also [291–293]).

Extensions of recursion relations

Attempts have been made to generalize various forms of recursion relations beyond gauge theory and gravity, for example to string theory amplitudes [294, 295] and to non-linear sigma models [6, 296]. Recursion relations can also be utilized to obtain rational functions that appear at 1-loop [297]. A review of various recursion relations is given in [298].

Triality: Wilson loop, correlation function, amplitude

In Section 5.3 we discussed the emergence of dual superconformal symmetry in planar $\mathcal{N} = 4$ SYM. It states that the scattering amplitude in the dual coordinates is superconformal covariant. Could we define $\mathcal{N} = 4$ SYM directly in these new coordinates? In dual coordinates, the kinematic setup for the on-shell momenta is a polygon with null edges. A similar physical quantity is a Wilson-loop specialized to a null-polygon contour. So could the amplitude in momentum space be dual to a null-polygon Wilson-loop in the dual space? Indeed it is. This duality has been established at strong coupling by Alday and Maldacena [107] as well as at weak coupling by Drummond, Korchemsky, and Sokatchev [299].

At strong coupling, the duality can be understood as a consequence of T-duality in string theory [300, 301]. At weak coupling, evidence for such duality was first reported by [299, 302]. It was later proven by defining the action of $\mathcal{N} = 4$ SYM directly in supertwistor space [303]. Remarkably, using the duality, the first computation of the 6-point 2 loop MHV amplitude was done by computing the 6-edged Wilson loop [110]. The amplitude/Wilson-loop duality was first established between the bosonic Wilson-loop and the ratio of the

MHV scattering amplitude, divided by the MHV tree amplitude. It can be generalized to N^KMHV amplitudes by supersymmetrizing the bosonic Wilson-loop [303–305].

Another extension is the realization that the super Wilson-loop in $\mathcal{N} = 4$ SYM is also dual to the correlation function of operators with light-like separation [306]. This can also be proven in supertwistor space [307]. For a review on the (MHV)amplitude/(bosonic)Wilson-loop duality see [308–310], for the general amplitude/super-Wilson loop duality in the framework of supertwistor space, see [311].

Based on an operator-product-expansion approach first developed in the perturbative computation of null Wilson-loops [312], a non-perturbative formulation of the S-matrix/Wilson-loop for planar $\mathcal{N} = 4$ SYM has been proposed in [313–315]. Its perturbative expansion gives important predictions for the explicit loop amplitude, and as a result the integrated functional form of the 6-point 3- and 4-loop MHV amplitude for planar $\mathcal{N} = 4$ SYM [316, 317] was found.

Twistors
Standard reviews of twistor space include [318–320]. For amplitude-friendly reviews, we suggest [321] as well as [54, 60].

What's next?
We hope you have found this book useful. We look forward to seeing what the future will bring for studies of scattering amplitudes.

Appendix Conventions for 4d spinor helicity formalism

The conventions of these notes follow those in Srednicki's QFT textbook [2].

We use a "mostly-plus" metric, $\eta_{\mu\nu} = \text{diag}(-1, +1, +1, +1)$ and define

$$(\sigma^\mu)_{a\dot{b}} = (1, \sigma^i)_{a\dot{b}}, \qquad (\bar{\sigma}^\mu)^{\dot{a}b} = (1, -\sigma^i)^{\dot{a}b}, \tag{A.1}$$

with Pauli matrices

$$\sigma^1 = \begin{pmatrix} 0 & 1 \\ 1 & 0 \end{pmatrix}, \quad \sigma^2 = \begin{pmatrix} 0 & -i \\ i & 0 \end{pmatrix}, \quad \sigma^3 = \begin{pmatrix} 1 & 0 \\ 0 & -1 \end{pmatrix}. \tag{A.2}$$

Two-index spinor indices are raised/lowered using

$$\varepsilon^{ab} = \varepsilon^{\dot{a}\dot{b}} = \begin{pmatrix} 0 & 1 \\ -1 & 0 \end{pmatrix} = -\varepsilon_{ab} = -\varepsilon_{\dot{a}\dot{b}}, \tag{A.3}$$

which obey $\varepsilon_{ab}\varepsilon^{bc} = \delta_a{}^c$.

We list the following properties

$$(\bar{\sigma}^\mu)^{\dot{a}a} = \varepsilon^{ab}\varepsilon^{\dot{a}\dot{b}}(\sigma^\mu)_{b\dot{b}}, \tag{A.4}$$

$$(\sigma^\mu)_{a\dot{a}}(\sigma_\mu)_{b\dot{b}} = -2\varepsilon_{ab}\varepsilon_{\dot{a}\dot{b}}, \tag{A.5}$$

$$\left(\sigma^\mu\bar{\sigma}^\nu + \sigma^\nu\bar{\sigma}^\mu\right)_a{}^b = -2\eta^{\mu\nu}\delta_a{}^b, \tag{A.6}$$

$$\text{Tr}(\sigma^\mu\bar{\sigma}^\nu) = \text{Tr}(\bar{\sigma}^\mu\sigma^\nu) = -2\eta^{\mu\nu}. \tag{A.7}$$

Define γ-matrices:

$$\gamma^\mu = \begin{pmatrix} 0 & (\sigma^\mu)_{a\dot{b}} \\ (\bar{\sigma}^\mu)^{\dot{a}b} & 0 \end{pmatrix}, \qquad \{\gamma^\mu, \gamma^\nu\} = -2\eta^{\mu\nu}, \tag{A.8}$$

and

$$\gamma_5 \equiv i\gamma^0\gamma^1\gamma^2\gamma^3 = \begin{pmatrix} -1 & 0 \\ 0 & 1 \end{pmatrix}, \qquad P_L = \frac{1}{2}(1 - \gamma_5), \quad P_R = \frac{1}{2}(1 + \gamma_5). \tag{A.9}$$

For a momentum 4-vector $p^\mu = (p^0, p^i) = (E, p^i)$ with $p^\mu p_\mu = -m^2$, we define momentum bi-spinors

$$p_{a\dot{b}} \equiv p_\mu (\sigma^\mu)_{a\dot{b}}, \qquad p^{\dot{a}b} \equiv p_\mu (\bar{\sigma}^\mu)^{\dot{a}b}. \tag{A.10}$$

For example,

$$p_{a\dot{b}} = \begin{pmatrix} -p^0 + p^3 & p^1 - ip^2 \\ p^1 + ip^2 & -p^0 - p^3 \end{pmatrix}. \tag{A.11}$$

Taking the determinant of this 2×2 matrix gives

$$\det p = -p^\mu p_\mu = m^2 \,. \tag{A.12}$$

The Dirac conjugate $\overline{\Psi}$ is defined as

$$\overline{\Psi} = \Psi^\dagger \beta \,, \qquad \beta = \begin{pmatrix} 0 & \delta^{\dot{a}}{}_{\dot{b}} \\ \delta_a{}^b & 0 \end{pmatrix} \,. \tag{A.13}$$

The 4×4 matrix β is the same as γ^0 but has a different index structure.

For convenience, we collect here some useful spinor helicity identities

$$
\begin{aligned}
[p|^a = \epsilon^{ab} |p]_b \,, & \qquad |p]_a = \epsilon_{ab} [p|^b \,, \\
|p\rangle^{\dot{a}} = \epsilon^{\dot{a}\dot{b}} \langle p|_{\dot{b}} & \qquad \langle p|_{\dot{a}} = \epsilon_{\dot{a}\dot{b}} |p\rangle^{\dot{b}} \,, \\
p_{a\dot{b}} = -|p]_a \langle p|_{\dot{b}} \,, & \qquad p^{\dot{a}b} = -|p\rangle^{\dot{a}} [p|^b \,, \\
\langle p\,q \rangle = \langle p|_{\dot{a}} |q\rangle^{\dot{a}} \,, & \qquad [p\,q] = [p|^a |q]_a \,, \\
\langle pq \rangle [pq] = 2\,p \cdot q = (p+q)^2 \,, & \\
[k|\gamma^\mu|p\rangle = \langle p|\gamma^\mu|k] \,, & \qquad [k|\gamma^\mu|p\rangle^* = [p|\gamma^\mu|k\rangle \quad \text{for real momenta} \,, \\
\langle p|P|k] = \langle p|_{\dot{a}} P^{\dot{a}b} |k]_b \,, & \qquad \langle p|y_1.y_2|k\rangle = \langle p|_{\dot{a}} (y_1)^{\dot{a}b} (y_2)_{b\dot{c}} |k\rangle^{\dot{c}} \,, \\
\langle p|q|k] = -\langle pq \rangle [qk] \,, & \qquad \langle 1|\gamma^\mu|2] \langle 3|\gamma_\mu|4] = 2\langle 13 \rangle [24] \,.
\end{aligned}
\tag{A.14}
$$

We also use the analytic continuation

$$|-p\rangle = -|p\rangle \,, \qquad |-p] = +|p] \,. \tag{A.15}$$

The angle and square spinor are related by complex conjugation:

$$[p|^a = \left(|p\rangle^{\dot{a}} \right)^* \,, \qquad \langle p|_{\dot{a}} = \left(|p]_a \right)^* \,, \tag{A.16}$$

if the momentum p^μ is real.

These identities are used in multiple places in the text and exercises.

References

[1] M. L. Mangano and S. J. Parke, "Multiparton amplitudes in gauge theories," *Phys. Rept.* **200**, 301 (1991) [hep-th/0509223].

[2] M. Srednicki, *Quantum Field Theory*, Cambridge, UK: Cambridge University Press (2007).

[3] L. J. Dixon, "Calculating scattering amplitudes efficiently," In *Boulder 1995, QCD and beyond*, 539–582 [hep-ph/9601359].

[4] L. J. Dixon, "Scattering amplitudes: the most perfect microscopic structures in the universe," *J. Phys.* A **44**, 454001 (2011) [arXiv:1105.0771 [hep-th]].

[5] S. J. Parke and T. R. Taylor, "An amplitude for n gluon scattering," *Phys. Rev. Lett.* **56**, 2459 (1986).

[6] K. Kampf, J. Novotny, and J. Trnka, "Tree-level amplitudes in the nonlinear sigma model," *JHEP* **1305**, 032 (2013) [arXiv:1304.3048 [hep-th]].

[7] R. Kleiss and H. Kuijf, "Multi-gluon cross-sections and five jet production at hadron colliders," *Nucl. Phys.* B **312**, 616 (1989).

[8] V. Del Duca, L. J. Dixon, and F. Maltoni, "New color decompositions for gauge amplitudes at tree and loop level," *Nucl. Phys.* B **571**, 51 (2000) [hep-ph/9910563].

[9] Z. Bern, J. J. M. Carrasco, and H. Johansson, "New relations for gauge-theory amplitudes," *Phys. Rev. D* **78**, 085011 (2008) [arXiv:0805.3993 [hep-ph]].

[10] C. F. Berger, V. Del Duca, and L. J. Dixon, "Recursive construction of Higgs-plus-multiparton loop amplitudes: The last of the phi-nite loop amplitudes," *Phys. Rev. D* **74**, 094021 (2006) [Erratum-*ibid.* D **76**, 099901 (2007)] [hep-ph/0608180];
S. D. Badger, E. W. N. Glover, and K. Risager, "One-loop phi-MHV amplitudes using the unitarity bootstrap," *JHEP* **0707**, 066 (2007) [arXiv:0704.3914 [hep-ph]];
L. J. Dixon and Y. Sofianatos, "Analytic one-loop amplitudes for a Higgs boson plus four partons," *JHEP* **0908**, 058 (2009) [arXiv:0906.0008 [hep-ph]];
S. Badger, E. W. Nigel Glover, P. Mastrolia, and C. Williams, "One-loop Higgs plus four gluon amplitudes: Full analytic results," *JHEP* **1001**, 036 (2010) [arXiv:0909.4475 [hep-ph]].

[11] L. J. Dixon, "A brief introduction to modern amplitude methods," [arXiv:1310.5353 [hep-ph]].

[12] F. A. Berends and W. T. Giele, "Recursive calculations for processes with n gluons," *Nucl. Phys.* B **306**, 759 (1988).
F. A. Berends, W. T. Giele, and H. Kuijf, "Exact and approximate expressions for multi-gluon scattering," *Nucl. Phys.* B **333**, 120 (1990).

[13] R. Britto, F. Cachazo, and B. Feng, "New recursion relations for tree amplitudes of gluons," *Nucl. Phys.* B **715**, 499 (2005) [hep-th/0412308].

[14] R. Britto, F. Cachazo, B. Feng, and E. Witten, "Direct proof of tree-level recursion relation in Yang–Mills theory," *Phys. Rev. Lett.* **94**, 181602 (2005) [hep-th/0501052].

[15] F. Cachazo, P. Svrcek, and E. Witten, "MHV vertices and tree amplitudes in gauge theory," *JHEP* **0409**, 006 (2004) [hep-th/0403047].

[16] B. Feng, J. Wang, Y. Wang, and Z. Zhang, "BCFW recursion relation with nonzero boundary contribution," *JHEP* **1001**, 019 (2010) [arXiv:0911.0301 [hep-th]].

[17] E. Conde and S. Rajabi, "The twelve-graviton next-to-MHV amplitude from Risager's construction," *JHEP* **1209**, 120 (2012) [arXiv:1205.3500 [hep-th]].

[18] N. Arkani-Hamed and J. Kaplan, "On tree amplitudes in gauge theory and gravity," *JHEP* **0804**, 076 (2008) [arXiv:0801.2385 [hep-th]].

[19] C. Cheung, "On-shell recursion relations for generic theories," *JHEP* **1003**, 098 (2010) [arXiv:0808.0504 [hep-th]].

[20] P. Benincasa, C. Boucher-Veronneau, and F. Cachazo, "Taming tree amplitudes in general relativity," *JHEP* **0711**, 057 (2007) [hep-th/0702032 [HEP-TH]].

[21] H. Kawai, D. C. Lewellen, and S. H. H. Tye, "A relation between tree amplitudes of closed and open strings," *Nucl. Phys. B* **269**, 1 (1986).

[22] S. Sannan, "Gravity as the limit of the type II superstring theory," *Phys. Rev. D* **34**, 1749 (1986).

[23] N. Arkani-Hamed, F. Cachazo, C. Cheung, and J. Kaplan, "A duality for the S matrix," *JHEP* **1003**, 020 (2010) [arXiv:0907.5418 [hep-th]].

[24] A. Hodges, "Eliminating spurious poles from gauge-theoretic amplitudes," *JHEP* **1305**, 135 (2013) [arXiv:0905.1473 [hep-th]].

[25] M. Spradlin, A. Volovich, and C. Wen, "Three applications of a bonus relation for gravity amplitudes," *Phys. Lett. B* **674**, 69 (2009) [arXiv:0812.4767 [hep-th]].

[26] T. Cohen, H. Elvang, and M. Kiermaier, "On-shell constructibility of tree amplitudes in general field theories," *JHEP* **1104**, 053 (2011) [arXiv:1010.0257 [hep-th]].

[27] H. Elvang, D. Z. Freedman, and M. Kiermaier, "Recursion relations, generating functions, and unitarity sums in N = 4 SYM theory," *JHEP* **0904**, 009 (2009) [arXiv:0808.1720 [hep-th]].

[28] K. Risager, "A direct proof of the CSW rules," *JHEP* **0512**, 003 (2005) [hep-th/0508206].

[29] H. Elvang, D. Z. Freedman, and M. Kiermaier, "Proof of the MHV vertex expansion for all tree amplitudes in N = 4 SYM theory," *JHEP* **0906**, 068 (2009) [arXiv:0811.3624 [hep-th]].

[30] L. J. Dixon, E. W. N. Glover, and V. V. Khoze, "MHV rules for Higgs plus multi-gluon amplitudes," *JHEP* **0412**, 015 (2004) [hep-th/0411092].

[31] S. D. Badger, E. W. N. Glover, and V. V. Khoze, "MHV rules for Higgs plus multi-parton amplitudes," *JHEP* **0503**, 023 (2005) [hep-th/0412275].

[32] A. Brandhuber, B. Spence, and G. Travaglini, "Tree-level formalism," *J. Phys. A* **44**, 454002 (2011) [arXiv:1103.3477 [hep-th]].

[33] A. Gorsky and A. Rosly, "From Yang–Mills Lagrangian to MHV diagrams," *JHEP* **0601**, 101 (2006) [hep-th/0510111];
P. Mansfield, "The Lagrangian origin of MHV rules," *JHEP* **0603**, 037 (2006) [hep-th/0511264];

J. H. Ettle and T. R. Morris, "Structure of the MHV-rules Lagrangian," *JHEP* **0608**, 003 (2006) [hep-th/0605121];

J. H. Ettle, C.-H. Fu, J. P. Fudger, P. R. W. Mansfield, and T. R. Morris, "S-matrix equivalence theorem evasion and dimensional regularisation with the canonical MHV Lagrangian," *JHEP* **0705**, 011 (2007) [hep-th/0703286];

H. Feng and Y.-t. Huang, "MHV Lagrangian for N = 4 super Yang–Mills," *JHEP* **0904**, 047 (2009) [hep-th/0611164].

[34] R. Boels, L. J. Mason, and D. Skinner, "From twistor actions to MHV diagrams," *Phys. Lett. B* **648**, 90 (2007) [hep-th/0702035].

[35] N. E. J. Bjerrum-Bohr, D. C. Dunbar, H. Ita, W. B. Perkins, and K. Risager, "MHV-vertices for gravity amplitudes," *JHEP* **0601**, 009 (2006) [hep-th/0509016].

[36] M. Bianchi, H. Elvang, and D. Z. Freedman, "Generating tree amplitudes in N = 4 SYM and N = 8 SG," *JHEP* **0809**, 063 (2008) [arXiv:0805.0757 [hep-th]].

[37] J. Wess and J. Bagger, *Supersymmetry and Supergravity*, Princeton, USA: University Press (1992).

[38] M. T. Grisaru and H. N. Pendleton, "Some properties of scattering amplitudes in supersymmetric theories," *Nucl. Phys. B* **124**, 81 (1977);

M. T. Grisaru, H. N. Pendleton, and P. van Nieuwenhuizen, "Supergravity and the S matrix," *Phys. Rev. D* **15**, 996 (1977).

[39] D. Z. Freedman and A. Van Proeyen, Supergravity, Cambridge, UK: Cambridge University Press (2012).

[40] L. Brink, J. H. Schwarz, and J. Scherk, "Supersymmetric Yang–Mills theories," *Nucl. Phys. B* **121**, 77 (1977).

[41] A. Ferber, "Supertwistors and conformal supersymmetry," *Nucl. Phys. B* **132**, 55 (1978).

[42] H. Elvang, D. Z. Freedman, and M. Kiermaier, "Solution to the Ward identities for superamplitudes," *JHEP* **1010**, 103 (2010) [arXiv:0911.3169 [hep-th]].

[43] H. Elvang, D. Z. Freedman, and M. Kiermaier, "SUSY Ward identities, super-amplitudes, and counterterms," *J. Phys. A* **44**, 454009 (2011) [arXiv:1012.3401 [hep-th]].

[44] M. Kiermaier and S. G. Naculich, "A super MHV vertex expansion for N = 4 SYM theory," *JHEP* **0905**, 072 (2009) [arXiv:0903.0377 [hep-th]].

[45] N. Arkani-Hamed, "What is the simplest QFT?," talk given at the Paris Workshop *Wonders of Gauge Theory and Supergravity*, 24 June 2008.

[46] A. Brandhuber, P. Heslop, and G. Travaglini, "A note on dual superconformal symmetry of the N = 4 super Yang–Mills S-matrix," *Phys. Rev. D* **78**, 125005 (2008) [arXiv:0807.4097 [hep-th]].

[47] N. Arkani-Hamed, F. Cachazo, and J. Kaplan, "What is the simplest quantum field theory?," *JHEP* **1009**, 016 (2010) [arXiv:0808.1446 [hep-th]].

[48] Z. Bern, J. J. M. Carrasco, H. Ita, H. Johansson, and R. Roiban, "On the structure of supersymmetric sums in multi-loop unitarity cuts," *Phys. Rev. D* **80**, 065029 (2009) [arXiv:0903.5348 [hep-th]].

[49] J. M. Drummond and J. M. Henn, "All tree-level amplitudes in N = 4 SYM," *JHEP* **0904**, 018 (2009) [arXiv:0808.2475 [hep-th]].

[50] J. M. Drummond, J. Henn, G. P. Korchemsky, and E. Sokatchev, "Dual superconformal symmetry of scattering amplitudes in N = 4 super-Yang–Mills theory," *Nucl. Phys. B* **828**, 317 (2010) [arXiv:0807.1095 [hep-th]].

[51] J. M. Drummond, M. Spradlin, A. Volovich, and C. Wen, "Tree-level amplitudes in N = 8 supergravity," *Phys. Rev. D* **79**, 105018 (2009) [arXiv:0901.2363 [hep-th]].

[52] J. M. Drummond, "Hidden simplicity of gauge theory amplitudes," *Class. Quant. Grav.* **27**, 214001 (2010) [arXiv:1010.2418 [hep-th]].

[53] R. Penrose, "Twistor algebra," *J. Math. Phys.* **8**, 345 (1967).

[54] E. Witten, "Perturbative gauge theory as a string theory in twistor space," *Commun. Math. Phys.* **252**, 189 (2004) [hep-th/0312171].

[55] P. A. M. Dirac, "Wave equations in conformal space," *Annals Math.* **37**, 429 (1936).

[56] W. Siegel, "Embedding versus 6D twistors," [arXiv:1204.5679 [hep-th]].

[57] W. Siegel, "Fields," [hep-th/9912205].

[58] J. M. Drummond, J. Henn, V. A. Smirnov, and E. Sokatchev, "Magic identities for conformal four-point integrals," *JHEP* **0701**, 064 (2007) [hep-th/0607160].

[59] J. M. Drummond, J. M. Henn, and J. Plefka, "Yangian symmetry of scattering amplitudes in N = 4 super Yang–Mills theory," *JHEP* **0905**, 046 (2009) [arXiv:0902.2987 [hep-th]].

[60] L. J. Mason and D. Skinner, "Dual superconformal invariance, momentum twistors and grassmannians," *JHEP* **0911**, 045 (2009) [arXiv:0909.0250 [hep-th]].
N. Arkani-Hamed, F. Cachazo, and C. Cheung, "The Grassmannian origin of dual superconformal invariance," *JHEP* **10033**, 036 (2010) [arXiv: 0909.0483 [hep-th]].
H. Elvang, Y.-t. Huang, C. Keeler, *et al.*, "Grassmannians for scattering amplitudes in 4d N =4 SYM and 3d ABJM," 2014 [arXiv: 1410.0621 [hep.th]].

[61] R. Roiban, "Review of AdS/CFT integrability, Chapter V.1: Scattering amplitudes – a brief introduction," *Lett. Math. Phys.* **99**, 455 (2012) [arXiv:1012.4001 [hep-th]].

[62] Lorenzo Magnea, *Lecture notes on Perturbative QCD* at the National School of Theoretical Physics of the University of Parma (2008). http://personalpages.to.infn.it/~magnea/QCD.pdf

[63] L. V. Bork, D. I. Kazakov, G. S. Vartanov, and A. V. Zhiboedov, "Construction of infrared finite observables in N = 4 super Yang–Mills theory," *Phys. Rev. D* **81**, 105028 (2010) [arXiv:0911.1617 [hep-th]].

[64] Z. Bern, G. Chalmers, L. J. Dixon, and D. A. Kosower, "One loop N gluon amplitudes with maximal helicity violation via collinear limits," *Phys. Rev. Lett.* **72**, 2134 (1994) [hep-ph/9312333].

[65] G. Mahlon, "Multi-gluon helicity amplitudes involving a quark loop," *Phys. Rev. D* **49**, 4438 (1994) [hep-ph/9312276].

[66] Z. Bern, L. J. Dixon, and D. A. Kosower, "Dimensionally regulated pentagon integrals," *Nucl. Phys. B* **412**, 751 (1994) [hep-ph/9306240].

[67] Z. Bern, L. J. Dixon, D. C. Dunbar, and D. A. Kosower, "One loop n point gauge theory amplitudes, unitarity and collinear limits," *Nucl. Phys. B* **425**, 217 (1994) [hep-ph/9403226];
Z. Bern, L. J. Dixon, D. C. Dunbar, and D. A. Kosower, "Fusing gauge theory tree amplitudes into loop amplitudes," *Nucl. Phys. B* **435**, 59 (1995) [hep-ph/9409265].

[68] J. J. M. Carrasco and H. Johansson, "Generic multiloop methods and application to N = 4 super-Yang–Mills," *J. Phys. A* **44**, 454004 (2011) [arXiv:1103.3298 [hep-th]].

[69] Z. Bern and Y.-t. Huang, "Basics of generalized unitarity," *J. Phys. A* **44**, 454003 (2011) [arXiv:1103.1869 [hep-th]].

[70] H. Ita, "Susy theories and QCD: Numerical approaches," *J. Phys. A* **44**, 454005 (2011) [arXiv:1109.6527 [hep-th]].

[71] R. Britto, "Loop amplitudes in gauge theories: Modern analytic approaches," *J. Phys. A* **44**, 454006 (2011) [arXiv:1012.4493 [hep-th]].

[72] W. L. van Neerven and J. A. M. Vermaseren, "Large loop integrals," *Phys. Lett. B* **137**, 241 (1984).

[73] Z. Bern, L. J. Dixon, and D. A. Kosower, "Dimensionally regulated one loop integrals," *Phys. Lett. B* **302**, 299 (1993) [Erratum-*ibid. B* **318**, 649 (1993)] [hep-ph/9212308].

[74] Z. Bern, L. J. Dixon, and D. A. Kosower, "Dimensionally regulated pentagon integrals," *Nucl. Phys. B* **412**, 751 (1994) [hep-ph/9306240].

[75] L. M. Brown and R. P. Feynman, "Radiative corrections to Compton scattering," *Phys. Rev.* **85**, 231 (1952);
G. 't Hooft and M. J. G. Veltman, "Scalar one loop integrals," *Nucl. Phys. B* **153**, 365 (1979).

[76] G. Passarino and M. J. G. Veltman, "One loop corrections for e+ e− annihilation into mu+ mu− in the Weinberg model," *Nucl. Phys. B* **160**, 151 (1979).

[77] R. K. Ellis, Z. Kunszt, K. Melnikov, and G. Zanderighi, "One-loop calculations in quantum field theory: from Feynman diagrams to unitarity cuts," *Phys. Rept.* **518**, 141 (2012) [arXiv:1105.4319 [hep-ph]].

[78] H. Johansson, D. A. Kosower, and K. J. Larsen, "An overview of maximal unitarity at two loops," *PoS LL* **2012**, 066 (2012) [arXiv:1212.2132 [hep-th]].

[79] C. Anastasiou, R. Britto, B. Feng, Z. Kunszt, and P. Mastrolia, "D-dimensional unitarity cut method," *Phys. Lett. B* **645**, 213 (2007) [hep-ph/0609191];
R. Britto and B. Feng, "Integral coefficients for one-loop amplitudes," *JHEP* **0802**, 095 (2008) [0711.4284 [hep-ph]];
R. Britto and B. Feng, "Unitarity cuts with massive propagators and algebraic expressions for coefficients," *Phys. Rev. D* **75**, 105006 (2007) [hep-ph/0612089];
G. Ossola, C. G. Papadopoulos, and R. Pittau, "Reducing full one-loop amplitudes to scalar integrals at the integrand level," *Nucl. Phys. B* **763**, 147 (2007) [hep-ph/0609007];
R. Britto, B. Feng, and P. Mastrolia, "Closed-form decomposition of one-loop massive amplitudes," *Phys. Rev. D* **78**, 025031 (2008) [0803.1989 [hep-ph]];
D. Forde, "Direct extraction of one-loop integral coefficients," *Phys. Rev. D* **75**, 125019 (2007) [0704.1835 [hep-ph]];

[80] Z. Bern and A. G. Morgan, "Massive loop amplitudes from unitarity," *Nucl. Phys. B* **467**, 479 (1996) [hep-ph/9511336];
Z. Bern, L. J. Dixon, and D. A. Kosower, "Progress in one-loop QCD computations," *Ann. Rev. Nucl. Part. Sci.* **46**, 109 (1996) [hep-ph/9602280].

[81] S. D. Badger, "Direct extraction of one loop rational terms," *JHEP* **0901**, 049 (2009) [0806.4600 [hep-ph]].

[82] M. B. Green, J. H. Schwarz, and L. Brink, "N = 4 Yang–Mills and N = 8 supergravity as limits of string theories," *Nucl. Phys. B* **198**, 474 (1982).

[83] Z. Bern, N. E. J. Bjerrum-Bohr, and D. C. Dunbar, "Inherited twistor-space structure of gravity loop amplitudes," *JHEP* **0505**, 056 (2005) [hep-th/0501137];
N. E. J. Bjerrum-Bohr, D. C. Dunbar, H. Ita, W. B. Perkins, and K. Risager, "The no-triangle hypothesis for N = 8 supergravity," *JHEP* **0612**, 072 (2006) [hep-th/0610043];
N. E. J. Bjerrum-Bohr and P. Vanhove, "Absence of triangles in maximal supergravity amplitudes," *JHEP* **0810**, 006 (2008) [arXiv:0805.3682 [hep-th]].

[84] Z. Bern, J. J. Carrasco, D. Forde, H. Ita, and H. Johansson, "Unexpected cancellations in gravity theories," *Phys. Rev. D* **77**, 025010 (2008) [arXiv:0707.1035 [hep-th]].

[85] S. Lal and S. Raju, "The next-to-simplest quantum field theories," *Phys. Rev. D* **81**, 105002 (2010) [arXiv:0910.0930 [hep-th]].

[86] D. C. Dunbar, J. H. Ettle, and W. B. Perkins, "Perturbative expansion of $N < 8$ supergravity," *Phys. Rev. D* **83**, 065015 (2011) [arXiv:1011.5378 [hep-th]].

[87] H. Elvang, Y.-t. Huang, and C. Peng, "On-shell superamplitudes in $N < 4$ SYM," *JHEP* **1109**, 031 (2011) [arXiv:1102.4843 [hep-th]].

[88] Y.-t. Huang, D. A. McGady, and C. Peng, "One-loop renormalization and the S-matrix," [arXiv:1205.5606 [hep-th]].

[89] N. Marcus, "Composite anomalies in supergravity," *Phys. Lett. B* **157**, 383 (1985).

[90] P. di Vecchia, S. Ferrara, and L. Girardello, "Anomalies of hidden local chiral symmetries in sigma models and extended supergravities," *Phys. Lett. B* **151**, 199 (1985).

[91] J. M. Drummond, J. Henn, G. P. Korchemsky, and E. Sokatchev, "Generalized unitarity for N = 4 super-amplitudes," *Nucl. Phys. B* **869**, 452 (2013) [arXiv:0808.0491 [hep-th]].

[92] A. Brandhuber, P. Heslop, and G. Travaglini, "One-loop amplitudes in N = 4 super Yang–Mills and anomalous dual conformal symmetry," *JHEP* **0908**, 095 (2009) [arXiv:0905.4377 [hep-th]].

[93] H. Elvang, D. Z. Freedman, and M. Kiermaier, "Dual conformal symmetry of 1-loop NMHV amplitudes in N = 4 SYM theory," *JHEP* **1003**, 075 (2010) [arXiv:0905.4379 [hep-th]].

[94] G. P. Korchemsky and E. Sokatchev, "Symmetries and analytic properties of scattering amplitudes in N = 4 SYM theory," *Nucl. Phys. B* **832**, 1 (2010) [arXiv:0906.1737 [hep-th]].

[95] J. Gluza, K. Kajda, and D. A. Kosower, "Towards a basis for planar two-loop integrals," *Phys. Rev. D* **83**, 045012 (2011) [arXiv:1009.0472 [hep-th]].

[96] D. A. Kosower and K. J. Larsen, "Maximal unitarity at two loops," *Phys. Rev. D* **85**, 045017 (2012) [arXiv:1108.1180 [hep-th]];
H. Johansson, D. A. Kosower, and K. J. Larsen, "Two-loop maximal unitarity with external masses," *Phys. Rev. D* **87**, 025030 (2013) [arXiv:1208.1754 [hep-th]].

[97] S. Badger, H. Frellesvig, and Y. Zhang, "Hepta-cuts of two-loop scattering amplitudes," *JHEP* **1204**, 055 (2012) [arXiv:1202.2019 [hep-ph]].

[98] Y. Zhang, "Integrand-level reduction of loop amplitudes by computational algebraic geometry methods," *JHEP* **1209**, 042 (2012) [arXiv:1205.5707 [hep-ph]].

[99] M. Sgaard, "Global residues and two-loop hepta-cuts," *JHEP* **1309**, 116 (2013) [arXiv:1306.1496 [hep-th]].

[100] A. V. Smirnov and A. V. Petukhov, "The number of master integrals is finite," *Lett. Math. Phys.* **97**, 37 (2011) [arXiv:1004.4199 [hep-th]].

[101] C. Anastasiou, Z. Bern, L. J. Dixon, and D. A. Kosower, "Planar amplitudes in maximally supersymmetric Yang–Mills theory," *Phys. Rev. Lett.* **91**, 251602 (2003) [hep-th/0309040].

[102] Z. Bern, L. J. Dixon, and V. A. Smirnov, "Iteration of planar amplitudes in maximally supersymmetric Yang–Mills theory at three loops and beyond," *Phys. Rev. D* **72**, 085001 (2005) [hep-th/0505205].

[103] Z. Bern, J. S. Rozowsky, and B. Yan, "Two loop four gluon amplitudes in N = 4 superYang–Mills," *Phys. Lett. B* **401**, 273 (1997) [hep-ph/9702424].

[104] Z. Bern, M. Czakon, D. A. Kosower, R. Roiban, and V. A. Smirnov, "Two-loop iteration of five-point N = 4 super-Yang–Mills amplitudes," *Phys. Rev. Lett.* **97**, 181601 (2006) [hep-th/0604074].

[105] F. Cachazo, M. Spradlin, and A. Volovich, "Iterative structure within the five-particle two-loop amplitude," *Phys. Rev. D* **74**, 045020 (2006) [hep-th/0602228].

[106] L. F. Alday and J. Maldacena, "Comments on gluon scattering amplitudes via AdS/CFT," *JHEP* **0711**, 068 (2007) [arXiv:0710.1060 [hep-th]].

[107] L. F. Alday and J. M. Maldacena, "Gluon scattering amplitudes at strong coupling," *JHEP* **0706**, 064 (2007) [arXiv:0705.0303 [hep-th]].

[108] Z. Bern, L. J. Dixon, D. A. Kosower, *et al.*, "The two-loop six-gluon MHV amplitude in maximally supersymmetric Yang–Mills theory," *Phys. Rev. D* **78** (2008) 045007 [arXiv:0803.1465 [hep-th]].

[109] F. Cachazo, M. Spradlin, and A. Volovich, "Leading singularities of the two-loop six-particle MHV amplitude," *Phys. Rev. D* **78**, 105022 (2008) [arXiv:0805.4832 [hep-th]].

[110] V. Del Duca, C. Duhr, and V. A. Smirnov, "An analytic result for the two-loop hexagon Wilson loop in N = 4 SYM," *JHEP* **1003**, 099 (2010) [arXiv:0911.5332 [hep-ph]].

[111] V. Del Duca, C. Duhr, and V. A. Smirnov, "The two-loop hexagon Wilson loop in N = 4 SYM," *JHEP* **1005**, 084 (2010) [arXiv:1003.1702 [hep-th]].

[112] A. B. Goncharov, M. Spradlin, C. Vergu, and A. Volovich, "Classical polylogarithms for amplitudes and Wilson loops," *Phys. Rev. Lett.* **105**, 151605 (2010) [arXiv:1006.5703 [hep-th]].

[113] E. W. Nigel Glover and C. Williams, "One-loop gluonic amplitudes from single unitarity cuts," *JHEP* **0812**, 067 (2008) [0810.2964 [hep-th]];
I. Bierenbaum, S. Catani, P. Draggiotis, and G. Rodrigo, "A tree-loop duality relation at two loops and beyond," *JHEP* **1010**, 073 (2010) [arXiv:1007.0194 [hep-ph]];
H. Elvang, D. Z. Freedman, and M. Kiermaier, "Integrands for QCD rational terms and N = 4 SYM from massive CSW rules," *JHEP* **1206**, 015 (2012) [arXiv:1111.0635 [hep-th]].

[114] S. Caron-Huot, "Loops and trees," *JHEP* **1105**, 080 (2011) [arXiv:1007.3224 [hep-ph]].

[115] N. Arkani-Hamed, J. L. Bourjaily, F. Cachazo, S. Caron-Huot, and J. Trnka, "The all-loop integrand for scattering amplitudes in planar N = 4 SYM," *JHEP* **1101**, 041 (2011) [arXiv:1008.2958 [hep-th]].

[116] R. H. Boels, "On BCFW shifts of integrands and integrals," *JHEP* **1011**, 113 (2010) [arXiv:1008.3101 [hep-th]].

[117] Z. Bern, J. J. M. Carrasco, H. Johansson, and D. A. Kosower, "Maximally supersymmetric planar Yang–Mills amplitudes at five loops," *Phys. Rev. D* **76**, 125020 (2007) [0705.1864 [hep-th]].

[118] R. Britto, F. Cachazo, and B. Feng, "Generalized unitarity and one-loop amplitudes in N = 4 super-Yang–Mills," *Nucl. Phys. B* **725**, 275 (2005) [hep-th/0412103];
E. I. Buchbinder and F. Cachazo, "Two-loop amplitudes of gluons and octa-cuts in N = 4 super Yang–Mills," *JHEP* **0511**, 036 (2005) [hep-th/0506126].

[119] N. Arkani-Hamed, J. L. Bourjaily, F. Cachazo, and J. Trnka, "Local integrals for planar scattering amplitudes," *JHEP* **1206**, 125 (2012) [arXiv:1012.6032 [hep-th]].
For extensions, see J. L. Bourjaily, S. Caron-Huot, and J. Trnka, "Dual-conformal regularization of infrared loop divergences and the chiral box expansion," [arXiv:1303.4734 [hep-th]].

[120] N. Arkani-Hamed, J. L. Bourjaily, F. Cachazo, *et al.*, "Scattering amplitudes and the positive Grassmannian," [arXiv:1212.5605 [hep-th]].

[121] F. Cachazo, "Sharpening the leading singularity," [arXiv:0803.1988 [hep-th]].

[122] J. M. Drummond and L. Ferro, "Yangians, Grassmannians and T-duality," *JHEP* **1007**, 027 (2010) [arXiv:1001.3348 [hep-th]].

[123] J. M. Drummond and L. Ferro, "The Yangian origin of the Grassmannian integral," *JHEP* **1012**, 010 (2010) [arXiv:1002.4622 [hep-th]];
G. P. Korchemsky and E. Sokatchev, "Superconformal invariants for scattering amplitudes in N = 4 SYM theory," *Nucl. Phys. B* **839**, 377 (2010) [arXiv:1002.4625 [hep-th]].

[124] R. Roiban, M. Spradlin, and A. Volovich, "On the tree level S matrix of Yang–Mills theory," *Phys. Rev. D* **70**, 026009 (2004) [hep-th/0403190].

[125] M. Spradlin and A. Volovich, "From twistor string theory to recursion relations," *Phys. Rev. D* **80**, 085022 (2009) [arXiv:0909.0229 [hep-th]].

[126] N. Arkani-Hamed, J. Bourjaily, F. Cachazo, and J. Trnka, "Unification of residues and Grassmannian dualities," *JHEP* **1101**, 049 (2011) [arXiv:0912.4912 [hep-th]].

[127] J. L. Bourjaily, J. Trnka, A. Volovich, and C. Wen, "The Grassmannian and the twistor string: Connecting all trees in N = 4 SYM," *JHEP* **1101**, 038 (2011) [arXiv:1006.1899 [hep-th]].

[128] M. Bullimore, L. J. Mason, and D. Skinner, "Twistor-strings, Grassmannians and leading singularities," *JHEP* **1003**, 070 (2010) [arXiv:0912.0539 [hep-th]].

[129] L. Dolan and P. Goddard, "Complete equivalence between gluon tree amplitudes in twistor string theory and in gauge theory," *JHEP* **1206**, 030 (2012) [arXiv:1111.0950 [hep-th]].

[130] N. Arkani-Hamed, J. L. Bourjaily, F. Cachazo, A. Hodges, and J. Trnka, "A note on polytopes for scattering amplitudes," *JHEP* **1204**, 081 (2012) [arXiv:1012.6030 [hep-th]].

[131] M. Bullimore, L. J. Mason, and D. Skinner, "MHV diagrams in momentum twistor space," *JHEP* **1012**, 032 (2010) [arXiv:1009.1854 [hep-th]].

[132] N. Arkani-Hamed and J. Trnka, "The amplituhedron," [arXiv:1312.2007 [hep-th]]; N. Arkani-Hamed and J. Trnka, "Into the amplituhedron," [arXiv:1312.7878 [hep-th]].

[133] L. Mason and D. Skinner, "Amplitudes at weak coupling as polytopes in AdS_5," *J. Phys. A* **44**, 135401 (2011) [arXiv:1004.3498 [hep-th]].

[134] H. Nastase and H. J. Schnitzer, "Twistor and polytope interpretations for subleading color one-loop amplitudes," *Nucl. Phys. B* **855**, 901 (2012) [arXiv:1104.2752 [hep-th]].

[135] R. Boels, "Covariant representation theory of the Poincaré algebra and some of its extensions," *JHEP* **1001**, 010 (2010) [arXiv:0908.0738 [hep-th]].

[136] S. Caron-Huot and D. O'Connell, "Spinor helicity and dual conformal symmetry in ten dimensions," *JHEP* **1108**, 014 (2011) [arXiv:1010.5487 [hep-th]].

[137] R. H. Boels and D. O'Connell, "Simple superamplitudes in higher dimensions," *JHEP* **1206**, 163 (2012) [arXiv:1201.2653 [hep-th]].

[138] S. Davies, "One-loop QCD and Higgs to partons processes using six-dimensional helicity and generalized unitarity," *Phys. Rev. D* **84**, 094016 (2011) [arXiv:1108.0398 [hep-ph]].

[139] Z. Bern, J. J. Carrasco, T. Dennen, Y.-t. Huang, and H. Ita, "Generalized unitarity and six-dimensional helicity," *Phys. Rev. D* **83**, 085022 (2011) [arXiv:1010.0494 [hep-th]].

[140] N. Craig, H. Elvang, M. Kiermaier, and T. Slatyer, "Massive amplitudes on the Coulomb branch of N = 4 SYM," *JHEP* **1112**, 097 (2011) [arXiv:1104.2050 [hep-th]].

[141] C. Cheung and D. O'Connell, "Amplitudes and spinor-helicity in six dimensions," *JHEP* **0907**, 075 (2009) [arXiv:0902.0981 [hep-th]].

[142] T. Dennen, Y.-t. Huang, and W. Siegel, "Supertwistor space for 6D maximal super Yang–Mills," *JHEP* **1004**, 127 (2010) [arXiv:0910.2688 [hep-th]].

[143] A. Brandhuber, D. Korres, D. Koschade, and G. Travaglini, "One-loop amplitudes in six-dimensional (1,1) theories from generalised unitarity," *JHEP* **1102**, 077 (2011) [arXiv:1010.1515 [hep-th]]; C. Saemann, R. Wimmer, and M. Wolf, "A twistor description of six-dimensional N = (1,1) super Yang–Mills theory," *JHEP* **1205**, 020 (2012) [arXiv:1201.6285 [hep-th]].

[144] T. Dennen and Y.-t. Huang, "Dual conformal properties of six-dimensional maximal super Yang–Mills amplitudes," *JHEP* **1101**, 140 (2011) [arXiv:1010.5874 [hep-th]].

[145] T. Chern, "Superconformal field theory in six dimensions and supertwistor," [arXiv:0906.0657 [hep-th]]; M. Chiodaroli, M. Gunaydin, and R. Roiban, "Superconformal symmetry, and maximal supergravity in various dimensions," *JHEP* **1203**, 093 (2012) [arXiv:1108.3085 [hep-th]]; L. J. Mason, R. A. Reid-Edwards, and A. Taghavi-Chabert, "Conformal field theories in six-dimensional twistor space," *J. Geom. Phys.* **62**, 2353 (2012) [arXiv:1111.2585 [hep-th]];

C. Saemann and M. Wolf, "On twistors and conformal field theories from six dimensions," *J. Math. Phys.* **54**, 013507 (2013) [arXiv:1111.2539 [hep-th]].

[146] B. Czech, Y.-t. Huang, and M. Rozali, "Amplitudes for multiple M5 branes," *JHEP* **1210**, 143 (2012) [arXiv:1110.2791 [hep-th]].

[147] C. Saemann and M. Wolf, "Non-Abelian tensor multiplet equations from twistor space," [arXiv:1205.3108 [hep-th]].

[148] Y.-t. Huang and A. E. Lipstein, "Amplitudes of 3D and 6D maximal superconformal theories in supertwistor space," *JHEP* **1010**, 007 (2010) [arXiv:1004.4735 [hep-th]].

[149] L. F. Alday, J. M. Henn, J. Plefka, and T. Schuster, "Scattering into the fifth dimension of N = 4 super Yang–Mills," *JHEP* **1001**, 077 (2010) [arXiv:0908.0684 [hep-th]].

[150] A. Agarwal, N. Beisert, and T. McLoughlin, "Scattering in mass-deformed $N \geq 4$ Chern–Simons models," *JHEP* **0906**, 045 (2009) [arXiv:0812.3367 [hep-th]].

[151] A. Gustavsson, "Algebraic structures on parallel M2-branes," *Nucl. Phys. B* **811**, 66 (2009) [arXiv:0709.1260 [hep-th]].

[152] J. Bagger and N. Lambert, "Gauge symmetry and supersymmetry of multiple M2-branes," *Phys. Rev. D* **77**, 065008 (2008) [arXiv:0711.0955 [hep-th]].

[153] M. A. Bandres, A. E. Lipstein, and J. H. Schwarz, "N = 8 superconformal Chern–Simons theories," *JHEP* **0805**, 025 (2008) [arXiv:0803.3242 [hep-th]].

[154] J. Gomis, G. Milanesi, and J. G. Russo, "Bagger-Lambert theory for general Lie algebras," *JHEP* **0806**, 075 (2008) [arXiv:0805.1012 [hep-th]].

[155] S. Benvenuti, D. Rodriguez-Gomez, E. Tonni, and H. Verlinde, "N = 8 superconformal gauge theories and M2 branes," *JHEP* **0901**, 078 (2009) [arXiv:0805.1087 [hep-th]].

[156] P. -M. Ho, Y. Imamura, and Y. Matsuo, "M2 to D2 revisited," *JHEP* **0807**, 003 (2008) [arXiv:0805.1202 [hep-th]].

[157] T. Bargheer, F. Loebbert, and C. Meneghelli, "Symmetries of tree-level scattering amplitudes in N = 6 superconformal Chern–Simons theory," *Phys. Rev. D* **82**, 045016 (2010) [arXiv:1003.6120 [hep-th]].

[158] O. Aharony, O. Bergman, D. L. Jafferis, and J. Maldacena, "N = 6 superconformal Chern–Simons-matter theories, M2-branes and their gravity duals," *JHEP* **0810**, 091 (2008) [arXiv:0806.1218 [hep-th]].

[159] M. Benna, I. Klebanov, T. Klose, and M. Smedback, "Superconformal Chern–Simons theories and AdS(4)/CFT(3) correspondence," *JHEP* **0809**, 072 (2008) [arXiv:0806.1519 [hep-th]].

[160] M. A. Bandres, A. E. Lipstein, and J. H. Schwarz, "Studies of the ABJM theory in a formulation with manifest SU(4) R-symmetry," *JHEP* **0809**, 027 (2008) [arXiv:0807.0880 [hep-th]].

[161] A. Gustavsson, "Selfdual strings and loop space Nahm equations," *JHEP* **0804**, 083 (2008) [arXiv:0802.3456 [hep-th]];
J. Bagger and N. Lambert, "Three-algebras and N = 6 Chern–Simons gauge theories," *Phys. Rev. D* **79**, 025002 (2009) [arXiv:0807.0163 [hep-th]].

[162] D. Gang, Y.-t. Huang, E. Koh, S. Lee, and A. E. Lipstein, "Tree-level recursion relation and dual superconformal symmetry of the ABJM theory," *JHEP* **1103**, 116 (2011) [arXiv:1012.5032 [hep-th]].

[163] Y.-t. Huang and A. E. Lipstein, "Dual superconformal symmetry of N = 6 Chern–Simons theory," *JHEP* **1011**, 076 (2010) [arXiv:1008.0041 [hep-th]].

[164] T. Bargheer, N. Beisert, F. Loebbert, and T. McLoughlin, "Conformal anomaly for amplitudes in $\mathcal{N} = 6$ superconformal Chern–Simons theory," *J. Phys. A* **45**, 475402 (2012) [arXiv:1204.4406 [hep-th]].

[165] M. S. Bianchi, M. Leoni, A. Mauri, S. Penati, and A. Santambrogio, "One loop amplitudes In ABJM," *JHEP* **1207**, 029 (2012) [arXiv:1204.4407 [hep-th]].

[166] A. Brandhuber, G. Travaglini, and C. Wen, "All one-loop amplitudes in N = 6 superconformal Chern–Simons theory," *JHEP* **1210**, 145 (2012) [arXiv:1207.6908 [hep-th]].

[167] W. -M. Chen and Y.-t. Huang, "Dualities for loop amplitudes of N = 6 Chern–Simons matter theory," *JHEP* **1111**, 057 (2011) [arXiv:1107.2710 [hep-th]];

[168] A. Brandhuber, G. Travaglini, and C. Wen, "A note on amplitudes in N = 6 superconformal Chern–Simons theory," *JHEP* **1207**, 160 (2012) [arXiv:1205.6705 [hep-th]].

[169] M. S. Bianchi, M. Leoni, A. Mauri, S. Penati, and A. Santambrogio, "Scattering amplitudes/Wilson loop duality in ABJM theory," *JHEP* **1201**, 056 (2012) [arXiv:1107.3139 [hep-th]].

[170] S. Caron-Huot and Y.-t. Huang, "The two-loop six-point amplitude in ABJM theory," *JHEP* **1303**, 075 (2013) [arXiv:1210.4226 [hep-th]].

[171] S. Lee, "Yangian invariant scattering amplitudes in supersymmetric Chern–Simons theory," *Phys. Rev. Lett.* **105**, 151603 (2010) [arXiv:1007.4772 [hep-th]].

[172] Y. -t. Huang, C. Wen, and D. Xie, "The positive orthogonal Grassmannian and loop amplitudes of ABJM," [arXiv:1402.1479 [hep-th]].

[173] Y.-t. Huang and S. Lee, "A new integral formula for supersymmetric scattering amplitudes in three dimensions," *Phys. Rev. Lett.* **109**, 191601 (2012) [arXiv:1207.4851 [hep-th]].

[174] O. T. Engelund and R. Roiban, "A twistor string for the ABJ(M) theory," [arXiv:1401.6242 [hep-th]].

[175] S. Weinberg, *Gravitation and Cosmology: Principles and Applications of the General Theory of Relativity*, USA: John Wiley & Sons (1972).

[176] R. M. Wald, *General Relativity*, Chicago, USA: University Press (1984).

[177] S. M. Carroll, *Spacetime and Geometry: An Introduction to General Relativity*, San Francisco, USA: Addison-Wesley (2004).

[178] H. Elvang and D. Z. Freedman, unpublished notes (2007).

[179] B. S. DeWitt, "Quantum theory of gravity. 2. The manifestly covariant theory," *Phys. Rev.* **162**, 1195 (1967);

B. S. DeWitt, "Quantum theory of gravity. 3. Applications of the covariant theory," *Phys. Rev.* **162**, 1239 (1967);

M. J. G. Veltman, "Quantum theory of gravitation," *Conf. Proc. C* **7507281**, 265 (1975).

[180] F. A. Berends, W. T. Giele, and H. Kuijf, "On relations between multi-gluon and multigraviton scattering," *Phys. Lett. B* **211**, 91 (1988).

[181] J. Bedford, A. Brandhuber, B. J. Spence, and G. Travaglini, "A recursion relation for gravity amplitudes," *Nucl. Phys. B* **721**, 98 (2005) [hep-th/0502146].

[182] H. Elvang and D. Z. Freedman, "Note on graviton MHV amplitudes," *JHEP* **0805**, 096 (2008) [arXiv:0710.1270 [hep-th]].

[183] D. Nguyen, M. Spradlin, A. Volovich, and C. Wen, "The tree formula for MHV graviton amplitudes," *JHEP* **1007**, 045 (2010) [arXiv:0907.2276 [hep-th]].

[184] Z. Bern, L. J. Dixon, M. Perelstein, and J. S. Rozowsky, "Multileg one loop gravity amplitudes from gauge theory," *Nucl. Phys. B* **546**, 423 (1999) [hep-th/9811140].

[185] Z. Bern and A. K. Grant, "Perturbative gravity from QCD amplitudes," *Phys. Lett. B* **457**, 23 (1999) [hep-th/9904026].

[186] Z. Bern, L. J. Dixon, D. C. Dunbar, *et al.*, "On perturbative gravity and gauge theory," *Nucl. Phys. Proc. Suppl.* **88**, 194 (2000) [hep-th/0002078].

[187] W. Siegel, "Two vierbein formalism for string inspired axionic gravity," *Phys. Rev. D* **47**, 5453 (1993) [hep-th/9302036].

[188] Z. Bern, "Perturbative quantum gravity and its relation to gauge theory," *Living Rev. Rel.* **5**, 5 (2002) [gr-qc/0206071].

[189] D. Z. Freedman, "Some beautiful equations of mathematical physics," In *ICTP (ed.): The Dirac Medals of the ICTP 1993* 25–53, and CERN Geneva - TH.-7367 (94/07,rec.Sep.) [hep-th/9408175].

[190] S. J. Gates, M. T. Grisaru, M. Rocek, and W. Siegel, "Superspace or one thousand and one lessons in supersymmetry," *Front. Phys.* **58**, 1 (1983) [hep-th/0108200].

[191] B. de Wit and D. Z. Freedman, "On SO(8) extended supergravity," *Nucl. Phys. B* **130**, 105 (1977).

[192] E. Cremmer and B. Julia, "The N = 8 supergravity theory. 1. The Lagrangian," *Phys. Lett. B* **80**, 48 (1978);
E. Cremmer and B. Julia, "The SO(8) supergravity," *Nucl. Phys. B* **159**, 141 (1979).

[193] B. de Wit and H. Nicolai, "N = 8 supergravity," *Nucl. Phys. B* **208**, 323 (1982).

[194] A. Hodges, "A simple formula for gravitational MHV amplitudes," [arXiv:1204.1930 [hep-th]].

[195] F. Cachazo, L. Mason, and D. Skinner, "Gravity in twistor space and its Grassmannian formulation," [arXiv:1207.4712 [hep-th]].

[196] S. He, "A link representation for gravity amplitudes," [arXiv:1207.4064 [hep-th]].

[197] F. Cachazo and Y. Geyer, "A 'twistor string' inspired formula for tree-level scattering amplitudes in N = 8 SUGRA," [arXiv:1206.6511 [hep-th]].

[198] D. Skinner, "Twistor strings for N = 8 supergravity," [arXiv:1301.0868 [hep-th]].

[199] F. Cachazo, S. He, and E. Y. Yuan, "Scattering equations and KLT orthogonality," [arXiv:1306.6575 [hep-th]].

[200] F. Cachazo, S. He, and E. Y. Yuan, "Scattering of massless particles in arbitrary dimension," [arXiv:1307.2199 [hep-th]].

[201] S. L. Adler, "Consistency conditions on the strong interactions implied by a partially conserved axial vector current," *Phys. Rev.* **137**, B1022 (1965).

[202] S. R. Coleman, "Secret symmetry: An introduction to spontaneous symmetry breakdown and gauge fields," *Subnucl. Ser.* **11**, 139 (1975).

[203] M. Kiermaier, "The Coulomb-branch S-matrix from massless amplitudes," [arXiv:1105.5385 [hep-th]].

[204] G. 't Hooft and M. J. G. Veltman, "One loop divergencies in the theory of gravitation," *Annales Poincaré Phys. Theor. A* **20**, 69 (1974).

[205] M. H. Goroff and A. Sagnotti, "Quantum gravity at two loops," *Phys. Lett. B* **160**, 81 (1985).

[206] A. E. M. van de Ven, "Two loop quantum gravity," *Nucl. Phys. B* **378**, 309 (1992).

[207] S. Deser and P. van Nieuwenhuizen, "One loop divergences of quantized Einstein–Maxwell fields," *Phys. Rev. D* **10**, 401 (1974).

[208] M. T. Grisaru, P. van Nieuwenhuizen, and J. A. M. Vermaseren, "One loop renormalizability of pure supergravity and of Maxwell–Einstein theory in extended supergravity," *Phys. Rev. Lett.* **37**, 1662 (1976).

[209] M. T. Grisaru, "Two loop renormalizability of supergravity," *Phys. Lett. B* **66**, 75 (1977).

[210] E. Tomboulis, "On the two loop divergences of supersymmetric gravitation," *Phys. Lett. B* **67**, 417 (1977).

[211] S. Deser, J. H. Kay, and K. S. Stelle, "Renormalizability properties of supergravity," *Phys. Rev. Lett.* **38**, 527 (1977).

[212] Z. Bern, L. J. Dixon, and R. Roiban, "Is N = 8 supergravity ultraviolet finite?," *Phys. Lett. B* **644**, 265 (2007) [hep-th/0611086].

[213] Z. Bern, J. J. Carrasco, L. J. Dixon, *et al.*, "Three-loop superfiniteness of N = 8 supergravity," *Phys. Rev. Lett.* **98**, 161303 (2007) [hep-th/0702112].

[214] Z. Bern, J. J. M. Carrasco, L. J. Dixon, H. Johansson, and R. Roiban, "Manifest ultraviolet behavior for the three-loop four-point amplitude of N = 8 supergravity," *Phys. Rev. D* **78**, 105019 (2008) [arXiv:0808.4112 [hep-th]].

[215] P. S. Howe and K. S. Stelle, "Supersymmetry counterterms revisited," *Phys. Lett. B* **554**, 190 (2003) [hep-th/0211279].

[216] Z. Bern, J. J. Carrasco, L. J. Dixon, H. Johansson, and R. Roiban, "The ultraviolet behavior of N = 8 supergravity at four loops," *Phys. Rev. Lett.* **103**, 081301 (2009) [arXiv:0905.2326 [hep-th]].

[217] Z. Bern, J. J. M. Carrasco, L. J. Dixon, H. Johansson, and R. Roiban, "The complete four-loop four-point amplitude in N = 4 super-Yang–Mills theory," *Phys. Rev. D* **82**, 125040 (2010) [arXiv:1008.3327 [hep-th]].

[218] J. Bjornsson and M. B. Green, "5 loops in 24/5 dimensions," *JHEP* **1008**, 132 (2010) [arXiv:1004.2692 [hep-th]].

[219] H. Elvang, D. Z. Freedman, and M. Kiermaier, "A simple approach to counterterms in N = 8 supergravity," *JHEP* **1011**, 016 (2010) [arXiv:1003.5018 [hep-th]].

[220] H. Elvang and M. Kiermaier, "Stringy KLT relations, global symmetries, and $E_{7(7)}$ violation," *JHEP* **1010**, 108 (2010) [arXiv:1007.4813 [hep-th]].

[221] N. Beisert, H. Elvang, D. Z. Freedman, *et al.*, "E7(7) constraints on counterterms in N = 8 supergravity," *Phys. Lett. B* **694**, 265 (2010) [arXiv:1009.1643 [hep-th]].

[222] P. van Nieuwenhuizen and C. C. Wu, "On integral relations for invariants constructed from three riemann tensors and their applications in quantum gravity," *J. Math. Phys.* **18**, 182 (1977).

[223] G. Bossard, C. Hillmann, and H. Nicolai, "E7(7) symmetry in perturbatively quantised N = 8 supergravity," *JHEP* **1012**, 052 (2010) [arXiv:1007.5472 [hep-th]].

[224] D. Z. Freedman and E. Tonni, "The $D^{2k}R^4$ invariants of $N = 8$ supergravity," *JHEP* **1104**, 006 (2011) [arXiv:1101.1672 [hep-th]].

[225] S. Deser and J. H. Kay, "Three loop counterterms for extended supergravity," *Phys. Lett. B* **76**, 400 (1978).

[226] J. M. Drummond, P. J. Heslop, and P. S. Howe, "A note on N = 8 counterterms," [arXiv:1008.4939 [hep-th]].

[227] G. Bossard and H. Nicolai, "Counterterms vs. dualities," *JHEP* **1108**, 074 (2011) [arXiv:1105.1273 [hep-th]].

[228] R. Kallosh and T. Kugo, "The footprint of E(7(7)) amplitudes of N = 8 supergravity," *JHEP* **0901**, 072 (2009) [arXiv:0811.3414 [hep-th]];
R. Kallosh, "$E_{7(7)}$ symmetry and finiteness of N = 8 supergravity," *JHEP* **1203**, 083 (2012) [arXiv:1103.4115 [hep-th]];
R. Kallosh, "N = 8 counterterms and $E_{7(7)}$ current conservation," *JHEP* **1106**, 073 (2011) [arXiv:1104.5480 [hep-th]];
R. Kallosh and T. Ortin, "New E77 invariants and amplitudes," *JHEP* **1209**, 137 (2012) [arXiv:1205.4437 [hep-th]];
M. Gunaydin and R. Kallosh, "Obstruction to $E_{7(7)}$ deformation in N = 8 supergravity," [arXiv:1303.3540 [hep-th]];
J. J. M. Carrasco and R. Kallosh, "Hidden supersymmetry may imply duality invariance," [arXiv:1303.5663 [hep-th]].

[229] S. Stieberger and T. R. Taylor, "Complete six-gluon disk amplitude in superstring theory," *Nucl. Phys. B* **801**, 128 (2008) [arXiv:0711.4354 [hep-th]].

[230] J. Broedel and L. J. Dixon, "R**4 counterterm and E(7)(7) symmetry in maximal supergravity," *JHEP* **1005**, 003 (2010) [arXiv:0911.5704 [hep-th]].

[231] N. Berkovits, "New higher-derivative R**4 theorems," *Phys. Rev. Lett.* **98**, 211601 (2007) [arXiv:hep-th/0609006];
M. B. Green, J. G. Russo, and P. Vanhove, "Non-renormalisation conditions in type II string theory and maximal supergravity," *JHEP* **0702**, 099 (2007) [arXiv:hep-th/0610299];
M. B. Green, J. G. Russo, and P. Vanhove, "Ultraviolet properties of maximal supergravity," *Phys. Rev. Lett.* **98**, 131602 (2007) [arXiv:hep-th/0611273];
M. B. Green, J. G. Russo, and P. Vanhove, "Modular properties of two-loop maximal supergravity and connections with string theory," *JHEP* **0807**, 126 (2008) [arXiv:0807.0389 [hep-th]];
N. Berkovits, M. B. Green, J. G. Russo, and P. Vanhove, "Non-renormalization conditions for four-gluon scattering in supersymmetric string and field theory," *JHEP* **0911**, 063 (2009) [arXiv:0908.1923 [hep-th]];
P. Vanhove, "The critical ultraviolet behaviour of N = 8 supergravity amplitudes," [arXiv:1004.1392 [hep-th]].

[232] Z. Bern, J. J. M. Carrasco, L. J. Dixon, H. Johansson, and R. Roiban, "Simplifying multiloop integrands and ultraviolet divergences of gauge theory and gravity amplitudes," *Phys. Rev. D* **85**, 105014 (2012) [arXiv:1201.5366 [hep-th]].

[233] P. S. Howe and U. Lindstrom, "Higher order invariants in extended supergravity," *Nucl. Phys. B* **181**, 487 (1981).

[234] R. E. Kallosh, "Counterterms in extended supergravities," *Phys. Lett. B* **99**, 122 (1981).

[235] G. Bossard, P. S. Howe, K. S. Stelle, and P. Vanhove, "The vanishing volume of D = 4 superspace," *Class. Quant. Grav.* **28**, 215005 (2011) [arXiv:1105.6087 [hep-th]].

[236] N. Berkovits, "Super Poincaré covariant quantization of the superstring," *JHEP* **0004**, 018 (2000) [hep-th/0001035].

[237] M. B. Green, H. Ooguri, and J. H. Schwarz, "Nondecoupling of maximal supergravity from the superstring," *Phys. Rev. Lett.* **99**, 041601 (2007) [arXiv:0704.0777 [hep-th]].

[238] T. Banks, "Arguments against a finite N = 8 supergravity," [arXiv:1205.5768 [hep-th]].

[239] M. Bianchi, S. Ferrara, and R. Kallosh, "Perturbative and non-perturbative N = 8 supergravity," *Phys. Lett. B* **690**, 328 (2010) [arXiv:0910.3674 [hep-th]].

[240] Z. Bern, S. Davies, T. Dennen, Y.-t. Huang, and J. Nohle, "Color-kinematics duality for pure Yang–Mills and gravity at one and two loops," [arXiv:1303.6605 [hep-th]].

[241] Z. Bern, S. Davies, T. Dennen, A. V. Smirnov, and V. A. Smirnov, "The ultraviolet properties of N = 4 supergravity at four loops," *Phys. Rev. Lett.* **111**, 231302 (2013) [arXiv:1309.2498 [hep-th]].

[242] Z. Bern, L. J. Dixon, D. C. Dunbar, M. Perelstein, and J. S. Rozowsky, "On the relationship between Yang–Mills theory and gravity and its implication for ultraviolet divergences," *Nucl. Phys. B* **530**, 401 (1998) [hep-th/9802162].

[243] D. C. Dunbar, B. Julia, D. Seminara, and M. Trigiante, "Counterterms in type I supergravities," *JHEP* **0001**, 046 (2000) [hep-th/9911158].

[244] Z. Bern, S. Davies, T. Dennen, and Y.-t. Huang, "Ultraviolet cancellations in half-maximal supergravity as a consequence of the double-copy structure," *Phys. Rev. D* **86**, 105014 (2012) [arXiv:1209.2472 [hep-th]].

[245] Z. Bern, S. Davies, T. Dennen, and Y.-t. Huang, "Absence of three-loop four-point divergences in N = 4 supergravity," *Phys. Rev. Lett.* **108**, 201301 (2012) [arXiv:1202.3423 [hep-th]];

[246] M. Fischler, "Finiteness calculations for O(4) through O(8) extended supergravity and O(4) supergravity coupled to selfdual O(4) matter," *Phys. Rev. D* **20**, 396 (1979).

[247] Z. Bern, S. Davies, and T. Dennen, "The ultraviolet structure of half-maximal supergravity with matter multiplets at two and three loops," [arXiv:1305.4876 [hep-th]].

[248] G. Bossard, P. S. Howe, and K. S. Stelle, "Invariants and divergences in half-maximal supergravity theories," [arXiv:1304.7753 [hep-th]].

[249] D. Vaman and Y.-P. Yao, "Constraints and generalized gauge transformations on tree-level gluon and graviton amplitudes," *JHEP* **1011**, 028 (2010) [arXiv:1007.3475 [hep-th]].

[250] R. H. Boels and R. S. Isermann, "On powercounting in perturbative quantum gravity theories through color-kinematic duality," *JHEP* **1306**, 017 (2013) [arXiv:1212.3473].

[251] Z. Bern, T. Dennen, Y.-t. Huang, and M. Kiermaier, "Gravity as the square of gauge theory," *Phys. Rev. D* **82**, 065003 (2010) [arXiv:1004.0693 [hep-th]].

[252] M. Kiermaier, Talk at *Amplitudes 2010*, May 2010 at QMUL, London, UK. http://www.strings.ph.qmul.ac.uk/~theory/Amplitudes2010/

[253] N. E. J. Bjerrum-Bohr, P. H. Damgaard, T. Sondergaard, and P. Vanhove, "The momentum kernel of gauge and gravity theories," *JHEP* **1101**, 001 (2011) [arXiv:1010.3933 [hep-th]].

[254] C. R. Mafra, O. Schlotterer, and S. Stieberger, "Explicit BCJ numerators from pure spinors," *JHEP* **1107**, 092 (2011) [arXiv:1104.5224 [hep-th]];
C.-H. Fu, Y.-J. Du, and B. Feng, "An algebraic approach to BCJ numerators," *JHEP* **1303**, 050 (2013) [arXiv:1212.6168 [hep-th]].

[255] N. E. J. Bjerrum-Bohr, P. H. Damgaard, and P. Vanhove, "Minimal basis for gauge theory amplitudes," *Phys. Rev. Lett.* **103**, 161602 (2009) [0907.1425 [hep-th]];
S. Stieberger, "Open & closed vs. pure open string disk amplitudes," [arXiv: 0907.2211 [hep-th]];
C. R. Mafra and O. Schlotterer, "The structure of n-point one-loop open superstring amplitudes," [arXiv:1203.6215 [hep-th]];
O. Schlotterer and S. Stieberger, "Motivic multiple zeta values and superstring amplitudes," [arXiv:1205.1516 [hep-th]];
J. Broedel, O. Schlotterer, and S. Stieberger, "Polylogarithms, multiple zeta values and superstring amplitudes," [arXiv:1304.7267 [hep-th]].

[256] S. H. Henry Tye and Y. Zhang, "Dual identities inside the gluon and the graviton scattering amplitudes," *JHEP* **1006**, 071 (2010) [Erratum-*ibid.* **1104**, 114 (2011)] [arXiv:1003.1732 [hep-th]].

[257] B. Feng, R. Huang, and Y. Jia, "Gauge amplitude identities by on-shell recursion relation in s-matrix program," *Phys. Lett. B* **695**, 350 (2011) [arXiv:1004.3417 [hep-th]].

[258] F. Cachazo, "Fundamental BCJ relation in N = 4 SYM from the connected formulation," [arXiv:1206.5970 [hep-th]].

[259] N. E. J. Bjerrum-Bohr, P. H. Damgaard, R. Monteiro, and D. O'Connell, "Algebras for amplitudes," *JHEP* **1206**, 061 (2012) [arXiv:1203.0944 [hep-th]].

[260] R. Monteiro and D. O'Connell, "The kinematic algebra from the self-dual sector," *JHEP* **1107**, 007 (2011) [arXiv:1105.2565 [hep-th]].

[261] M. Tolotti and S. Weinzierl, "Construction of an effective Yang–Mills Lagrangian with manifest BCJ duality," [arXiv:1306.2975 [hep-th]].

[262] Z. Bern, J. J. M. Carrasco, and H. Johansson, "Perturbative quantum gravity as a double copy of gauge theory," *Phys. Rev. Lett.* **105**, 061602 (2010) [arXiv:1004.0476 [hep-th]].

[263] J. J. Carrasco and H. Johansson, "Five-point amplitudes in N = 4 super-Yang–Mills theory and N = 8 supergravity," *Phys. Rev. D* **85**, 025006 (2012) [arXiv:1106.4711 [hep-th]].

[264] N. E. J. Bjerrum-Bohr, T. Dennen, R. Monteiro, and D. O'Connell, "Integrand oxidation and one-loop colour-dual numerators in N = 4 gauge theory," [arXiv:1303.2913 [hep-th]].

[265] R. H. Boels, R. S. Isermann, R. Monteiro, and D. O'Connell, "Colour-kinematics duality for one-loop rational amplitudes," *JHEP* **1304**, 107 (2013) [arXiv:1301.4165 [hep-th]].

[266] J. J. M. Carrasco, M. Chiodaroli, M. Günaydin, and R. Roiban, "One-loop four-point amplitudes in pure and matter-coupled $N \leq 4$ supergravity," *JHEP* **1303**, 056 (2013) [arXiv:1212.1146 [hep-th]].

[267] M. Chiodaroli, Q. Jin, and R. Roiban, "Color/kinematics duality for general abelian orbifolds of N = 4 super Yang–Mills theory," *JHEP* **1401**, 152 (2014) [arXiv:1311.3600 [hep-th]].

[268] J. Nohle, "Color-kinematics duality in one-loop four-gluon amplitudes with matter," [arXiv:1309.7416 [hep-th]].

[269] Z. Bern, C. Boucher-Veronneau, and H. Johansson, "$N \geq 4$ supergravity amplitudes from gauge theory at one loop," *Phys. Rev. D* **84**, 105035 (2011) [arXiv:1107.1935 [hep-th]];
C. Boucher-Veronneau and L. J. Dixon, "$N \geq 4$ supergravity amplitudes from gauge theory at two loops," *JHEP* **1112**, 046 (2011) [arXiv:1110.1132 [hep-th]].

[270] M. T. Grisaru and W. Siegel, "Supergraphity. 2. Manifestly covariant rules and higher loop finiteness," *Nucl. Phys. B* **201**, 292 (1982) [Erratum-*ibid. B* **206**, 496 (1982)].

[271] S. Ferrara, R. Kallosh, and A. Van Proeyen, "Conjecture on hidden superconformal symmetry of N = 4 supergravity," *Phys. Rev. D* **87**, 025004 (2013) [arXiv:1209.0418 [hep-th]].

[272] J. Broedel and L. J. Dixon, "Color-kinematics duality and double-copy construction for amplitudes from higher-dimension operators," *JHEP* **1210**, 091 (2012) [arXiv:1208.0876 [hep-th]].

[273] R. H. Boels, B. A. Kniehl, O. V. Tarasov, and G. Yang, "Color-kinematic duality for form factors," *JHEP* **1302**, 063 (2013) [arXiv:1211.7028 [hep-th]];

[274] T. Bargheer, S. He, and T. McLoughlin, "New relations for three-dimensional supersymmetric scattering amplitudes," *Phys. Rev. Lett.* **108**, 231601 (2012) [arXiv:1203.0562 [hep-th]].

[275] Y.-t. Huang and H. Johansson, "Equivalent D = 3 supergravity amplitudes from double copies of three-algebra and two-algebra gauge theories," [arXiv:1210.2255 [hep-th]].

[276] A. Zee, "Quantum field theory in a nutshell," Princeton, USA: Princeton University Press (2010).

[277] M. D. Schwartz, *Quantum Field Theory and the Standard Model*, Cambridge, UK: Cambridge University Press (2013).

[278] J. M. Henn and J. C. Plefka, "Scattering amplitudes in gauge theories," *Lecture Notes in Physics 883* Heidelbera: Springer (2014).

[279] M. Wolf, "A first course on twistors, integrability and gluon scattering amplitudes," *J. Phys. A* **43**, 393001 (2010) [arXiv:1001.3871 [hep-th]].

[280] Z. Bern, L. J. Dixon, and D. A. Kosower, "On-shell methods in perturbative QCD," *Annals Phys.* **322**, 1587 (2007) [arXiv:0704.2798 [hep-ph]].

[281] M. E. Peskin, "Simplifying multi-jet QCD computation," [arXiv:1101.2414 [hep-ph]].

[282] J. F. Donoghue, "Introduction to the effective field theory description of gravity," gr-qc/9512024.

[283] L. J. Dixon, "Ultraviolet behavior of N = 8 supergravity," [arXiv:1005.2703 [hep-th]].

[284] Z. Bern, P. Gondolo, and M. Perelstein, "Neutralino annihilation into two photons," *Phys. Lett. B* **411**, 86 (1997) [hep-ph/9706538];
Z. Bern, A. De Freitas, and L. J. Dixon, "Two loop helicity amplitudes for gluon–gluon scattering in QCD and supersymmetric Yang–Mills theory," *JHEP* **0203**, 018 (2002) [hep-ph/0201161];
Z. Bern, A. De Freitas, and L. J. Dixon, "Two loop helicity amplitudes for quark gluon scattering in QCD and gluino gluon scattering in supersymmetric Yang–Mills theory," *JHEP* **0306**, 028 (2003) [hep-ph/0304168].

[285] S. J. Bidder, N. E. J. Bjerrum-Bohr, D. C. Dunbar, and W. B. Perkins, "One-loop gluon scattering amplitudes in theories with N < 4 supersymmetries," *Phys. Lett. B* **612**, 75 (2005) [hep-th/0502028].

[286] R. Britto, E. Buchbinder, F. Cachazo, and B. Feng, "One-loop amplitudes of gluons in SQCD," *Phys. Rev. D* **72**, 065012 (2005) [hep-ph/0503132].

[287] S. Lal and S. Raju, "Rational terms in theories with matter," *JHEP* **1008**, 022 (2010) [arXiv:1003.5264 [hep-th]].

[288] S. Dittmaier, "Weyl-van der Waerden formalism for helicity amplitudes of massive particles," *Phys. Rev. D* **59**, 016007 (1998) [hep-ph/9805445].

[289] R. Boels and C. Schwinn, "CSW rules for massive matter legs and glue loops," *Nucl. Phys. Proc. Suppl.* **183**, 137 (2008) [arXiv:0805.4577 [hep-th]].

[290] R. H. Boels, "No triangles on the moduli space of maximally supersymmetric gauge theory," *JHEP* **1005**, 046 (2010) [arXiv:1003.2989 [hep-th]].

[291] P. Ferrario, G. Rodrigo, and P. Talavera, "Compact multigluonic scattering amplitudes with heavy scalars and fermions," *Phys. Rev. Lett.* **96**, 182001 (2006) [hep-th/0602043].

[292] D. Forde and D. A. Kosower, "All-multiplicity amplitudes with massive scalars," *Phys. Rev. D* **73**, 065007 (2006) [hep-th/0507292].

[293] G. Rodrigo, "Multigluonic scattering amplitudes of heavy quarks," *JHEP* **0509**, 079 (2005) [hep-ph/0508138].

[294] C. Cheung, D. O'Connell, and B. Wecht, "BCFW recursion relations and string theory," *JHEP* **1009**, 052 (2010) [arXiv:1002.4674 [hep-th]].

[295] R. H. Boels, D. Marmiroli, and N. A. Obers, "On-shell recursion in string theory," *JHEP* **1010**, 034 (2010) [arXiv:1002.5029 [hep-th]].

[296] K. Kampf, J. Novotny, and J. Trnka, "Recursion relations for tree-level amplitudes in the SU(N) non-linear sigma model," *Phys. Rev. D* **87**, 081701 (2013) [arXiv:1212.5224 [hep-th]].

[297] Z. Bern, L. J. Dixon, and D. A. Kosower, "The last of the finite loop amplitudes in QCD," *Phys. Rev. D* **72**, 125003 (2005) [hep-ph/0505055].

[298] B. Feng and M. Luo, "An introduction to on-shell recursion relations," [arXiv:1111.5759 [hep-th]].

[299] J. M. Drummond, G. P. Korchemsky, and E. Sokatchev, "Conformal properties of four-gluon planar amplitudes and Wilson loops," *Nucl. Phys. B* **795**, 385 (2008) [arXiv:0707.0243 [hep-th]].

[300] N. Berkovits and J. Maldacena, "Fermionic T-duality, dual superconformal symmetry, and the amplitude/Wilson loop connection," *JHEP* **0809**, 062 (2008) [arXiv:0807.3196 [hep-th]].

[301] N. Beisert, R. Ricci, A. A. Tseytlin, and M. Wolf, "Dual superconformal symmetry from AdS(5) x S**5 superstring integrability," *Phys. Rev. D* **78**, 126004 (2008) [arXiv:0807.3228 [hep-th]].

[302] A. Brandhuber, P. Heslop, and G. Travaglini, "MHV amplitudes in N = 4 super Yang–Mills and Wilson loops," *Nucl. Phys. B* **794**, 231 (2008) [arXiv:0707.1153 [hep-th]].

[303] L. J. Mason and D. Skinner, "The complete planar S-matrix of N = 4 SYM as a Wilson loop in twistor space," *JHEP* **1012**, 018 (2010) [arXiv:1009.2225 [hep-th]].

[304] S. Caron-Huot, "Notes on the scattering amplitude / Wilson loop duality," *JHEP* **1107**, 058 (2011) [arXiv:1010.1167 [hep-th]].

[305] B. Eden, P. Heslop, G. P. Korchemsky, and E. Sokatchev, "The super-correlator/super-amplitude duality: Part I," *Nucl. Phys. B* **869**, 329 (2013) [arXiv:1103.3714 [hep-th]]; Part II, *Nucl. Phys. B* **869**, 378 (2013) [arXiv:1103.4353 [hep-th]].

[306] L. F. Alday, B. Eden, G. P. Korchemsky, J. Maldacena, and E. Sokatchev, "From correlation functions to Wilson loops," *JHEP* **1109**, 123 (2011) [arXiv:1007.3243 [hep-th]].

[307] T. Adamo, M. Bullimore, L. Mason, and D. Skinner, "A proof of the supersymmetric correlation function / Wilson loop correspondence," *JHEP* **1108**, 076 (2011) [arXiv:1103.4119 [hep-th]].

[308] L. F. Alday and R. Roiban, "Scattering amplitudes, Wilson loops and the string/gauge theory correspondence," *Phys. Rept.* **468**, 153 (2008) [arXiv:0807.1889 [hep-th]].

[309] R. M. Schabinger, "One-loop N = 4 super Yang–Mills scattering amplitudes in d dimensions, relation to open strings and polygonal Wilson loops," *J. Phys. A* **44**, 454007 (2011) [arXiv:1104.3873 [hep-th]].

[310] J. M. Henn, "Duality between Wilson loops and gluon amplitudes," *Fortsch. Phys.* **57**, 729 (2009) [arXiv:0903.0522 [hep-th]].

[311] T. Adamo, M. Bullimore, L. Mason, and D. Skinner, "Scattering amplitudes and Wilson loops in twistor space," *J. Phys. A* **44**, 454008 (2011) [arXiv:1104.2890 [hep-th]].

[312] L. F. Alday, D. Gaiotto, J. Maldacena, A. Sever, and P. Vieira, "An operator product expansion for polygonal null Wilson loops," *JHEP* **1104**, 088 (2011) [arXiv:1006.2788 [hep-th]].

[313] B. Basso, A. Sever, and P. Vieira, "Space-time S-matrix and flux-tube S-matrix at finite coupling," *Phys. Rev. Lett.* **111**, 091602 (2013) [arXiv:1303.1396 [hep-th]].

[314] B. Basso, A. Sever, and P. Vieira, "Space-time S-matrix and flux tube S-matrix II. Extracting and matching data," *JHEP* **1401**, 008 (2014) [arXiv:1306.2058 [hep-th]].

[315] B. Basso, A. Sever, and P. Vieira, "Space-time S-matrix and flux-tube S-matrix III. The two-particle contributions," [arXiv:1402.3307 [hep-th]].

[316] L. J. Dixon, J. M. Drummond, M. von Hippel, and J. Pennington, "Hexagon functions and the three-loop remainder function," *JHEP* **1312**, 049 (2013) [arXiv:1308.2276 [hep-th]].

[317] L. J. Dixon, J. M. Drummond, C. Duhr, and J. Pennington, "The four-loop remainder function and multi-Regge behavior at NNLLA in planar N = 4 super-Yang–Mills theory," [arXiv:1402.3300 [hep-th]].

[318] R. Penrose and W. Rindler, *Spinors and Space-Time*, vol. 2, Cambridge: Cambridge University Press (1986).

[319] R. Ward and R. Wells, *Twistor Geometry and Field Theory*, Cambridge: Cambridge University Press (1990).

[320] S. Huggett and P. Tod, *An Introduction to Twistor Theory*, Student Texts 4, London: London Mathematical Society (1985).

[321] F. Cachazo and P. Svrcek, "Lectures on twistor strings and perturbative Yang–Mills theory," *PoS RTN* **2005**, 004 (2005) [hep-th/0504194].

Index

Printed in the United States
By Bookmasters